Phase Estimation
in
Optical
Interferometry

Phase Estimation
in
Optical
Interferometry

Edited by
Pramod Rastogi
EPFL - Swiss Federal Institute of Technology
Lausanne, Switzerland
Erwin Hack
Empa - Swiss Federal Laboratories for Materials Science and Technology
Dubendorf, Switzerland

CRC Press
Taylor & Francis Group
Boca Raton London New York

CRC Press is an imprint of the
Taylor & Francis Group, an **informa** business

CRC Press
Taylor & Francis Group
6000 Broken Sound Parkway NW, Suite 300
Boca Raton, FL 33487-2742

First issued in paperback 2016

Version Date: 20140922

ISBN 13: 978-1-138-03365-8 (pbk)
ISBN 13: 978-1-4665-9831-7 (hbk)

Library of Congress Cataloging-in-Publication Data

Phase estimation in optical interferometry / [edited by] Pramod Rastogi, Erwin Hack.
    pages cm
    Summary: "This tutorial-style book reviews experimental approaches for the analysis of solid mechanics based on interference-optical methods, including coherent-optical techniques and photoelastic methods. Suitable for advanced students, young researchers, and practicing metrologists and engineers, it describes the theory and scope of application in everyday practice. Takes a practical approach, the book familiarizes readers with various techniques and explains how and when to apply them"-- Provided by publisher.
    Includes bibliographical references and index.
    ISBN 978-1-4665-9831-7 (hardback)
    1. Interferometry--Mathematics. 2. Optical measurements. 3. Solids--Optical properties. 4. Image analysis. I. Rastogi, Pramod K., 1951- editor. II. Hack, Erwin, editor.

QC367.P43 2015
535'.470287--dc23   2014015082

Visit the Taylor & Francis Web site at
http://www.taylorandfrancis.com

and the CRC Press Web site at
http://www.crcpress.com

Contents

KIERAN G. LARKIN

Preface

The advent of the laser as an intense monochromatic light source has remorselessly dwarfed the experimental possibilities of the spontaneous emission lines from gas lamps. The emergence of interferometry using laser light could be traced to the experiments performed independently by Yuri Denisyuk and the team of Emmett Leith and Juris Upatnieks in 1962 with the objective of marrying coherent light obtained from lasers with holography, a method discovered by Dennis Gabor in 1948. The result was a novel interference methodology that opened the way to a greatly expanded range of applications. A significant contribution of such laser-based techniques like holographic and speckle interferometry has been to bring rough objects within the reach of interferometry. In early experiments, photographic plates were used for recording, and tedious fringe-counting methods were necessary for the evaluation of the interferograms produced. Although measurement accuracy in fringe-counting methods could be enhanced by interpolating interference fringes by purposely shifting the relative phase of reference and object beams between successive exposures, a quantitative analysis of these interferograms remained prohibitively labor intensive.

With the output results in most forms of interferometry displayed in the form of fringe patterns, a wide range of efforts went into decoding as accurately as possible the information on important physical quantities that had been encoded in these patterns during their formation. These methods for fringe analysis have included the visual photographic identification of the regions of minimum and maximum intensity, counting of the number of fringes, fringe interpolation, and fringe skeletonizing. The hard-pressed optical scientists and engineers working in the area were thus in for a sea change as new approaches adroitly rode the computer wave and rich developments in the electronic recording medium started trickling into their research domain. In their wake, they brought a variety of rapid and flexible ways to handle data that were becoming ever more massive when obtained from interferometers in quest of higher performance and new challenges.

Interferometry techniques, either in their avatar of plane wave or rough scattered wave, have each contributed significantly to various fields of engineering and science, and in the majority of the cases, the phase of the fringe patterns is known to carry the key information. Although the first instance of the use of phase-stepping methods for estimating phase distribution in interferometry dates to 1964 when Carré suggested imparting four arbitrary but controlled phase steps to a piezoelectric device employed in one arm of the interferometer, it was only in the mid-1970s that these methods started implanting and integrating themselves in interferometry. Although standard methods such as the four-frame algorithm were quick to be embraced by the scientific community, they still left ample room for development and improvement. Especially, the robustness of the algorithms started becoming an issue in view of the experimental imperfections or the dependence of their sensitivity to a multitude of factors, such as phase shifter nonlinearity, additive noise, or higher harmonics, to name just a few. Added to these were new challenges arising from experimental developments, such as high-speed imaging, use of double reference beams, digital holography, and fringe projection.

This book presents the essentials of modern methods of phase-stepping algorithms used in interferometry and pseudointerferometric techniques with the aim to offer the basic concepts and mathematics needed for the understanding of the phase estimation methods in vogue today. With the real push for the deployment of novel technologies and fast-evolving techniques having become more than a reality, we hope that the book—in addition to providing a framework of understanding—will also assist the user in obtaining a comparative view of the performance and limitations of these various approaches.

The first four chapters in this book treat the subject of phase retrieval from image transforms using a single frame. The classical Fourier transform method uses the addition of carrier fringes in an interference pattern and filtered back-transformation to obtain the phase map of the original intensity fringe pattern. Although the windowed Fourier transform method in Chapter 2 is seen to augment robustness against higher-order harmonics, the approach based on wavelet transforms presented in Chapter 3 is ideally suited for fringe patterns with irregular carrier frequency content. In spite of these developments, interference patterns with closed fringes have long evaded solution using these evaluation methods. The shortcoming was redeemed with the introduction of the spiral phase transform, discussed in Chapter 4, which finally allows retrieving phase from fringe patterns containing closed fringes. Chapter 5 on fringe regularization looks into the local environment of a fringe pattern, an important aspect to be taken into account in both spatial and temporal phase measurements. Chapter 6 paints a broad picture of the phase estimation approach based on local polynomial phase modeling; Chapter 7 covers temporal high-resolution phase evaluation methods relying on a sequence of image values in a single pixel and that are seen to exhibit the capability of minimizing several systematic and random errors arising during phase measurements. Methods of phase unwrapping are described in Chapter 8, in which a continuous total phase is calculated from the wrapped (modulo 2π) phase image, which then can be scaled using an appropriate calibration of the phase values. The book winds up by giving in its final chapter an overview of experimental imperfections that are liable to adversely influence the accuracy of phase measurements. The phase measurement uncertainty is calculated for a single-pixel-based as well as for single-image-based phase measurement techniques.

We are extremely thankful to the contributors, all of whom are internationally renowned experts in their respective fields, for their warm collegiality and for the chance we have had to work with them as many belong to the category of founders of the approaches covered in this book.

Pramod Rastogi
Erwin Hack
Switzerland

About the Editors

Professor Pramod Rastogi received his MTech degree from the Indian Institute of Technology Delhi and doctorate degree from the University of Franche Comté in France. He started his research at the EPFL in Switzerland 1978. He is the author or coauthor of over 150 scientific papers published in peer-reviewed archival journals. Prof. Rastogi is also the author of encyclopedia articles and has edited several books in the field of optical metrology. He is the 2014 recipient of the SPIE  Dennis Gabor Award and is a member of the Swiss Academy of Engineering Sciences. He is a Fellow of the Society of the Photo-Optical Instrumentation Engineers (1995) and of the Optical Society of America (1993). He is also a recipient of the "Hetényi Award" for the most significant research paper published in experimental mechanics in the year 1982. Prof. Rastogi is the co-editor-in-chief of the *International Journal of Optics and Lasers in Engineering*, Elsevier.

Dr. Erwin Hack holds a diploma in theoretical physics and a PhD in physical chemistry (1990) from the University of Zurich, Switzerland. He is senior scientist at EMPA, the Swiss Federal Laboratories for Materials Science and Technology, and a member of its research commission. His research interest includes THz imaging, digital speckle pattern interferometry and thermography. He coordinated a European research project on the validation of dynamic models using integrated simulation and experimentation (ADVISE) and has been partner in

national and international research projects on optical full-field techniques. Dr. Hack has authored or coauthored more than 80 papers in peer-reviewed journals and conferences and co-edited the book *Optical Methods in Solid Mechanics*. He is lecturer at ETH Zurich, associate editor of *Optics and Lasers in Engineering*, vice-chair of CEN WS71 on Validation of computational solid mechanics models using strain fields from calibrated measurement (VANESSA), vice-president of the Swiss Society for Non-destructive Testing, and a member of EOS and OSA.

List of Contributors

David R. Burton
Liverpool John Moores University
General Engineering Research
    Institute
Liverpool, United Kingdom
E-mail:
    d.r.burton@ljmu.ac.uk

Sai Siva Gorthi
Indian Institute of Science (IISc)
Department of Instrumentation
    and Applied Physics
Bangalore, India
E-mail:
    saisiva.gorthi@isu.iisc.ernet.in

Erwin Hack
Electronics/Metrology/Reliability
    Laboratory
EMPA—Materials Science and
    Technology
Dubendorf, Switzerland
E-mail: erwin.hack@empa.ch

Rajesh Langoju
GE Global Research Center
Bangalore, India
E-mail: rajesh.langoju@ge.com

Kieran G. Larkin
Nontrivialzeros Research
Sydney, Australia
E-mail:
    spizralphase@yahoo.com.au

Moises Padilla
Centro de Investigaciones en
    Optica
Leon Guanajuato, Mexico
E-mail: moises@cio.mx

Abhijit Patil
John F. Welch Technology
    Centre
GE Global Research
EPIP—Phase 2, Hoodi Village
Bangalore, India
E-mail: abhijit.patil1@ge.com

Gannavarpu Rajshekhar
Beckman Institute
Urbana, Illinois, USA
E-mail: gshekhar@illinois.edu

Pramod Rastogi
EPFL—Swiss Federal Institute of
 Technology
Applied Computing and
 Mechanics Laboratory
Lausanne, Switzerland
E-mail: pramod.rastogi@epfl.ch

Manuel Servin
Centro de Investigaciones en Optica
Leon Guanajuato, Mexico
E-mail: mservin@cio.mx

Mitsuo Takeda
Utsunomiya University
Center for Optical Research and
 Education (CORE)
Utsunomiya, Japan
E-mail: takeda@opt.
 utsunomiya-u.ac.jp

Lionel R. Watkins
Department of Physics
University of Auckland
Auckland, New Zealand
E-mail: l.watkins@auckland.ac.nz

Jiawen Weng
College of Science
South China Agricultural
 University
Guangzhou, China
E-mail: weng-jw@163.com

Jingang Zhong
Department of Optoelectronic
 Engineering
Jinan University
Guangzhou, China
E-mail: tzjg@jnu.edu.cn

Abbreviations

1-D	One dimensional
2-D	Two dimensional
3-D	Three dimensional
4-D	Four dimensional
ADC	Analog-to-digital conversion
AM	Amplitude modulation
AWFT	Adaptive windowed Fourier transform
AWGN	Additive white Gaussian noise
CCD	Charge-coupled device
CMOS	Complementary metal-oxide semiconductor
CPF	Cubic phase function
CRB	Cramér-Rao bound
CWT	Continuous wavelet transform
DC	Direct current
DFT	Discrete Fourier transform
DHI	Digital holographic interferometry
DoG	Derivative of Gaussian
DSPI	Digital speckle pattern interferometry
ESPRIT	Estimation of signal parameters by rotational invariance technique
FFA	Fourier fringe analysis
FFT	Fast Fourier transform
FM	Frequency modulation
FT	Fourier transform
FTF	Frequency transfer function
FTM	Fourier transform method
GL	Gray level
GMW	Generalized Morse wavelet
GPU	Graphics processing unit
GUM	*Guide to the Expression of Uncertainty in Measurement*
GWT	Gabor wavelet transform
HAF	High-order ambiguity function
HIM	High-order instantaneous moment

IFEIF	Iterative frequency estimation by interpolation on Fourier coefficients
IFFT	Inverse fast Fourier transform
IFR	Instantaneous frequency rate
LoG	Laplacian-of-Gaussian or Mexican hat function
LPE	Linear predictive extrapolation
LPF	Low-pass filter
LS	Least squares
LSB	Least-significant bit
LSI	Lateral shearing interferometry
ML	Maximum likelihood
MLE	Maximum likelihood estimation
MRI	Magnetic resonance imaging
MSE	Mean squared error
MUSIC	Multiple signal classification
NOF	Number of frequencies
OSA	Optical Society of America
PGSL	Probabilistic global search Lausanne
PSA	Phase-stepping (or shifting) algorithm
PZT	Lead zirconate titanate, a piezoelectric material
rms	Root mean square
RPT	Regularized phase tracker
SAR	Synthetic aperture radar
SNR	Signal-to-noise ratio
SPIDER	Spectral phase interferometry for direct electric-field reconstruction
SPIE	Society of Photo-Optical Instrumentation Engineers
SPT	Spiral phase transform
SRFT	Superresolution Fourier transform
UV	Ultraviolet spectral range
WFF	Windowed Fourier filtering
WFR	Windowed Fourier ridges
WFT	Windowed Fourier transform
WT	Wavelet transform
yawtb	"Yet another wavelet toolbox"

CHAPTER 1

Fourier Fringe Demodulation

Mitsuo Takeda

Utsunomiya University
Center for Optical Research and Education (CORE)
Utsunomiya, Japan

1.1 INTRODUCTION

How to extract the desired object information, with the highest possible precision and speed, from a temporally or spatially modulated quasi-periodic fringe signal has been a critical issue common to all kinds of sensing and metrology. Various techniques of fringe generation and analysis have been developed [1–4] to meet different requirements in diverse applications. They are categorized into two types: the temporal carrier technique and the spatial carrier technique. The temporal carrier technique detects the phase and the amplitude from the periodic temporal variation of fringe signals (called temporal carrier fringes), as typified by heterodyne interferometry [5], in which the temporal carrier is introduced by the optical frequency beat generated by interference between two mutually frequency-shifted beams, and the phase shift technique [6], in which the temporal carrier is introduced by varying the optical path difference between the interfering beams. In the spatial carrier techniques, the phase information is encoded into quasi-periodic spatial variation of fringe signals, which is created, for example, by the interference between two beams with mutually tilted wavefronts. To characterize their features in short, the temporal carrier technique gains spatial independence of the probing points at the cost of time-sequential data acquisitions; the spatial carrier technique gains spatial parallelism for instantaneous data acquisitions at the cost of losing the spatial independence of the probing points.

This chapter is devoted to a review of Fourier fringe demodulation techniques, which were initially developed for spatial carrier fringe analysis and later came to be applied for a wider class of spatiotemporal fringe signals. Among others, the focus is on the technique for fringe analysis, known by the name Fourier fringe analysis (FFA) or the Fourier transform method (FTM). In the early 1980s, the generic FTM was proposed and experimentally demonstrated [7,8], initially as a means for demodulating a fringe pattern with a spatial carrier frequency. Since then, through the three decades of active participation of many scientists and engineers all over the world, the FTM has been critically analyzed, continuously improved, and refined and has found new application areas [9–12]. Starting from a brief review of the principle of the generic FTM, the features, the strength, and the weakness of the FTM as compared with other techniques are described. Then, some typical applications of the FTM are introduced, and it is shown how the advantages of the FTM are exploited in these practical applications. Part of this chapter is based on my review articles published elsewhere [11,12].

1.2 PRINCIPLE OF THE GENERIC FTM FOR FRINGE DEMODULATION

The principle of the FTM was inspired by the demodulation scheme in communication theory. For ease of understanding, let us first start with a one-dimensional (1-D) spatial fringe signal detected by a line sensor:

$$g(x) = a(x) + b(x)\cos[2\pi f_0 x + \phi(x)], \tag{1.1}$$

which resembles a 1-D time signal in traditional communication theory. In Equation (1.1), $a(x)$ is unwanted background intensity, $b(x)$ is amplitude, and $\phi(x)$ is phase, all of which are assumed to vary much slower than the spatial carrier frequency f_0. A typical example of the fringe signal $g(x)$ and its amplitude $b(x)$ and phase $\phi(x)$ are illustrated, respectively, in the illustration in the a–c parts of the Box 1.1 figure with specified parameters. For convenience of explanation, we rewrite Equation (1.1) as

$$g(x) = a(x) + \frac{1}{2}b(x)\exp[i\phi(x)]\exp(2\pi i f_0 x)$$

$$+ \frac{1}{2}b(x)\exp[-i\phi(x)]\exp(-2\pi i f_0 x) \tag{1.2}$$

$$= a(x) + c(x)\exp(2\pi i f_0 x) + c^*(x)\exp(-2\pi i f_0 x),$$

BOX 1.1 NUMERICAL EXAMPLES OF 1-D GENERIC FTM

To give an idea about the principle of the Fourier transform method (FTM), a specific numerical example is presented for a fringe signal:

$$g(x) = a(x) + b(x)\cos[2\pi f_0 x + \phi(x)] \quad (-0.5 \le x \le 0.5)$$

with

$$f_0 = 32 \text{ [lines/(unit length)]},$$

$$a(x) = 0.25 \times [1 + 0.01 \times (9 + \cos(2\pi x))^2],$$

$$b(x) = 0.1 \times (9 + \cos(2\pi x)),$$

and

$$\phi(x) = 8 \times \exp[-85 \times (x - 0.25)^2] - 5 \times \exp[-50 \times (x + 0.2)^2].$$

The (a) portion of the Box 1.1 figure shows a fringe signal $g(x)$, of which the amplitude $b(x)$ and the phase $\phi(x)$ are modulated by the functions specified and shown in the (b) and (c) portions of the figure, respectively. The modulus of the Fourier spectrum of the fringe signal $G(f)$ is shown by a solid line in the d portion of the figure. The signal spectra are separated from the background by the spatial carrier frequency, and the spectrum $C(f - f_0)$ on the carrier frequency f_0 is selected by the band-pass filter window shown by a broken line in the (d) portion. The filtered spectrum is shifted down to the origin to remove the carrier frequency f_0 and to have $C(f)$ as shown in the (e) part, which is then inverse Fourier transformed to obtain $c(x)$. In the (f) section of the figure, the solid line and the open circles show, respectively, the amplitude detected by FTM and the original amplitude $b(x)$. The (g) part shows the wrapped phase obtained from Equation (1.5) in the text. In the (h) portion, the solid line and the open circles show the unwrapped phase and the original phase $\phi(x)$, respectively.

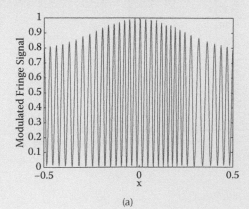

(a)

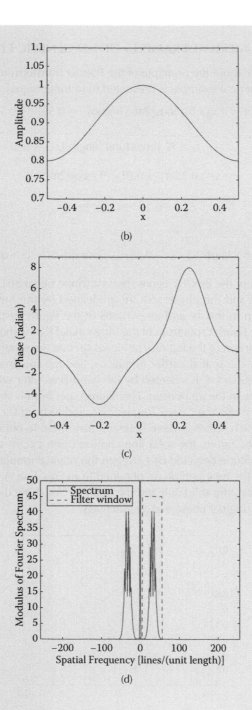

(b)

(c)

(d)

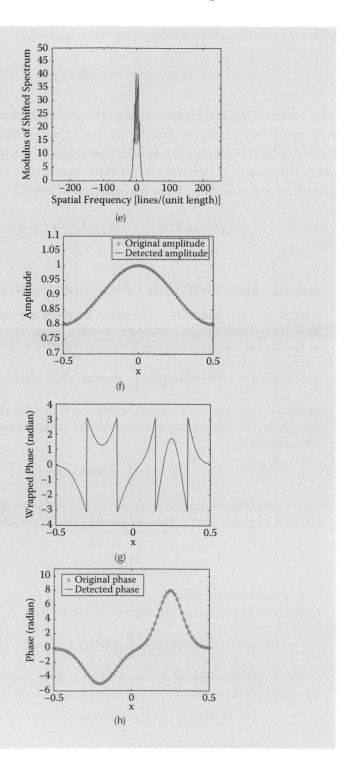

(e)

(f)

(g)

(h)

where $c(x,y)$ is the complex fringe amplitude defined by

$$c(x) = \frac{1}{2}b(x)\exp[i\phi(x)], \tag{1.3}$$

and * denotes the complex conjugate. The principle of the generic FTM is shown schematically in Box 1.1 for the typical fringe signal created numerically. We compute the Fourier transform of the spatial fringe signal in Equation (1.2). Referring to the shift theorem [13] in the list of basic properties of Fourier transform summarized in Box 1.2, we have

$$G(f) = A(f) + C(f - f_0) + C^*[-(f + f_0)], \tag{1.4}$$

BOX 1.2   BASIC PROPERTIES OF FOURIER TRANSFORM

Some of the basic properties of the Fourier transform relevant to the Fourier transform method (FTM) are summarized; for more details, refer to Reference 13.

The Fourier transform of a signal $g(x)$ is defined by

$$G(f) = F[g(x)] = \int_{-\infty}^{\infty} g(x)\exp(-2\pi i f x)\, dx,$$

which gives the spectral frequency components that constitute the signal $g(x)$. Conversely, the signal $g(x)$ is given by the inverse Fourier transform of the spectrum $G(f)$:

$$g(x) = F^{-1}[G(f)] = \int_{-\infty}^{\infty} G(f)\exp(2\pi i f x)\, df,$$

which states that the signal can be reconstructed by the superposition of complex harmonic components weighted by the Fourier spectrum.

A. Shift theorem
$$F[g(x)\exp(2\pi i f_0 x)] = G(f - f_0)$$

B. Convolution theorem
Convolution of two signals $g(x)$ and $h(x)$ is defined by

$$g(x) \otimes h(x) = \int_{-\infty}^{\infty} g(\tilde{x})h(x - \tilde{x})d\tilde{x}.$$

The Fourier transform of the convolution of the two signals is given by the product of the Fourier spectra of the two signals such that

$$F[g(x) \otimes h(x)] = F[g(x)] \times F[h(x)] = G(f)H(f)$$

From the inverse Fourier transform, we have

$$g(x) \otimes h(x) = F^{-1}[G(f)H(f)] \tag{1.21}$$

Regarding $G(f)$ as the Fourier spectrum of the fringe signal and $H(f)$ as the filter function, one can see that the filtering operation of the FTM in the Fourier spectrum domain is equivalent to the convolution operation in signal domain. Whereas the Fourier transform is a global operation applied to the entire fringe data, the convolution can be a local operation if the impulse response $h(x) = F^{-1}[H(f)]$ has a finite spread because the operation is performed only on the data within the range of the impulse response.

where the uppercase letters stand for the Fourier spectra of the signals denoted by the corresponding lowercase letters. Because of the high carrier frequency, the Fourier spectrum of this fringe signal splits into three spectrum components separated from each other, as shown by a solid line in the d portion of the figure in Box 1.1. This permits selective filtering of only the second spectrum component $C(f - f_0)$ on the carrier frequency f_0 (using a filter window indicated by a broken line in the d part of the figure in Box 1.1); shift it down to the origin to remove its carrier frequency, and obtain $C(f)$ as shown in the Box 1.1 illustration, part e. The inverse Fourier transform of the filtered spectrum $C(f)$, with its carrier frequency now removed, gives the complex fringe amplitude defined by Equation (1.3). Computing a complex logarithm of Equation (1.3), we have

$$\log[c(x)] = \log\left[\frac{1}{2}b(x)\right] + i\phi(x), \tag{1.5}$$

where the real and the imaginary parts give, respectively, the log amplitude and the phase of the fringe signal. The solid line in the Box 1.1 figure, part f, shows the amplitude $b(x)$ obtained from the retrieved log amplitude, and the open circles show the original numerical value of the amplitude. The phase obtained from the imaginary part of Equation (1.5) is wrapped into the principal value $[-\pi, \pi)$ as shown in part g of the Box 1.1 figure. The wrapped phase is unwrapped using a suitable phase-unwrapping algorithm [14–16] that gives the desired phase map. In the Box 1.1 figure, part h, the solid line and the open circles show, respectively, the retrieved phase and the original numerical phase value.

The traditional communication theory for a 1-D time signal was naturally extended to deal with a two-dimensional (2-D) spatial fringe signal (referred to as a fringe pattern or an interferogram) in such a manner that one can detect phase and amplitude from a single fringe pattern by introducing a 2-D spatial carrier frequency (f_{0X}, f_{0Y}):

$$g(x, y) = a(x, y) + b(x, y)\cos[2\pi(f_{0X}x + f_{0Y}y) + \phi(x, y)]. \qquad (1.6)$$

An example of such a 2-D interferogram for $f_{Y0} = 0$ and a phase with a peak and a valley created by the standard MATLAB peaks function is shown in the a part of the illustration in Box 1.3a. Similarly to the 1-D case, the 2-D Fourier transform of the interferogram in Equation (1.6) is computed, which gives

$$G(f_X, f_Y) = A(f_X, f_Y) + C(f_X - f_{X0}, f_Y - f_{Y0})$$
$$+ C^*[-(f_X + f_{X0}), -(f_Y + f_{Y0})], \qquad (1.7)$$

where $C(f_{0X}, f_{0Y})$ is the Fourier transform of the 2-D complex amplitude

$$c(x, y) = \frac{1}{2}b(x, y)\exp[i\phi(x, y)]. \qquad (1.8)$$

The three spectra in Equation (1.7) are separated from each other by the spatial carrier frequency as shown in part b of the figure in Box 1.3. This permits one to selectively filter only the spectrum component $C(f_X - f_{X0}, f_Y - f_{Y0})$ on the carrier frequency (f_{X0}, f_{Y0}), shift it down to the origin to remove its carrier frequency, and obtain $C(f_X, f_Y)$ as shown in the c part of the Box 1.3 figure. The inverse Fourier transform of the filtered spectrum gives the complex fringe amplitude $c(x, y)$ defined by Equation (1.8). The process of the amplitude and phase detection is the same as the 1-D case. The phase obtained from the imaginary part of Equation (1.8) is wrapped into the principal value $[-\pi, \pi)$ as shown in part d of the Box 1.3 figure. The wrapped phase is unwrapped using a suitable 2-D phase unwrapping algorithm [14–16], which gives the final phase map as shown in part e of the Box 1.3 illustration.

BOX 1.3 NUMERICAL EXAMPLES OF 2-D GENERIC FTM

To demonstrate how 2-D FTM works, a numerical example is created. Shown in the (a) portion of the illustration for Box 1.3 is a two-dimensional (2-D) inter-ferogram $g(x,y)$ for a phase distribution $\phi(x,y)$ with a peak and a valley created by the standard MATLAB peaks function. The spatial carrier frequency f_{X0} is introduced in the x direction, with $f_{Y0} = 0$. The (b) part shows the modulus of the 2-D Fourier transform of the interferogram. The spectra are separated from each other by the spatial carrier frequency f_{X0}, which permits selective filtering of the spectrum $C(f_X - f_{X0}, f_Y - f_{Y0})$. The filtered spectrum is shifted down to the origin to remove the carrier frequency, as shown in the (c) portion of the figure, and then inverse Fourier transformed. The (d) portion shows the retrieved 2-D phase map that is wrapped into the range $[-\pi,\pi)$, and the (e) part shows the final result obtained after unwrapping.

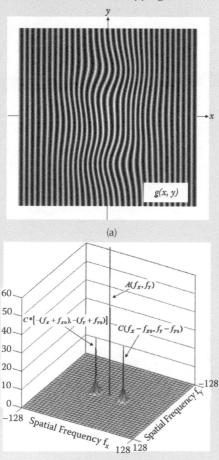

(a)

(b)

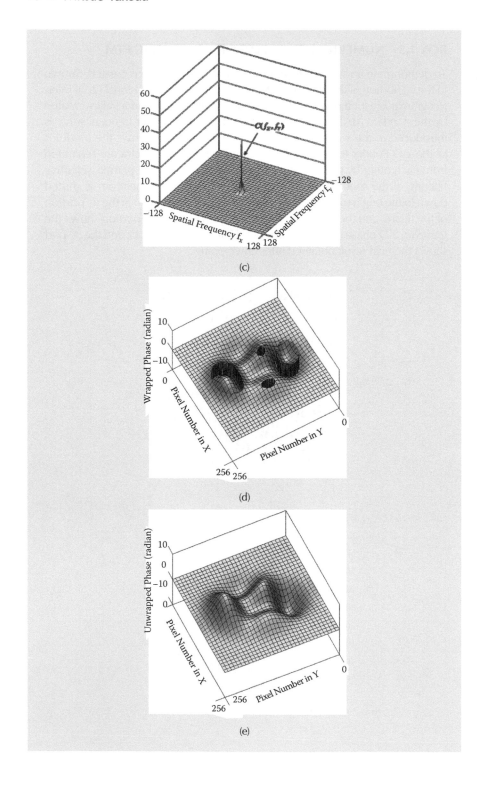

(c)

(d)

(e)

1.3 GENERAL FEATURES OF THE FTM

It should be noted that the application of the FTM is not restricted to the 2-D spatial fringe pattern but includes a general multidimensional spatiotemporal fringe signal $g(x,y,z;t)$ with a spatiotemporal carrier frequency vector $f_0 = (f_{X0}, f_{Y0}, f_{Z0}; f_{T0})$. The advantage of introducing a multidimensional spatial carrier frequency vector was first pointed out by Bone et al. [17], who demonstrated with a 2-D interferogram that aliasing errors caused by undersampling can be avoided by an appropriate choice of the direction of the 2-D carrier frequency vector. Shown in Figure 1.1 is a schematic illustration of the Fourier spectrum $G(f_X, f_Y)$ of a discretely sampled interferogram with a 1-D spatial carrier frequency in the x direction $f_0 = (f_{0X}, 0)$. The shaded square area in the center bounded by the Nyquist boundary is the maximum usable range of the spectrum set by the Shannon sampling theorem, and because of the discrete sampling, the

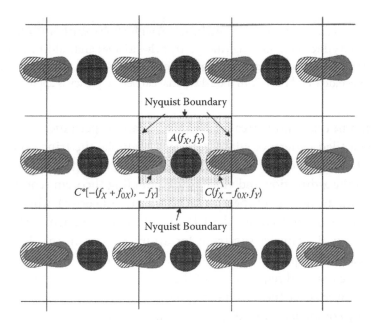

FIGURE 1.1 Fourier spectra of an interferogram with a 1-D spatial carrier frequency in the x direction. The spectrum has a large bandwidth in the x direction that violates the sampling theory requirement. Because of the aliasing error of the overlapped spectra, FTM with the 1-D spatial carrier frequency results in failure.

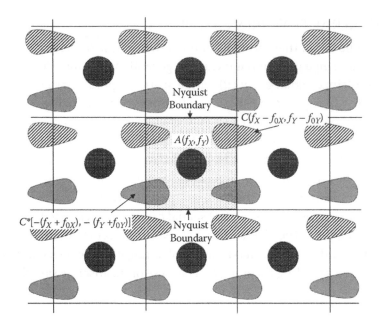

FIGURE 1.2   Fourier spectra of an interferogram with a 2-D spatial carrier frequency in both x and y directions. Although the spectrum has a large bandwidth in the x direction that violates the sampling theory requirement, the aliasing error of the overlapped spectra is avoided, and FTM with the 2-D spatial carrier frequency is applicable.

spectrum in this central area (or Nyquist square) is periodically repeated to form a tiled pattern of aliasing. In this example, the fringe spectrum $C(f_X - f_{0X}, f_Y)$ (and its Hermite conjugate spectrum) has a large bandwidth in the x direction and violates the sampling theorem requirement. This causes the overlap of spectra (called the aliasing error), which results in the failure of the FTM for the 1-D spatial carrier frequency. Shown in Figure 1.2 is the Fourier spectrum for an interferogram that has a 2-D spatial carrier frequency $f_0 = (f_{0X}, f_{0Y})$. Even though the interferogram violates the nominal requirement of the sampling theorem set for the 1-D case, the overlap of the spectra is avoided by the introduction of the y frequency component. By adequately combining the aliased spectra that fall within the area of the Nyquist square without overlapping, one can reconstruct the original spectrum $C(f_X - f_{X0}, f_Y)$ and retrieve the phase and the amplitude by the 2-D FTM.

Because the principle of the FTM has its origin in communication theory, the FTM inherits the merit of frequency multiplexing that multiple signals having different information can be accommodated simultaneously in a single fringe signal (or a single fringe pattern) by assigning a different carrier frequency to each signal:

$$g(\boldsymbol{u}) = a(\boldsymbol{u}) + \sum_{n=0}^{N} b_n(\boldsymbol{u}) \cos[2\pi f_n \cdot \boldsymbol{u} + \phi_n(\boldsymbol{u})] \qquad (1.9)$$

where $\boldsymbol{u} = (x, y, z; t)$ and $\boldsymbol{f}_n = (f_{Xn}, f_{Yn}, f_{Zn}; f_{Tn})$. Similarly to the 2-D case, the spectra of the $(N + 1)$-fold multiplexed signals are given by

$$G(\boldsymbol{f}) = A(\boldsymbol{f}) + \sum_{n=0}^{N} C_n(\boldsymbol{f} - \boldsymbol{f}_n) + \sum_{n=0}^{N} C_n^*[-(\boldsymbol{f} + \boldsymbol{f}_n)], \qquad (1.10)$$

where $\boldsymbol{f} = (f_X, f_Y, f_Z; f_T)$, and $C_n(\boldsymbol{f})$ is the Fourier spectrum of the complex amplitude

$$c_n(\boldsymbol{u}) = \frac{1}{2} b_n(\boldsymbol{u}) \exp[i\phi_n(\boldsymbol{u})]. \qquad (1.11)$$

The spectra of the multiplexed signals are separated from each other in the frequency domain so that any desired spectrum can be selected by filtering (or frequency tuning to the desired channel in terms of communication theory). The maximum number of signals that can be accommodated by frequency multiplexing can be increased using a multidimensional carrier frequency vector that spans the Fourier space of a higher dimension [18]. This frequency multiplexing is a unique function that differentiates the FTM from other fringe analysis techniques, such as heterodyne interferometry [5] and the phase shift technique [3,4,6], which assume a single-frequency pure sinusoid. Further in the chapter, applications are introduced in which spatial frequency multiplexing plays a key role.

Although the generic FTM is based on band-pass filtering in the Fourier spectrum domain, a mathematically equivalent operation can be performed by convolution in the signal domain [19–21] (see the convolution theorem in Box 1.2). Whereas filtering in the Fourier domain is an operation on global data, convolution in the signal domain is an operation on local data. The operation in the signal domain may be suitable for applications that need locally adaptive convolution operations. Details on the principle and various variants of the FTM as well as corresponding signal domain fringe analysis techniques can be found in Malacara, Servín, and

Malacara [2] and Su and Chen [10]. As is common to other fringe analysis techniques, the FTM has strength and weakness. The strength is that the phase and the amplitude can be detected from a single fringe pattern, and that the single fringe pattern can accommodate multiple sensing signals simultaneously by spatial carrier frequency multiplexing. Because of this unique feature, the major application areas of the FTM include instantaneous measurements of dynamic or high-speed phenomena. The weakness is that, to guarantee the separation of spectra in the Fourier domain, one needs to introduce a spatial carrier frequency that is much higher than the amplitude and phase variations. In other words, one has to use a high-resolution image sensor with a large number of pixels to avoid aliasing errors caused by undersampling the spatial carrier fringes, and the image sensor has to be free from image distortion because the deformation of the fringe pattern caused by the distortion of the image sensor (which was unavoidable in classical vacuum image tubes) gives rise to artifact spatial phase modulation.

Whereas the phase shift technique was successfully demonstrated by Bruning et al. [6] using a diode array camera with only 32 x 32 pixels, the FTM demanded by far higher spatial resolution and lower distortion than the image sensor technology at the time could offer (e.g., a Vidicon vacuum tube camera had high distortion, and the maximum resolution of a Reticon solid-state image sensor was limited to 100 x 100 pixels). This requirement of high performance of the image sensor was the major drawback of the FTM and its variant techniques that make use of the spatial carrier frequency. However, now the situation has been changed greatly by the rapid advance of solid-state image sensor technology.

Today's advanced image sensor technology offers, with significantly reduced costs, a virtually distortion-free image sensor with more than 120 megapixels and a high-speed image sensor that can capture images at more than 10,000 frames/s. The combination of advanced image sensors and high-speed computers has opened up a new possibility of the FTM for multidimensional fringe analysis with a spatiotemporal carrier frequency vector. Interestingly, the advanced image sensor technology also brought about a drastic change in the implementation of the phase shift technique for interferometry. The time-sequential acquisition of multiple phase-shifted fringe data is now replaced by space-parallel data acquisition using a polarization mask that introduces a pixel-level phase shift to circularly polarized light [22]. This has enabled the high-speed measurement that was not possible by the traditional time-sequential phase shift technique.

Remember that the FTM and its variant spatial carrier techniques are also based on the pixel-level spatial phase shift, which is introduced by the tilt of the wavefront and increases linearly in the direction normal to the carrier fringe. Despite the difference in the mechanism of the phase shift and the algorithm of fringe analysis, it is common that the spatial parallelism in the fringe detection plays a key role in the high-speed measurements of fast dynamic phenomena. Many other issues have been addressed about the weakness of the FTM, and various solutions have been proposed, among which are fringe extrapolation techniques to reduce the influence of fringe discontinuities at the support boundaries [23,24] and windowed FFA [25] that compromises the pixel-wise processing of the phase shift technique and global processing of the FTM.

Through the continuous improvements and refinements made by many researchers, the FTM has now come to have many variants. The term *Fourier fringe analysis* (FFA) or Fourier fringe demodulation appears to include these variants and encompass a wider family of fringe analysis techniques than the generic FTM. A new insight into the relation between FFA and the phase shift techniques has been gained by the Fourier interpretations and Fourier representations of the phase shift techniques [26–28], which has filled the gap between these two approaches.

1.4 APPLICATIONS OF FOURIER FRINGE DEMODULATION

Applications of the FTM cover wide areas of science and technology, as summarized in Figure 1.3 [12].

1.4.1 Vibration Mode Measurement

The branch of industrial applications includes high-speed dynamic three-dimensional (3-D) profilometry and interferometry [29]. As an illustrative example, we first introduce an application of the FTM to 3-D shape measurement of the vibration mode of a micromechanical device with a stroboscopic interferometric microscope by Petitgrand et al. [30]. Shown in Figure 1.4a is an initial static interferogram of a $100\mu m \times 8\mu m \times 0.6\mu m$ Al cantilever microbeam. To avoid the errors caused by the fringe discontinuities at object boundaries, an artificial fringe pattern that has continuous connections to the real fringe pattern at the boundaries was created by the fringe extrapolation technique proposed by Roddier [24], as shown in Figure 1.4b. Figure 1.4c shows the log-scaled modulus of the 2-D Fourier spectrum of the interferogram given by Equation (1.7). The three spectra

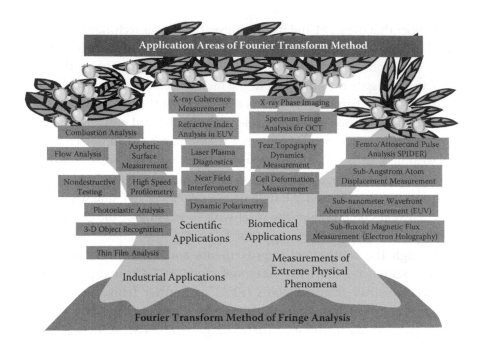

FIGURE 1.3   (See color insert.) Application areas of Fourier transform method. (After M. Takeda, *Proceedings of SPIE 8011* (2011): 80116S; *AIP Conference Proceedings 1236* (2010): 445–448 with permission from SPIE.)

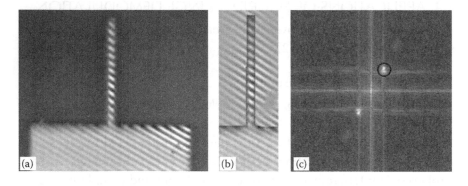

FIGURE 1.4   An interferogram recorded on a 100μm × 8μm × 0.6μm Al cantilever microbeam. (a) Initial static interferogram; (b) partial view after fringe extrapolation; and (c) Fourier spectrum separated by the spatial carrier frequency. The circled spectrum is filtered and then inverse Fourier transformed. (After Petitgrand, S., Yahiaoui, R., Danaie, K., Bosseboeuf, A., and Gilles, J. P., *Optics and Lasers in Engineering* **36** (2001): 77–101 with permission from Elsevier.)

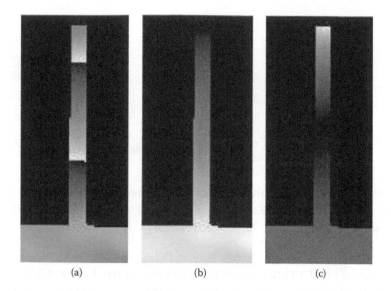

(a)   (b)   (c)

FIGURE 1.5 (a) Wrapped and (b) unwrapped static deflection phases; maximum static deflection is 0.63 μm. (c) Second flexural mode ($f = 296.1$ kHz) after static deflection subtraction; maximum vibration amplitude is 180 nm. (After Petitgrand, S., Yahiaoui, R., Danaie, K., Bosseboeuf, A., and Gilles, J. P., *Optics and Lasers in Engineering* **36** (2001): 77–101 with permission from Elsevier.)

are separated from each other by the spatial carrier frequency, which allows selective filtering of the desired spectrum indicated by the black circle.

Figures 1.5a and 1.5b show, respectively, the wrapped and unwrapped phase maps obtained for the static microbeam. Figure 1.5c shows the unwrapped phase map for the second vibration mode at a frequency of about 300 kHz, which is obtained after correcting the static displacement shown in Figure 1.5b.

1.4.2 Imaging Polarimetry

Another practical application of the FTM is found in a compact imaging polarimeter developed by Oka and Kaneko [31]. As shown in Figure 1.6, a beam from a He-Ne laser is spatially filtered and expanded by telescopic optics to pass through a polarizer. The expanded linearly polarized beam enters a sample plate with spatially varying birefringence. The light from the sample passes through an imaging lens that images the sample onto a charge-coupled device (CCD) image sensor. A block of polarimetric devices, consisting of four birefringent wedge prisms (PR$_1$, PR$_2$, PR$_3$, and PR$_4$ shown

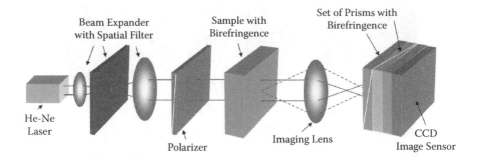

FIGURE 1.6 Schematic of the imaging polarimeter using birefringent wedge prisms. (After Oka, K., and Kaneko, K., *Optical Express* **11**(13) (2003): 1510–1519 with permission from OSA.)

in Figure 1.7) and an analyzer A, is placed just in front of the CCD image sensor. The fast axes of PR_1, PR_2, PR_3, and PR_4 are oriented at 0°, 90°, 45°, and −45°, respectively, relative to the x-axis, and the transmission axis of A is aligned with the x-axis. The planes of contact between PR_1 and PR_2 and between PR_3 and PR_4 are inclined with respect to the y- and x-axes with an angle α, whereas the other contact planes as well as the input face of PR_1 are parallel to the xy-plane. Figure 1.7 illustrates the configuration of the

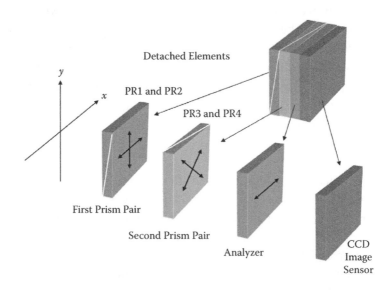

FIGURE 1.7 Configuration of the block of the polarimetric devices. (After Oka, K., and Kaneko, K., *Optical Express* **11**(13) (2003): 1510–1519 with permission from OSA.)

block of the polarimetric devices (shown detached). Mueller calculus of the optical system gives the detected intensity distribution as

$$g(x,y) = \frac{1}{2}S_0(x,y) + \frac{1}{2}S_1(x,y)\cos(2\pi f_0 x)$$

$$+ \frac{1}{4}|S_{23}(x,y)|\cos[2\pi f_0(x-y) + \phi_{23}(x,y)] \qquad (1.12)$$

$$- \frac{1}{4}|S_{23}(x,y)|\cos[2\pi f_0(x+y) - \phi_{23}(x,y)],$$

with

$$S_{23}(x,y) = |S_{23}(x,y)| \exp[i\phi_{23}(x,y)] = S_2(x,y) + iS_3(x,y), \qquad (1.13)$$

$$f_0 = \frac{2B}{\lambda}\tan\alpha, \qquad (1.14)$$

where $S_0(x,y), S_1(x,y), S_2(x,y)$, and $S_3(x,y)$ are Stokes parameters with spatial variations, and B and λ denote the birefringence of the prisms and the wavelength of the light, respectively.

Figure 1.8 shows an example of an intensity distribution detected for a birefringent object made of a 90° twisted nematic liquid crystal cell, part of which is covered with patterned transparent electrodes to apply electric fields across the cell so that a spatially modulated birefringence pattern of

FIGURE 1.8 Spatial frequency-multiplexed fringe pattern. (After Oka, K., and Kaneko, K., *Optical Express* **11**(13) (2003): 1510–1519 with permission from OSA.)

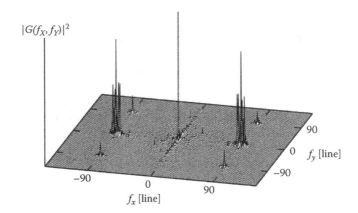

FIGURE 1.9 Power spectrum of the intensity pattern. Frequency-multiplexed signals are separated from each other by the different carrier frequencies in the spatial frequency domain. (After Oka, K., and Kaneko, K., *Optical Express* **11**(13) (2003): 1510–1519 with permission from OSA.)

the letter A is created. As expressed by Equation (1.12), a periodic structure with multiple spatial carrier frequencies is observed.

Shown in Figure 1.9 is the spatial frequency spectra obtained by applying the FTM to this intensity distribution. As intended, each term in Equation (1.12) is separated from others in the spectral domain by the spatial carrier frequency, which allows filtering the desired spectra and obtaining Stokes parameters.

Figures 1.10a and 1.10b show, respectively, the azimuth angle $\theta(x,y)$ and ellipticity angle $\varepsilon(x, y)$ obtained from the Stokes parameters. The advantage

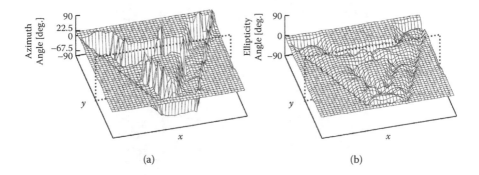

FIGURE 1.10 Spatial distribution of polarization states: (a) azimuth angle $\theta(x,y)$ and (b) ellipticity angle $\varepsilon(x,y)$. (After Oka, K., and Kaneko, K., *Optical Express* **11**(13) (2003): 1510–1519 with permission from OSA.)

of spatial frequency multiplexing of FTM is fully exploited in this system to enable imaging polarimetry of dynamic phenomena with a single-shot fringe recording without mechanical moving components.

1.4.3 Plasma Diagnosis

Important application areas of the FTM include characterization of laser plasma for interaction studies. Shown in Figure 1.11 is an experimental setup used by Gizzi et al. [32] and Borghesi et al. [33] for the production and characterization of long scale length expanding plasmas. The plasma was produced

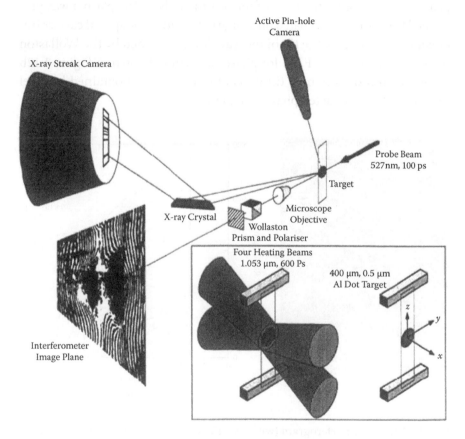

FIGURE 1.11 Experimental setup for the production and characterization of long scale length expanding plasmas. In the separate frame enclosed in the figure are the Al dot target, plastic substrate, and the target holder (right) and configuration of the four heating beams on target. (After Gizzi, L. A., Giulietti, D., Giulietti, A., Afshar-Rad, T., Biancalana, V., Chessa, P., Danson, C., Schifano, E., Viana, S. M., and Willi, O., *Physical Review E* **49**(6) (1994): 5628–5643 with permission from APS.)

by four 600-ps, 1.053-µm heating beams. The targets used were Al disks, alternatively coated onto 0.1-µm thick plastic foil support. The 1-ps chirped pulse amplification beam, frequency doubled to 0.53 µm, was used as an optical probe for interferometric measurements in a line of view parallel to the target plane. A modified Nomarski interferometer was employed to detect the phase changes undergone by the probe beam and to measure the electron density profiles. A confocal optical system imaged the plasma and recollimated the probe beam. The Wollaston prism was located close to this plane.

The interferogram shown in Figure 1.12a was obtained using a 1-ps probe pulse 2 ns after the peak of the heating pulses. The plasma was preformed from an Al disk coated on a large plastic foil. The spatial carrier frequency introduced by the tilt of the wavefront generated by the Wollaston prism allows use of the FTM for phase extraction. Shown in Figure 1.12b is the electron density map (in units of electrons/cm³) obtained by Abel inversion of the extracted phase distribution.

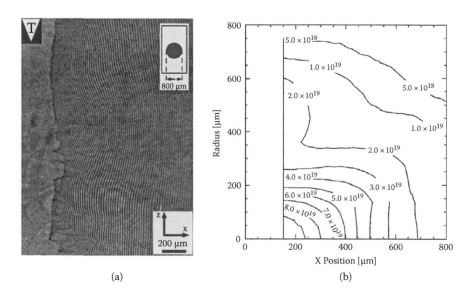

(a)   (b)

FIGURE 1.12   (a) Interferogram (with a spatial carrier frequency) taken 2 ns after the peak of the heating pulses using a 1-ps probe pulse. The target was an Al disk (800-µm diameter) on a large plastic foil. The heating irradiance was 7×10^{13} W/ cm² on each side of the target. (b) Density map obtained by Abel inversion of the phase map. The density is expressed in electrons/cm³. The radius is measured with respect to the symmetry axis. (After Borghesi, M., Giulietti, A., Giulietti, D., Gizzi, L. A., Macchi, A., and Willi, O., *Physical Review E* **54**(6) (1996): 6769–6773 with permission from APS.)

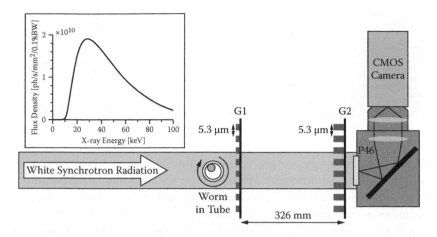

FIGURE 1.13 (See color insert.) Experimental setup (side view) of four-dimensional X-ray phase tomography with a Talbot interferometer consisting of a phase grating (G1) and an amplitude grating (G2) and white synchrotron radiation. The inset shows the calculated flux density at the sample position. (After Momose, A., Yashiro, W., Harasse, S, and Kuwabara, H., *Optics Express* **19**(9) (2011): 8424–8432 with permission from OSA.)

1.4.4 X-ray Phase Tomography

Next, as an example of successful applications of the FTM in the field of biomedical imaging, we introduce four-dimensional X-ray phase tomography demonstrated by Momose et al. [34]. A schematic of the experimental setup is shown in Figure 1.13. An X-ray beam from a white synchrotron radiation source passes through a three-dimensional test sample and enters a Talbot interferometer consisting of a phase grating (G1) and an amplitude grating (G2); the inset shows the flux density at the sample position. Both of the gratings have the same pitch of 5.3 μm, and the distance between the gratings was set to 326 mm, which is optimal for 28.8-keV X-rays and gives the highest contrast for the Talbot image. To apply the FTM, spatial carrier moiré fringes were introduced by inclining the gratings to each other around the optical axis. The sample was a living larva of *Nokona regalis* glued softly onto the inner wall of a tube of low-density polyethylene placed in front of the phase grating and was rotated for the computed tomographic (CT) scan. An X-ray image detector composed of a phosphor screen and a complementary metal-oxide semiconductor (CMOS) camera was operated at a frame rate of about 500 fps, and the exposure time per image was about 2 ms. The advantage of single-shot fringe pattern analysis of the FTM was exploited in this high-speed measurement.

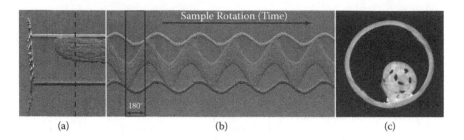

(a)   (b)   (c)

FIGURE 1.14 Procedure for the reconstruction of a four-dimensional phase tomogram movie: (a) differential phase image; (b) sinogram corresponding to the position indicated by the dashed line in (a); and (c) phase tomogram reconstructed from the portion indicated by the box in (b). A movie of the phase tomogram is generated by shifting the box along the sinogram. (After Momose, A., Yashiro, W., Harasse, S, and Kuwabara, H., *Optics Express* **19**(9) (2011): 8424–8432 with permission from OSA.)

Figure 1.14a shows a differential phase image of a worm in a tube obtained by FTM, and Figure 1.14b shows a sinogram on the position indicated by the dashed line in Figure 1.14a. A phase tomogram was reconstructed by extracting a 180° portion from the sinogram, as shown in Figure 1.14c. Momose et al. [34] succeeded in obtaining a volume movie of the dynamic sample that has never been acquired by any other techniques, which stimulates entomological interest.

1.4.5 Measurement of Ultrashort Optical Pulses

As the final example of the successful applications of the FTM, we introduce SPIDER (spectral phase interferometry for direct electric field reconstruction) developed by Iaconis and Walmsley [35] for the measurement of ultrashort optical pulses. SPIDER makes use of a spectral interferogram generated by interference of two replicas of the pulse to be measured. The two pulses are identical except that they are temporally delayed by τ in the time domain and spectrally shifted by Ω with respect to each other in the frequency spectrum domain. The spectrum interferogram is given by

$$S(\omega) = |\tilde{E}(\omega)|^2 + |\tilde{E}(\omega + \Omega)|^2 + 2|\tilde{E}(\omega)\tilde{E}(\omega + \Omega)| \cos[\phi(\omega + \Omega) - \phi(\omega) + \tau\omega],$$

$$(1.15)$$

where $\tilde{E}(\omega)$ is the Fourier spectrum of the pulse electric field, the frequency shift Ω serves as the spectral shear, and the temporal delay τ between the two replicas serves as a carrier frequency of the spectral fringe in spectrum domain. Applying the FTM to the spectrum interferogram $S(\omega)$ and assuming small spectral shear Ω, one can obtain the amplitude $|\tilde{E}(\omega)| \cong |\tilde{E}(\omega)\tilde{E}(\omega+\Omega)|^{1/2}$ and the derivative of the phase $d\phi/d\omega \cong [\phi(\omega+\Omega)-\phi(\omega)]/\Omega$ of the Fourier spectrum $\tilde{E}(\omega)$. From this information about the Fourier spectrum, the electric field $E(t)$ of the ultrashort pulse can be directly reconstructed.

Figure 1.15 shows an example of a specific implementation of SPIDER by Gallmann et al. [36]. Two pulse replicas with a delay of $\tau = 300$ fs were generated in a Michelson-type interferometer. The input pulse for the interferometer was the reflection from the surface of a 6.5-cm long

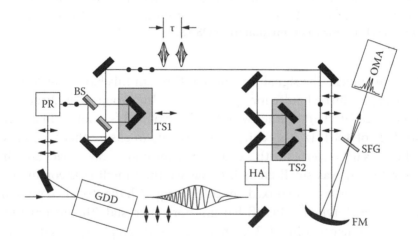

FIGURE 1.15 SPIDER setup. GDD, SF10 glass block; PR, periscope for polarization rotation; BS, beam splitters; TS1, translation stage for adjustment of delay τ; TS2, translation stage for adjustment of temporal overlap of the short pulse pair with the stretched pulse; HA, periscope for height adjustment; FM, focusing mirror (30-cm radius of curvature); SFG, upconversion crystal (30-μm thick type II b-barium borate [BBO]); OMA, optical multichannel analyzer. The filled shapes represent silver-coated mirrors, and the filled circles and arrows on the beam path display the polarization state of the beam. (After Gallmann, L., Sutter, D. H., Matuschek, N., Steinmeyer, G., Keller, U., Iaconis, C., and Walmsley, I. A., *Optics Letters* **24** (1999): 1314–1316 with permission from OSA.)

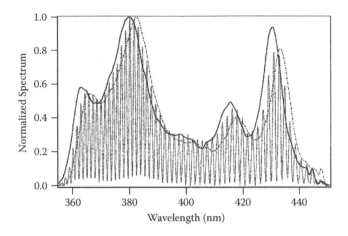

FIGURE 1.16   SPIDER spectral interferogram of a sub-6-fs pulse (dotted curve). In addition, the spectra of the individual upconverted pulses are shown (solid and dashed-dotted curves). (After Gallmann, L., Sutter, D. H., Matuschek, N., Steinmeyer, G., Keller, U., Iaconis, C., and Walmsley, I. A., *Optics Letters* **24** (1999): 1314–1316 with permission from OSA.)

SF10 glass block. The signal transmitted through this block was used as the strongly chirped pulse for upconversion. The group delay dispersion of the block (10^4 fs²) together with the delay of 300 fs resulted in a spectral shear of 4.8 THz. This chirp was adequate to ensure that each pulse replica was upconverted with a quasi-cw (continuous wave) field, and the spectral shear was small enough to satisfy the sampling theorem. The two delayed pulses were mixed noncollinearly with the stretched pulse in a 30-μm thick type II BBO (β-barium borate) crystal. The upconverted pulses were detected in a 0.3-m imaging spectrograph equipped with a 1,200-groove/mm grating and a 1,024 by 128 pixel ultraviolet-enhanced CCD array allowing for rapid acquisition.

Figure 1.16 shows the SPIDER interferogram of a pulse with a transform limit of 5.3 fs. For comparison, the individual spectra of the upconverted pulses are also displayed. Note that they are identical but shifted by the spectral shear Ω. The spectral phase reconstructed from the SPIDER trace is plotted in Figure 1.17, together with the independently measured pulse spectrum and the corresponding temporal intensity profile [36]. Information on recent advances in SPIDER with improved performance by multiple-shearing spectral interferometry can be found elsewhere [37].

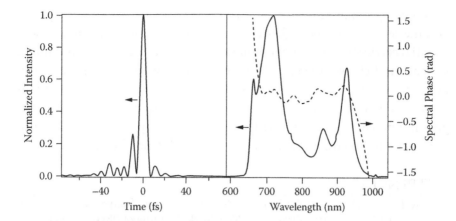

FIGURE 1.17   Reconstructed temporal intensity profile (left) and spectral phase (right; dashed curve). The independently measured power spectrum of the pulse (right; solid curve) has a transform limit of 5.3 fs. The solid curves are referenced to the left and the dashed curves to the right vertical axes. (After Gallmann, L., Sutter, D. H., Matuschek, N., Steinmeyer, G., Keller, U., Iaconis, C., and Walmsley, I. A., *Optics Letters* **24** (1999): 1314–1316 with permission from OSA.)

1.5 CONCLUSION

The principle, characteristics, and applications of Fourier phase demodulation have been reviewed with particular focus on the generic FTM of fringe analysis. The advantages of FTM are summarized as follows:

a. Phase and amplitude can be demodulated from a single fringe pattern (interferogram), which makes the FTM suitable for the measurement of high-speed phenomena.

b. Multiple signals with independent information can be encoded into and decoded from a single fringe signal by means of multidimensional carrier frequency multiplexing, which eases synchronization of measurement of independent physical phenomena.

The fundamental limitation arising from the principle of the FTM is that the spatiotemporal frequency bandwidth of the object signal is restricted to the range that guarantees that the signal spectrum is separated from other spectra in the frequency domain by the carrier frequency. In practice, however, this restriction on the signal bandwidth can be eased by adequately introducing a multidimensional carrier frequency so that the spectra become separated in the higher dimension. For example, one may choose a spatial carrier frequency for a signal that varies rapidly in time but slowly in space and choose a temporal carrier frequency for a signal that varies slowly in time but rapidly in space. Recent technological advances in high-resolution image sensors and dedicated high-speed processors such as the graphics processing unit (GPU) have paved the way to practical applications of multidimensional Fourier phase demodulation by the FTM.

REFERENCES

1. Robinson, D. W., and Reid, G. T., eds. *Interferogram Analysis: Digital Fringe Pattern Measurement Techniques*. New York: Institute of Physics, 1993.
2. Malacara, D., Servín, M., and Malacara, Z. *Interferogram Analysis for Optical Testing*. New York: Dekker, 1998.
3. Creath, K. Phase-measurement interferometry techniques. In *Progress in Optics*, vol. 26. Edited by E. Wolf. New York: Elsevier, 1988, 349–393.
4. Schwider, J. Advanced evaluation techniques in interferometry. In *Progress in Optics*, vol. 28. Edited by E. Wolf. New York: Elsevier, 1990, 271–359.
5. Massie, N. A., Nelson, R. D., and Holly, S. High-performance real-time heterodyne interferometry. *Applied Optics* **18** (1979): 1797–1803.
6. Bruning, J. H., Herriott, D. R., Gallagher, J. E., Rosenfeld, D. P., White, A. D., and Brangaccio, D. J. Digital wavefront measuring interferometer for testing optical surfaces and lenses. *Applied Optics* **13** (1974): 2693–2703.
7. Takeda, M., Ina, H., and Kobayashi, S. Fourier-transform method of fringe-pattern analysis for computer-based topography and interferometry. *Journal of the Optical Society of America* **72** (1982): 156–160.
8. Takeda, M., and Mutoh, K. Fourier transform profilometry for the automatic measurement of 3-D object shapes. *Applied Optics* **22** (1983): 3977–3982.
9. Takeda, M. Spatial-carrier fringe pattern analysis and its applications to precision interferometry and profilometry: An overview. *Industrial Metrology* **1** (1990): 79–99.
10. Su, X., and Chen, W. Fourier transform profilometry: a review. *Optics and Lasers in Engineering* **35** (2001): 263–284.
11. Takeda, M. Fourier fringe analysis and its application to metrology of extreme physical phenomena: A review [Invited]. *Applied Optics* **52**(1) (2013): 20–29.
12. Takeda, M. Measurement of extreme physical phenomena by Fourier fringe analysis: A review. *Proceedings of SPIE* **8011** (2011): 80116S; *AIP Conference Proceedings* **1236** (2010): 445–448.
13. Bracewell, R. N. *The Fourier Transform and Its Applications*. New York: Tata McGraw-Hill, 2003.
14. Judge, T.R., and Bryanston-Cross, P.J. A review of phase unwrapping techniques in fringe analysis. *Optics and Lasers in Engineering* **21** (1994): 199–239.
15. Takeda, M. Recent progress in phase unwrapping techniques. *Proceedings of SPIE* **2782** (1996): 334–343.

16. Su, X., and Chen, W. Reliability-guided phase unwrapping algorithm: a review. *Optics and Lasers in Engineering* **42** (2004): 245–261.
17. Bone, D. J., Bachor, H.-A., and Sandeman, J. Fringe-pattern analysis using a 2-D Fourier transform. *Applied Optics* **25** (1986): 1653–1660.
18. Takeda, M., and Kitoh, M. Spatiotemporal frequency multiplex hetero-dyne interferometry. *Journal of the Optical Society of America* **A9** (1992): 1607–1613.
19. Womack, K. H. Interferometric phase measurement using spatial synchro-nous detection. *Optical Engineering* **23** (1984): 391–395.
20. Mertz, L. Real-time fringe-pattern analysis. *Applied Optics* **22** (1983): 1535–1539.
21. Ichioka, Y., and Inuiya, M. Direct phase detecting system. *Applied Optics* **11** (1972): 1507–1514.
22. Novak, M., Millerd, J., Brock, N., North-Morris, M., Hayes, J., and Wyant, J. Analysis of a micropolarizer array-based simultaneous phase-shifting inter-ferometer. *Applied Optics* **44** (2005): 6861–6868.
23. Kujawinska, M., Spik, A., and Wojciak, J. Fringe pattern analysis using Fourier transform techniques. *Proceedings SPIE* **1121**, Interferometry '89, 130 (April 3, 1990); doi:10.1117/12.961260.
24. Roddier, C., and Roddier, F. Interferogram analysis using Fourier transform techniques. *Applied Optics* **26** (1987): 1668–1673.
25. Kemao, Q. Two-dimensional windowed Fourier transform for fringe pattern analysis: Principles, applications and implementation. *Optics and Lasers in Engineering* **45** (2007): 304–317.
26. Freischlad, K., and Koliopoulos, C. L. Fourier description of digital phase-measuring interferometry. *Journal of the Optical Society of America A* **7** (1990): 542–551.
27. Larkin, K. G., and Oreb, B. F. Design and assessment of symmetrical phase-shifting algorithms. *Journal of the Optical Society of America A* **9** (1992): 1740–1748.
28. Surrel, Y. Design of algorithms for phase measurements by the use of phase stepping. *Applied Optics* **35** (1996): 51–60.
29. Su, X., and Zhang, Q. Dynamic 3-D shape measurement method: A review. *Optics and Lasers in Engineering* **48** (2010): 191–204.
30. Petitgrand, S., Yahiaoui, R., Danaie, K., Bosseboeuf, A., and Gilles, J. P. 3D measurement of micromechanical devices vibration mode shapes with a stroboscopic interferometric microscope. *Optics and Lasers in Engineering* **36** (2001): 77–101.
31. Oka, K., and Kaneko, K. Compact complete imaging polarimeter using bire-fringent wedge prisms. *Optical Express* **11**(13) (2003): 1510–1519.
32. Gizzi, L. A., Giulietti, D., Giulietti, A., Afshar-Rad, T., Biancalana, V., Chessa, P., Danson, C., Schifano, E., Viana, S. M., and Willi, O. Characterization of laser plasma for interaction studies. *Physical Review E* **49**(6) (1994): 5628–5643.
33. Borghesi, M., Giulietti, A., Giulietti, D., Gizzi, L. A, Macchi, A., and Willi, O. Characterization of laser plasmas for interaction studies: Progress in time-resolved density mapping. *Physical Review E* **54**(6) (1996): 6769–6773.

34. Momose, A., Yashiro, W., Harasse, S, and Kuwabara, H. Four-dimensional X-ray phase tomography with Talbot interferometry and white synchrotron radiation: dynamic observation of a living worm. Optics Express **19**(9) (2011): 8424–8432.
35. Iaconis, C., and Walmsley, I. A. Spectral phase interferometry for direct electric-field reconstruction of ultrashort optical pulses. *Optics Letters* **23** (1998): 792–794.
36. Gallmann, L., Sutter, D. H., Matuschek, N., Steinmeyer, G., Keller, U., Iaconis, C., and Walmsley, I. A. Characterization of sub-6-fs optical pulses with spectral phase interferometry for direct electric-field reconstruction. *Optics Letters* **24** (1999): 1314–1316.
37. Austin, D. R., Witting, T., and Walmsley, I. A. High precision self-referenced phase retrieval of complex pulses with multiple-shearing spectral interferometry. *Journal of the Optical Society of America B* **26** (2009): 1818–1830.

CHAPTER 2

Windowed Fourier Transforms

Jingang Zhong

Department of Optoelectronic Engineering
Jinan University
Guangzhou, China

Jiawen Weng

College of Science
South China Agricultural University
Guangzhou, China

2.1 INTRODUCTION

Phase demodulation techniques, as automatic analysis techniques for fringe signals in optical metrology, have been widely used for applications such as three-dimensional (3-D) shape measurement [1,2], vibration analysis [3], and strain analysis [4]. Fourier transform (FT) methods [2] and phase-shifting methods [5,6] are typical techniques for phase demodulation. In practice, the fringe signals may be nonsinusoidal signals because of image distortion, a nonlinear response of the detector, or discrete sampling. Phase demodulation of a nonsinusoidal fringe signal using the phase-shifting method would provide results with large error [7]. Furthermore, it is not fit for dynamic analysis as it requires multiple fringe patterns for a phase-shifting calculation. Compared with the phase-shifting method, the FT method [2,8,9] proposed by Takeda et al. in 1982 is a fringe signal analysis technique that enables phase demodulation from a single fringe pattern and makes dynamic analysis possible. Because of its prominent ability to analyze dynamic events, the FT method has been applied to measure extreme physical phenomena [10]. For phase demodulation, the fundamental spectral component should be extracted accurately from the frequency domain by filtering in the FT method, so it must avoid being superposed by other spectral components. However, modulated fringe signals are best described as nonstationary signals with local instantaneous frequency changing in the spatial domain, and FT is a global operation with poor spatial localization. Therefore, when employing FT for analysis, the fundamental spectral component within an area in the spatial domain may be superposed by higher-order spectral components or the zero component within other areas. In this case, the FT method will provide results with large error. In addition, when the signal is polluted by the noise within an area, the spectrum of the noise would have an effect on the whole spectrum of the signal, and errors would be introduced into the analysis of this whole signal.

Various spatial frequency analysis methods, such as the wavelet transform (WT) method [11–15], that excel at spatial localization, have been developed for the analysis of these fringe signals in recent years. Using the WT method, the phase of a fringe signal can be obtained directly from the WT coefficients at the ridge. Because of the spatial localization ability, this WT method can limit the large errors locally in the neighborhood of the error points. However, for the WT method, local linear approximation of the phase is necessary [12,15]; otherwise, the accuracy of phase demodulation can be low.

Windowed Fourier transform (WFT) is another excellent spatial frequency analysis method for nonstationary signal analysis. To overcome the deficiency of FT with poor spatial localization, Dennis Gabor adapted the FT to analyze only a small section of the signal at a time and proposed a technique called WFT or Gabor transform in 1946 [16]. The main idea of this technique involves introducing a window to the FT in the spatial domain to achieve spatial localization analysis. In recent years, WFT has been applied to the analysis of fringe signals. Jiawen Weng and Jingang Zhong have presented an approach for the analysis of fringe patterns using the WFT for 3-D profilometry [17]. Based on the WFT, windowed Fourier ridges (WFR) and windowed Fourier filtering (WFF) algorithms have been proposed by Qian Kemao et al. for fringe pattern analysis [18–22]. Hlubina et al. [23] proposed processing spectral interference signals using a method based on WFT applied in the wavelength domain. A two-dimensional WFT has been developed for fringe pattern analysis [24,25]. For the WFT method, an essential factor is the determination of an appropriate window size for the localization analysis. Qian Kemao [26] proposed that the window size should be determined according to the balance between the linear phase approximation error and the noise level. However, the WFT method with an invariable window size is not suitable for analyzing the fringe signal with widely spectral parameters because of the invariable resolution in the spatial and frequency domains.

To balance the spatial and frequency resolution, adaptive control of the size of the window should be introduced to the WFT method. Jingang Zhong and Jiawen Weng presented a dilating Gabor transform for fringe analysis [27], and Suzhen Zheng et al. presented a method by which the window size is determined by the instantaneous frequency extracted from the wavelet ridge [28]. Furthermore, Jingang Zhong et al. have presented an adaptive windowed Fourier transform (AWFT) for the phase demodulation of the fringe pattern [29,30]. For the AWFT method, the adaptive window size is determined by the length of the local stationary signal, which is defined as the local signal whose fundamental spectral component is separated from the other spectral components within the local signal area. In fact, the FT method can be considered a special case of the AWFT method with a maximum spatial window. The WFT method and the AWFT method are discussed in detail in this chapter. Furthermore, a comparison of phase demodulation with the FT, WFT, WT, and AWFT methods in numerical simulations and experiments is presented.

2.2 PHASE DEMODULATION BASED ON FOURIER TRANSFORM

The FT phase demodulation method, proposed by Mitsuo Takeda et al. [2], has been extensively studied since the 1980s and has found wide implementation in many fields.

Assume a one-dimensional (1-D) nonsinusoidal fringe signal $I(x)$ for analysis, described as follows:

$$I(x) = r(x) \sum_{n=-\infty}^{+\infty} A_n \cos[2\pi n f_0 x + n\varphi(x)] \tag{2.1}$$

where $r(x)$ is a slowly varying parameter, A_n are the weighting factors of the Fourier series, f_0 is the fundamental frequency of the signal, and $\varphi(x)$ is the phase modulation. A complex expression is employed to make the calculation more convenient, and it does not affect the physical analysis of signals:

$$I(x) = r(x) \sum_{n=-\infty}^{+\infty} A_n \exp\{j[2\pi n f_0 x + n\varphi(x)]\} \qquad (2.2)$$

where j is the imaginary unit. Applying 1-D FT for analysis, the corresponding Fourier spectrum $F(f)$ of the modulated fringe signal $I(x)$ (sinusoidal or nonsinusoidal) can be expressed as follows:

$$F(f) = \int_{-\infty}^{\infty} I(x) \exp(-j2\pi fx) dx \qquad (2.3)$$

With a suitable filter function, the fundamental spectral component $F(f_0)$ can be obtained. Applying the inverse FT to $F(f_0)$, a complex signal can be obtained as follows:

$$I_1(x) = \int_{-\infty}^{\infty} F(f_0) \exp(j2\pi fx) df \qquad (2.4)$$

The modulated phase $\varphi(x)$ can be obtained by

$$\varphi(x) = \arctan \frac{\text{Im}[I_1(x)]}{\text{Re}[I_1(x)]} - 2\pi f_0 x \qquad (2.5)$$

where $\text{Im}[I_1(x)]$ and $\text{Re}[I_1(x)]$ represent the imaginary and real part of $I_1(x)$, respectively.

For the FT method, accurate extraction of the fundamental spectral component is the key process for phase demodulation. To describe the deficiency of FT, analysis of a sinusoidal fringe signal and a nonsinusoidal fringe signal is employed for explication and comparison. Figures 2.1a and 2.1b show the intensity of 1-D modulated sinusoidal and nonsinusoidal fringe signals with 512 pixels. Figures 2.2a and 2.2b show the power spectra of the two fringe signals with the zero peak removed. The normalized

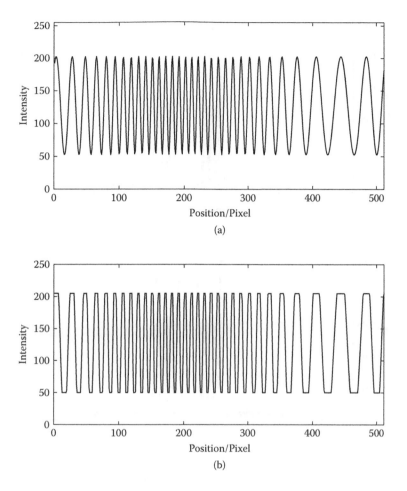

FIGURE 2.1 The 1-D modulated (a) sinusoidal and (b) nonsinusoidal fringe signals with 512 pixels.

frequency of a signal is its frequency expressed in units of cycles per sample. Therefore, for the numerical simulations, the units of the normalized frequency are cycles/pixel. From Figure 2.2a, we can see that the modulated sinusoidal signal has only one broad spectral component (i.e., the fundamental spectral component) in the frequency domain. Therefore, it would not be disturbed by other higher-order spectral components, and it can be extracted correctly with a wide enough filter. However, for the modulated nonsinusoidal fringe signal, as shown in Figure 2.2b, there are some other spectral components.

To better understand this point, local signals are extracted by Gaussian window for spatial localization analysis. Definition and properties of the

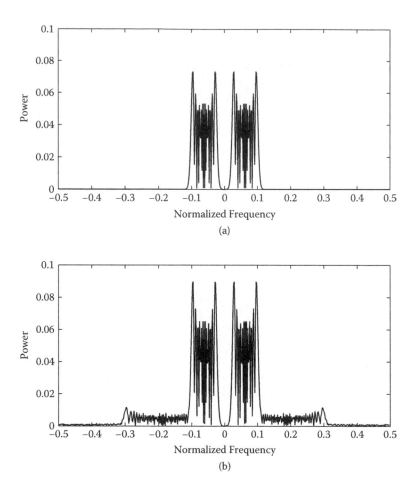

FIGURE 2.2 The power spectra of (a) sinusoidal and (b) nonsinusoidal fringe signals with the zero peak removed.

Gaussian function are given in Box 2.1. Figure 2.3 shows the Gaussian window (a = 30) centered at the 128th pixel (b = 128) and 384th pixel (b = 384). The power spectra of the local sinusoidal and nonsinusoidal fringe signals selected by the corresponding Gaussian window are carried out for comparative analysis, as shown in Figures 2.4a and 2.4b. For the nonsinusoidal fringe signal, the fundamental spectral component of the local signal centered at the 128th pixel is superposed by the higher-order spectral components of the local signal centered at the 384th pixel. However, this does not happen in the case of the sinusoidal fringe signal. Therefore, using the FT method, it is impossible to achieve correct filtering

BOX 2.1   THE GAUSSIAN FUNCTION

A one-dimensional (1-D) basic Gaussian function is defined as

$$gaus(x) = \frac{1}{\sqrt{2\pi}} \exp\left[-\frac{x^2}{2}\right].$$

By introducing a scale factor $a\,(>0)$ corresponding to the size of the Gaussian window and a shift factor b corresponding to the central position, the Gaussian function can be rewritten as $gaus_a^b(x) = |a|^{-1} gaus(\frac{x-b}{a})$. According to the normalized feature of the Gaussian function, the relationship $\int_{-\infty}^{\infty} gaus_a^b(x)dx = \int_{-\infty}^{\infty} gaus_a^b(x)db = 1$ can be obtained. The width of the Gaussian window $L_{window}(x)$ is defined as the full width at $1/e^2$ maximum. By setting $\exp[-\frac{x^2}{2a^2}] = 1/e^2$, we can obtain $x = \pm 2a$, so the width of the Gaussian window is $4a$.

in the frequency domain, and phase demodulation of such a nonsinusoidal fringe signal will provide results with errors.

From this discussion, for accurate extraction, the fundamental spectral component must be separated from all the other spectral components. The condition [2] that should be satisfied is

$$\begin{cases} f_0 + \dfrac{1}{2\pi}\varphi'(x)_{max} < nf_0 + \dfrac{n}{2\pi}\varphi'(x)_{min} & (n = 2, 3, \cdots) \\[2mm] f_b < f_0 + \dfrac{1}{2\pi}\varphi'(x)_{min} \end{cases} \tag{2.6}$$

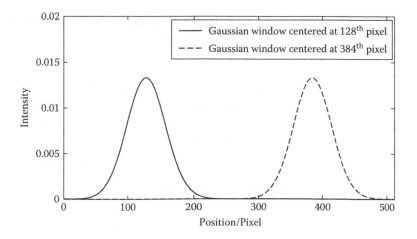

FIGURE 2.3   The Gaussian window centered at the 128th and 384th pixels.

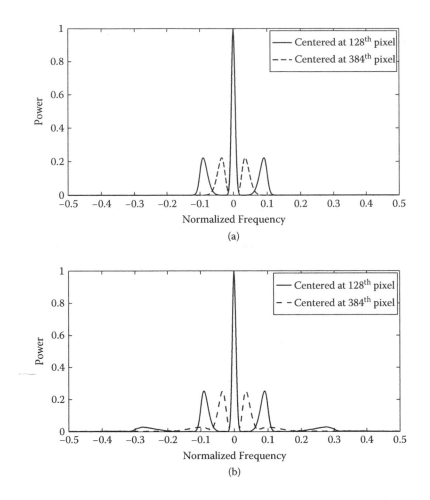

FIGURE 2.4 The power spectra of the local (a) sinusoidal and (b) nonsinusoidal fringe signals centered at the 128th and 384th pixels.

where f_0 is the fundamental frequency; f_b is the cutoff frequency of the background term; and $\varphi'(x)_{max}$ and $\varphi'(x)_{min}$ correspond to the maximum and minimum value of $\frac{d\varphi(x)}{dx}$, respectively. Figure 2.5 shows the positions of f_b, $(f_n)_{min} = nf_0 + \frac{n}{2\pi}\varphi'(x)_{min}$, and $(f_n)_{max} = nf_0 + \frac{n}{2\pi}\varphi'(x)_{max}$ for $n = 1, 2,$ and 3.

For the FT, f_b is much smaller than $f_0/2$ in most cases [2]. So, the condition expressed by Equation (2.6) can be rewritten as

$$f_0 + \frac{1}{2\pi}\varphi'(x)_{max} < 2f_0 + \frac{2}{2\pi}\varphi'(x)_{min} \quad (n = 2) \tag{2.7}$$

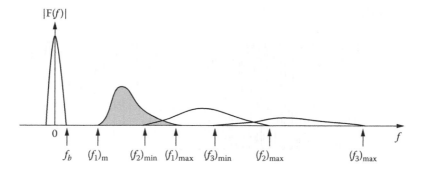

FIGURE 2.5 The distribution of the spectral components.

However, the phase $\varphi(x)$ is unknown before performing phase demodulation. Therefore, it cannot be predetermined whether the FT method would be valid. To determine the restrictive conditions for the FT method, the instantaneous frequency of the local signal is introduced, expressed as

$$f_{inst}(x) = f_0 + \frac{1}{2\pi}\varphi'(x) \tag{2.8}$$

The maximum and minimum of the instantaneous frequency of the signal are as follows:

$$\begin{cases} (f_{inst})_{max} = f_0 + \dfrac{1}{2\pi}\varphi'(x)_{max} \\[2mm] (f_{inst})_{min} = f_0 + \dfrac{1}{2\pi}\varphi'(x)_{min} \end{cases} \tag{2.9}$$

By employing the instantaneous frequency, Equation (2.7) can be rewritten as

$$(f_{inst})_{max} < 2(f_{inst})_{min} \tag{2.10}$$

This means that when $(f_{inst})_{max}$ and $(f_{inst})_{min}$ of the signal satisfy the condition expressed by Equation (2.10), the fundamental spectral component would not be disturbed by the higher-order spectral components and the FT method can achieve the phase demodulation correctly. Therefore, when performing the analysis of the signal, determination of the validity

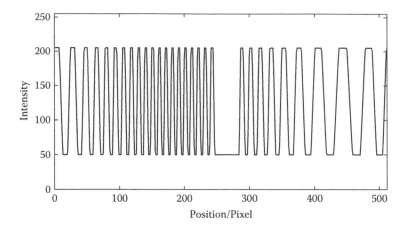

FIGURE 2.6   The 1-D modulated nonsinusoidal fringe signal with local signal loss.

of the FT method for the phase demodulation should be performed first; that is, the instantaneous frequency of the whole signal should be preestimated whether the condition for separating the fundamental spectral component is satisfied or not.

We also consider the effect of the local signal loss seen as noise, such as the projected shadow in projection profilometry. Figure 2.6 shows the intensity of a 1-D modulated nonsinusoidal fringe signal with 512 pixels for which the local signal within [250, 280] pixels is lost by an occlusion. Figure 2.7 shows the power spectra, with the zero peak removed, of the

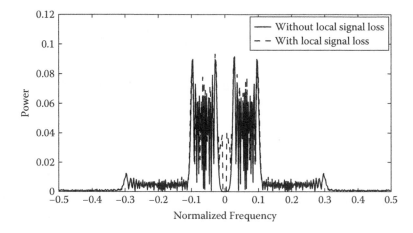

FIGURE 2.7   The power spectra of the nonsinusoidal fringe signals with the zero peak removed.

fringe signal according to Figure 2.1b, as well as that of the fringe signal according to Figure 2.6. Comparing the distribution of the spectra, it is evident that the noise has an effect on the whole spectrum of the signal, and it is difficult to filter out such a noise spectrum in the frequency domain.

2.3 PHASE DEMODULATION BASED ON WINDOWED FOURIER TRANSFORM

WFT is an important method for nonstationary signal analysis. To overcome the deficiency of FT with poor localization in the spatial frequency domains, Dennis Gabor adapted the FT in 1946 to analyze only a local section of the signal at a time and proposed a technique called WFT or Gabor transform [16]. The main idea of this technique involves introducing a window in the FT calculation of the spatial domain to achieve a more local analysis.

2.3.1 Principle of Windowed Fourier Transform for Phase Demodulation

By introducing the WFT for analysis, the spectra of different local fringe signals can be separated to some extent. The WFT of the modulated signal $I(x)$ is defined as follows:

$$WFT_a^b(f) = \int_{-\infty}^{\infty} [I(x)\exp(-j2\pi fx)]gaus_a^b(x)dx \qquad (2.11)$$

where $gaus_a^b(x)$ is the Gaussian function given in Box 2.1. $WFT_a^b(f)$ can be considered as the FT of the local signal selected by a Gaussian window in the spatial domain. Using a normalized Gaussian function, the relationship between $WFT_a^b(f)$ and $F(f)$ can be obtained as

$$\int_{-\infty}^{\infty} WFT_a^b(f)db = \int_{-\infty}^{\infty}\int_{-\infty}^{\infty} [I(x)\exp(-j2\pi fx)]gaus_a^b(x)dx\,db$$

$$= \int_{-\infty}^{\infty}\left\{[I(x)\exp(-j2\pi fx)]\int_{-\infty}^{\infty} gaus_a^b(x)db\right\}dx \quad (2.12)$$

$$= \int_{-\infty}^{\infty} [I(x)\exp(-j2\pi fx)]dx$$

$$= F(f)$$

Equation (2.12) indicates that the sum of the spectra obtained by the WFT for all the Gaussian windows is equal to the spectrum obtained by conventional FT. By shifting the central position of the Gaussian window point by point and computing the WFT of the signal, the fundamental spectral component of each local signal can be obtained. Then, by summing all these fundamental spectral components, a global fundamental spectral component can be reconstructed as follows:

$$\int_{-\infty}^{\infty} WFT_a^b(f_0)\,db = F(f_0) \tag{2.13}$$

Finally, applying the inverse FT to the fundamental spectral component, the phase $\varphi(x)$ can be obtained by Equations (2.4) and (2.5).

The WFT was calculated with $a = 30$ for the analysis of the fringe signal shown in Figure 2.1b. Figures 2.8a to 2.8c show the power spectra of different local fringe signals centered at the 50th, 200th, and 400th pixels, respectively. The fundamental spectral components of different local fringe signals can be extracted without superposition by the higher-order spectral components by performing WFT. Therefore, to some extent, by introducing the Gaussian window in the spatial domain for analysis, WFT can achieve local analysis.

2.3.2 Deficiency of Windowed Fourier Transform with an Invariable Window Size

To extract the fundamental spectral component in the frequency domain, high-frequency resolution is needed. On the other hand, to separate the spectra of different local signals, high spatial resolution is needed. However, according to the Heisenberg uncertainty theory [31], the spatial resolution δ_x and the frequency resolution δ_f satisfy the following equation:

$$\delta_x \delta_f \geq 1/4\pi \tag{2.14}$$

The minimal value of $\delta_x \delta_f$ is called the Heisenberg box. For WFT with a Gaussian window $gaus_a^b(x)$, the δ_x is approximately the width of the Gaussian window. That is, $\delta_x \approx 4a$. The FT of the Gaussian function is also a Gaussian function. The δ_f is approximately the width of the spectral

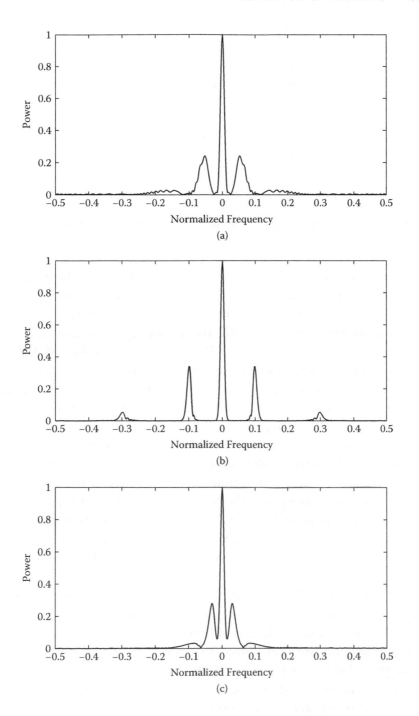

FIGURE 2.8 The power spectra of different local fringe signals centered at the (a) 50th pixel, (b) 200th pixel, and (c) 400th pixel.

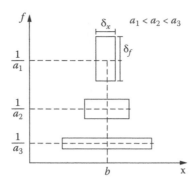

FIGURE 2.9   The spatial frequency windows of a Gaussian function.

Gaussian function *gaus(f)*. That is, $\delta_f \approx 2/(\pi a)$. Figure 2.9 shows that an enlargement of the spatial window width of the Gaussian window causes a reduction of the frequency window width and vice versa; that is, the spatial resolution increases while the frequency resolution decreases because of the constant value of the Heisenberg box. In other words, both spatial and frequency resolution cannot be maximized at the same time. To obtain higher-frequency resolution, a wider window with low spatial resolution should be employed and vice versa. For conventional FT, a unit rectangular function is employed as a window function, where δ_x is equal to the whole size of the analysis signal, resulting in the highest-frequency resolution but the lowest spatial resolution. To some extent, WFT achieves higher spatial resolution by introducing the Gaussian window in the spatial domain for analysis at the cost of the reduction of the frequency resolution. When the scale factor *a* of the Gaussian function $gaus_a^b(x)$ is large, the Gaussian window will be wide and high-frequency resolution can be obtained, while the spatial resolution decreases. For a nonstationary fringe signal, we should apply different size Gaussian windows for the analysis of different local signals to satisfy the different requirements of frequency resolution and spatial resolution. However, the window size of WFT is invariable, so there is only constant-frequency resolution and spatial resolution while analyzing. Therefore, the WFT method with an invariable window size has certain defects in the analysis of nonstationary fringe signals. To balance the spatial and frequency resolution, adaptive control of the size of the window should be introduced to the WFT method. Determination of the appropriate window size for optimal localization analysis is a key factor.

2.4 PHASE DEMODULATION BASED ON ADAPTIVE WINDOWED FOURIER TRANSFORM

The trade-off between spatial and frequency resolution can be partially remedied by optimizing the size of the analysis window with respect to the signal characteristics [30]. A wide spatial window should be employed for a local signal whose instantaneous frequency varies smoothly because a wide spatial window with high-frequency resolution contributes to a more concentrative spatial frequency representation [32]. A narrow spatial window should be employed for a local signal whose instantaneous frequency varies sharply because a narrow spatial window with high spatial resolution is better able to distinguish frequencies at different spatial positions.

In this section, the AWFT method [30] is presented for the phase demodulation of the fringe signal. The adaptive window size is determined by the length of the local stationary signal, which is defined as the local signal whose fundamental spectral component is separated from all the other spectral components within the corresponding spatial area. The instantaneous frequency of the local signal is introduced to determine whether the condition for separating the fundamental spectral component is satisfied.

2.4.1 Principle of Adaptive Windowed Fourier Transform

The AWFT can be considered as an extension of the WFT with the scale factor a of the Gaussian function variable. The AWFT of the modulated fringe signal $I(x)$ is

$$AWFT_a^b(f) = \int_{-\infty}^{\infty} [I(x)\exp(-j2\pi fx)]gaus_a^b(x)dx \qquad (2.15)$$

According to Equations (2.12) and (2.15), the following relationship can be obtained:

$$\int_{-\infty}^{\infty} AWFT_a^b(f)db = F(f) \qquad (2.16)$$

This means that the sum of the spectra obtained by the AWFT for all the Gaussian windows is equal to the spectrum obtained by the FT. By summing all the fundamental spectral components of each local stationary signal, the global fundamental spectral component can be reconstructed as Equation (2.13). Applying the inverse FT to the fundamental spectral component, the phase $\varphi(x)$ can be obtained by Equations (2.4) and (2.5).

2.4.2 Principle of the Determination of the Scale Factor for AWFT

The adaptive window size is determined by the length of the local stationary signal, which is defined as the local signal whose fundamental spectral component is separated from all the other spectral components within the corresponding spatial area. The instantaneous frequency of the local signal is introduced to determine whether the condition for separating the fundamental spectral component is satisfied. Generally, for the calculation, the instantaneous frequency of the local signal is detected by the ridge of the WT [33]. For a local signal of length $2\Delta x$ around the position x, that is, within $[x - \Delta x, x + \Delta x]$, when the maximum and minimum of the instantaneous frequency of the local signal satisfy the condition expressed by Equation (2.10), the fundamental spectral component of such a local signal will not be disturbed by higher-order spectral components. This means that the fundamental spectral component can be extracted accurately. Therefore, the corresponding local signal is considered to be a local stationary signal. Here, the maximal length $2(\Delta x)_{max}$ is defined as the length of the local stationary signal $L_{signal}(x)$:

$$L_{signal}(x) = 2(\Delta x)_{max} \tag{2.17}$$

The appropriate size of the window is determined according to the condition that the width of the window $L_{window}(x) = 4a$ is equal to the length of the local stationary signal. Therefore, the scale factor a of an appropriate Gaussian window for the local analysis can be determined by

$$a = \frac{L_{signal}(x)}{4} \tag{2.18}$$

For the second condition expressed by Equation (2.6), the limitation can be set as

$$f_b < (f_{inst})_{min} \tag{2.19}$$

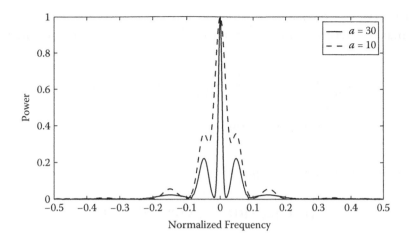

FIGURE 2.10 The power spectra of the local fringe signal centered at the 350th pixel with $a = 30$ and $a = 10$.

For the FT, f_b is much smaller than $f_0/2$ in most cases. However, for the WFT and the AWFT, because of the introduction of the window in the spatial domain, f_b spreads out. The narrower the window, the higher the f_b will be. Figure 2.10 shows the power spectra of the local fringe signals, according to Figure 2.1b, centered at the 350th pixel selected by the Gaussian window with $a = 30$ and $a = 10$. The fundamental spectral component is superposed by the zero component when employing $a = 10$. Therefore, for the AWFT method, the effect of f_b should be reconsidered according to the different size windows employed.

Computing the AWFT of the nonsinusoidal fringe signal expressed by Equation (2.2),

$$AWFT_a^b(f) = \frac{1}{\sqrt{2\pi}a} \int_{-\infty}^{\infty} r(x)A_0 \exp\left[-\frac{(x-b)^2}{2a^2}\right] \exp(-j2\pi fx)\,dx$$

$$+ \frac{1}{\sqrt{2\pi}a} \int_{-\infty}^{\infty} r(x) \left\{ \sum_{\substack{n=-\infty \\ n \neq 0}}^{+\infty} A_n \exp[j2\pi n f_0 x + jn\varphi(x)] \right\} \qquad (2.20)$$

$$\exp\left[-\frac{(x-b)^2}{2a^2}\right] \exp(-j2\pi fx)\,dx$$

where $r(x)$ is a slowly varying function, and the zero component ($n = 0$) is primarily determined by the first term. The first term can be calculated as

$$\frac{1}{\sqrt{2\pi a}} \int_{-\infty}^{\infty} r(x) A_0 \exp\left[-\frac{(x-b)^2}{2a^2}\right] \exp(-j2\pi fx) \, dx$$

$$= r(x) A_0 \exp(-j2\pi fb) \exp(-2\pi^2 a^2 f^2) \tag{2.21}$$

Therefore, the power spectrum is

$$|F(f)|_{n=0}^2 = [r(x) A_0]^2 \exp(-4\pi^2 a^2 f^2) \tag{2.22}$$

We define f_b as the frequency at which the power is $1/e^2$ of the maximum as

$$\exp(-4\pi^2 a^2 f_b^2) = \exp(-2) \tag{2.23}$$

We can obtain $f_b = \frac{1}{\sqrt{2\pi a}}$. Therefore, the limitation expressed by Equation (2.19) can be rewritten as

$$f_b = \frac{1}{\sqrt{2\pi a}} < (f_{inst})_{min} \tag{2.24}$$

It can be obtained that

$$a > \frac{1}{\sqrt{2\pi}(f_{inst})_{min}} \tag{2.25}$$

According to the conditions expressed by Equations (2.18) and (2.25), the scale factor of the Gaussian window can be determined as

$$\begin{cases} a = \dfrac{L_{signal}(x)}{4}, & \text{if } L_{signal}(x) > \dfrac{2\sqrt{2}}{\pi(f_{inst})_{min}} \\[3mm] a = \dfrac{1}{\sqrt{2\pi}(f_{inst})_{min}} & \text{if } L_{signal}(x) \le \dfrac{2\sqrt{2}}{\pi(f_{inst})_{min}} \end{cases} \tag{2.26}$$

Note that the $(f_{inst})_{min}$ of Equation (2.26) is the minimum of the instantaneous frequency for the local stationary signal.

2.5 NUMERICAL ANALYSIS

A numerical simulation is presented to illustrate the AWFT method and to compare it with the FT method and the WFT method with different scale factors. First, the original and modulated sinusoidal fringe signals with 512 pixels in the horizontal direction are given by the following expressions:

$$I_0(x) = 255 \times \{1 + \cos[2\pi n f_0 x]\}/2 \tag{2.27}$$

$$I(x) = 255 \times \{1 + \cos[2\pi n f_0 x + \varphi(x)]\}/2 \tag{2.28}$$

where $f_0 = 1/16(1/pixel)$, and $x (= 0, 1, 2, ..., 511)$. The modulated phase $\varphi(x)$ at each position is given by the following equation:

$$\varphi(x) = 7[1 - \xi]^2 \exp[-\xi^2 - 1] - 35\left[\frac{1}{5}\xi - \xi^3\right]\exp[-\xi^2] - \frac{7}{6}\exp\{-[\xi + 1]^2\}$$

$$\tag{2.29}$$

where $\xi = \frac{6}{512}x - 3$. Nonsinusoidal signals are considered for the cases of image distortion, nonlinear detector response, and discrete sampling. According to the signals given by Equations (2.27) and (2.28), the intensity values of the original and modulated nonsinusoidal fringe signals are set as 205 when greater than 205 and set as 50 when less than 50. Figures 2.11a and 2.11b show the original and the modulated 1-D nonsinusoidal fringe patterns, respectively.

2.5.1 Numerical Analysis by FT

First, the FT method is employed for the numerical analysis. Figure 2.12 shows the normalized power spectra of the original and modulated fringe signals; the dashed-dot line and dot line represent a narrow and a large rectangular filter employed for filtering in the frequency domain, respectively. Figure 2.13 shows the unwrapped demodulated phase. Figure 2.14 shows the phase errors. Figure 2.15 shows the distribution of the instantaneous frequency. In Figure 2.14, we can see that higher error

(a)

(b)

FIGURE 2.11   (a) The original and (b) modulated fringe patterns.

peaks appear when employing a narrow filter. When employing a larger filter, greater errors appear within local fringe signals with lower frequency than appear with signals of higher frequency. According to Equation (2.8), the instantaneous frequency of the whole signal is calculated for preestimating whether the condition for separating the fundamental spectral component is satisfied. We see that $(f_{inst})_{max}$ is about 0.11 and $(f_{inst})_{min}$ is about 0.036 for the whole fringe signal, which do not satisfy the condition $(f_{inst})_{max} < 2(f_{inst})_{min}$ expressed by Equation (2.10). The fundamental spectral component of such a signal would be disturbed by the higher-order spectral components, and it is impossible to find a filter that is applicable

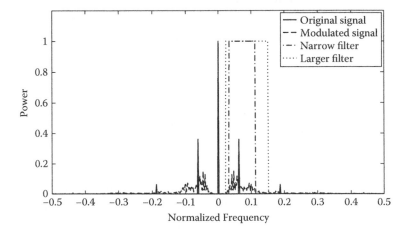

FIGURE 2.12   The power spectra.

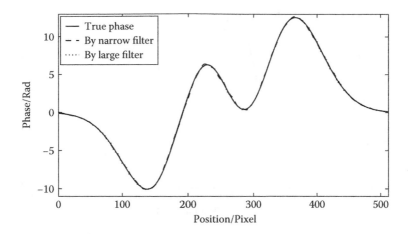

FIGURE 2.13   The unwrapped demodulated phase using FT.

and satisfactory for the whole fringe signal. Therefore, errors would be introduced to the phase demodulation by the FT method, as described in Section 2.2.

2.5.2 Numerical Analysis by WFT with an Invariable Window Size

For comparison, the WFT method is employed for analysis with different scale factors. Figure 2.16 shows the unwrapped demodulated phase with WFT employing the scale factors $a = 30$ and $a = 8$. Figure 2.17 shows the respective errors. When employing an inappropriate scale factor with

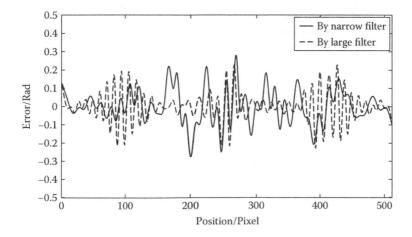

FIGURE 2.14   The errors from FT.

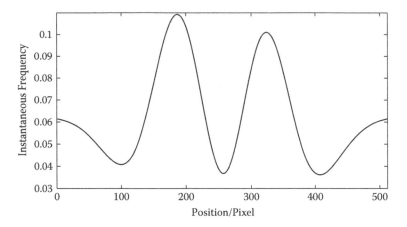

FIGURE 2.15   The instantaneous frequency.

a small value (e.g., $a = 8$), greater errors are introduced into the phase demodulation by the WFT method because the fundamental spectral component of the local fringe pattern is superposed by the zero component. The errors are limited by employing an appropriate scale factor. Therefore, it is clear that finding an appropriate scale factor for the Gaussian window in the localization analysis is the key for WFT. For the analysis of fringe signals with widely varying spectral parameters, WFT still has some deficiency because of the invariable resolution in the spatial and frequency

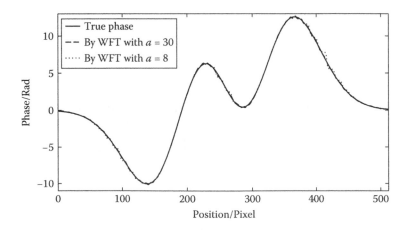

FIGURE 2.16   The unwrapped demodulated phase using WFT with $a = 30$ and $a = 8$.

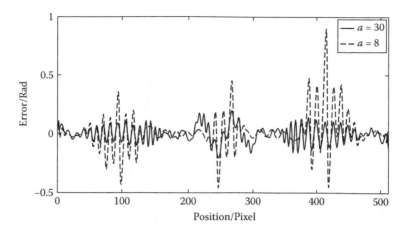

FIGURE 2.17 The errors using WFT with $a = 30$ and $a = 8$.

domains because of the fixed scale factor. To determine what spectral components exist at different positions and how they change, spatial frequency representation [32] is employed. Figure 2.18 is the spectrogram acquired by performing the WFT with a fixed scale factor of $a = 30$. The contour lines indicate the signal energy of a particular frequency at a particular position. The density of the contours reflects the spatial frequency resolution. The denser contours imply better spatial frequency resolution. As Figure 2.18 shows, a fixed scale factor of $a = 30$ results in poor spatial

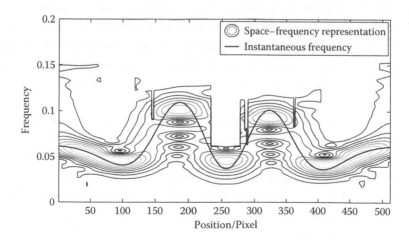

FIGURE 2.18 The spatial frequency representation of the modulated signal using WFT with $a = 30$.

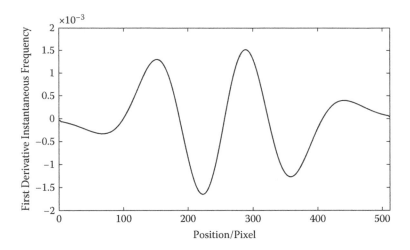

FIGURE 2.19   The first derivative of the instantaneous frequency.

frequency resolution. The energy is dispersed, and the contour line fails to match the true instantaneous frequency curve well.

2.5.3 Numerical Analysis by AWFT

Figure 2.19 shows the first derivative of the instantaneous frequency shown in Figure 2.15. It is clear that a wide spatial window should be employed for the analysis of the local signal centered at the 70th pixel because of the smooth variation of the instantaneous frequency, and a narrow spatial window should be employed for the analysis of the local signal centered at the 300th pixel because of the sharp variation of the instantaneous frequency. Therefore, to balance the spatial and frequency resolution, adaptive control of the size of the window is introduced to achieve variable localization analysis.

First, the instantaneous frequency of the whole signal is obtained by performing the WT [33]. By detecting the ridge of the WT, the instantaneous frequency of the modulated fringe signal at each position x can be obtained, as shown in Figure 2.20. Then, the maximal spatial domain, that is, $[x-(\Delta x)_{max}, x+(\Delta x)_{max}]$, is determined, where the maximum and minimum of the instantaneous frequency of the local signal around the position x satisfy the condition expressed by Equation (2.10). This means that the local stationary length around each position x can be obtained

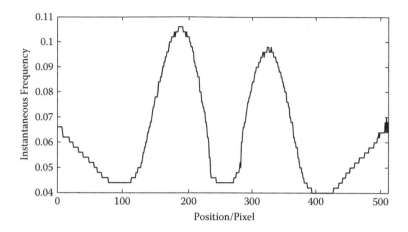

FIGURE 2.20 The instantaneous frequency detected by wavelet transform.

as $2(\Delta x)_{max}$ according to Equation (2.17). It should be noted that, for the initial boundary of the signal, the maximal spatial domain is considered to be $[0, x + (\Delta x)_{max}]$, and the local stationary length around the position x is defined as $x + (\Delta x)_{max}$. For the final boundary of the signal, the maximal spatial domain is considered to be $[x - (\Delta x)_{max}, 511]$, and the local stationary length around the position x is defined as $511 - x + (\Delta x)_{max}$.

Finally, according to Equation (2.26), the scale factors of each appropriate Gaussian window at different positions can be determined, as shown in Figure 2.21. Comparing Figure 2.15, Figure 2.19, and Figure 2.21 for the analysis of the local signal around position $x = 70$ with $f_{inst} = 0.047$, a wide Gaussian window with $a = 34.75$ is employed. In contrast, for the analysis of the local signal around position $x = 300$ with $f_{inst} = 0.084$, a narrow Gaussian window with $a = 10.75$ is employed. Figure 2.22 shows the local fringe signals centered at the 70th pixel and 300th pixel, which are selected by Gaussian windows with $a = 34.75$ and $a = 10.75$, respectively. Figure 2.23 shows the corresponding power spectra, for which the fundamental spectral component is separated from the other spectral components. Therefore, the fundamental spectral component can be extracted. Figure 2.24 shows the unwrapped demodulated phase using AWFT. Figure 2.25 shows the corresponding errors, and the errors at the borders are caused by undersampling within the Gaussian window. The analysis result shows that the large errors of the local fringe signal at low frequency

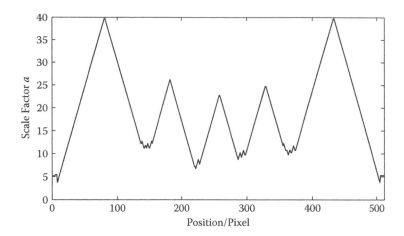

FIGURE 2.21   The scale factor of the Gaussian window.

by the FT are well eliminated in the AWFT. Compared with Figure 2.18, Figure 2.26 shows better spatial frequency resolution using AWFT, with the contour line concentrating most of the energy and matching the true instantaneous frequency curve well.

The WT is another spatial frequency analysis method for signal analysis. Here, the Gabor wavelet transform (GWT) method [11,12]

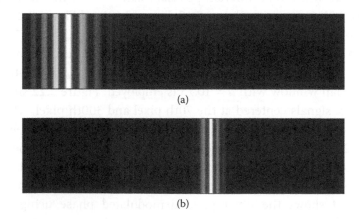

FIGURE 2.22   The local fringe patterns centered at the (a) 70th pixel and (b) 300th pixel selected by Gaussian windows $a = 34.75$ and $a = 10.75$, respectively.

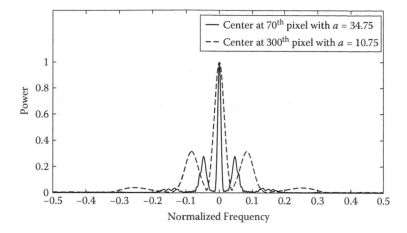

FIGURE 2.23 The power spectra of local fringe signals centered at the 70th pixel and 300th pixel.

is performed for comparison. Definition and properties of the GWT are given in Box 2.2. The central frequencies of the analyzing wavelets employed are within [0.01, 0.2] with the interval 0.002, and the scale factors employed are the inverse of the frequencies. Figure 2.27 and Figure 2.28 show the unwrapped demodulated phase and the corresponding errors. Considering the local linear approximation condition for the

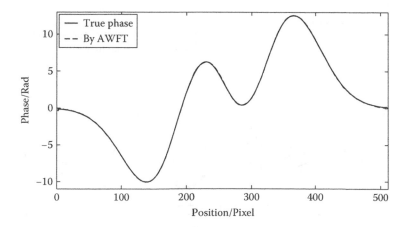

FIGURE 2.24 The unwrapped demodulated phase using AWFT.

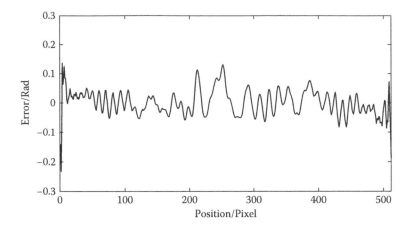

FIGURE 2.25   The errors from AWFT.

application of the GWT method, the second derivative of the modulated phase $\varphi''(x)$ is performed for analysis, as shown in Figure 2.29. It can be found that the error distribution curve by the WT has an analogy with the $\varphi''(x)$ curve. This shows that the phase errors using GWT are related to the local linearity of the modulated phase [15], whereas AWFT is free from this limitation. In short, the AWFT method with a series of automatically determined appropriate Gaussian windows can present a more accurate result for the phase demodulation of nonsinusoidal fringe signals.

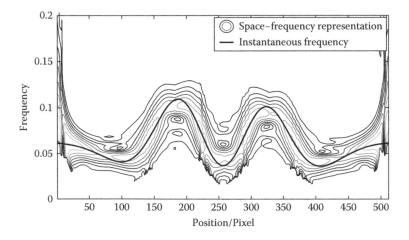

FIGURE 2.26   The spatial frequency representation of the modulated signal using AWFT.

BOX 2.2 THE GABOR WAVELET TRANSFORM

The continuous one-dimensional (1-D) Gabor wavelet transform (GWT) is defined as follows:

$$W(s,b) = \int_{-\infty}^{\infty} f(x)\psi_{s,b}^{*}(x)dx = \langle f(x), \psi_{s,b}(x) \rangle$$

$$\psi_{s,b}(x) = \frac{1}{s}\psi\left(\frac{x-b}{s}\right)$$

$$\psi(x) = \frac{1}{\sqrt[4]{\pi}}\sqrt{\frac{2\pi}{\gamma}}\exp\left[-\frac{(2\pi/\gamma)^2 x^2}{2} + j2\pi x\right]$$

where $\gamma = \pi\sqrt{2/\ln 2}$; $s > 0$ is the scale factor related to the frequency; b is the shift factor related to the position; $f(x)$ is the signal to be analyzed; $\psi_{s,b}(x)$ is the analyzing wavelet obtained by shifting and scaling the mother wavelet $\psi(x)$; and * indicates the complex conjugate. Computing the 1-D GWT, the modulus of the wavelet coefficients can be obtained as

$$|W(s,b)| = \sqrt{[imag(W(s,b))]^2 + [real(W(s,b))]^2}$$

where $imag(W(s,b))$ and $real(W(s,b))$ indicate the imaginary part and real part of $W(s,b)$, respectively. At position b, the ridge of the wavelet transform is defined as the location where $|W(s,b)|$ reaches its local maximum along the scale direction s.

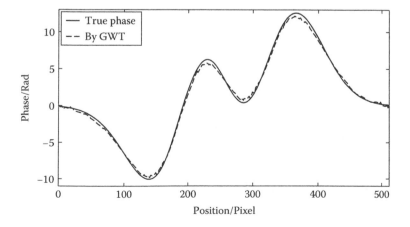

FIGURE 2.27 The unwrapped demodulated phase using GWT.

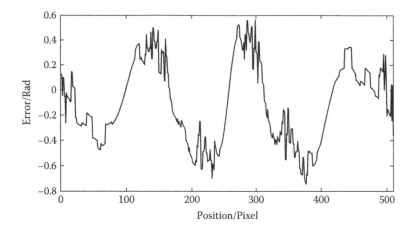

FIGURE 2.28　The errors from GWT.

2.6 EXPERIMENTAL ANALYSIS EXAMPLE

Optical profilometry of fringe projection [1,2] was considered for the experimental analysis. A 400-W digital projector was used to project a sinusoidal fringe pattern onto a real person's face. The modulated fringe pattern was detected by a charge-coupled device (CCD) camera. An analogue video output signal was converted into an 8-bit digital signal, stored in a frame memory, and then analyzed by the computer. Figure 2.30 shows a modulated fringe pattern detected by the CCD camera with saturated exposure. The pattern within the black rectangle box is 512 × 200

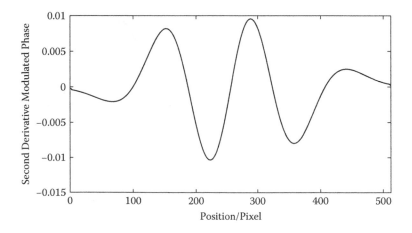

FIGURE 2.29　The second derivative of the modulated phase.

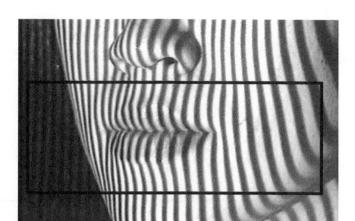

FIGURE 2.30 The modulated fringe pattern.

pixels and includes signals with high and low frequencies. Figure 2.31 shows the intensity of the modulated fringe pattern at the 100th row. This fringe pattern can be considered a nonsinusoidal signal because of the saturated exposure. Figure 2.32 shows the power spectra of the original and modulated fringe pattern at the 100th row. The fundamental spectral component of the modulated fringe pattern is superposed by the other higher-order spectral components, which makes it impossible to filter out the fundamental spectral component accurately. The instantaneous frequency of the fringe pattern at the 100th row was detected by GWT. It can

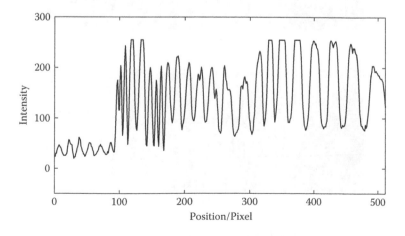

FIGURE 2.31 The modulated non-sinusoidal fringe pattern at the 100th row.

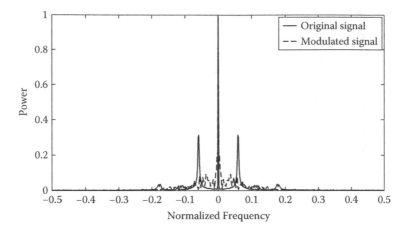

FIGURE 2.32   The power spectra.

be determined that $(f_{inst})_{max}$ is about 0.38 and $(f_{inst})_{min}$ is about 0.16, which do not satisfy the condition expressed by Equation (2.10); thus, the FT method is not applicable. For comparison, FT, WFT, GWT, and AWFT were performed. Figures 2.33 and 2.34 show the wrapped and unwrapped demodulated phase using FT, WFT with $a = 30$, GWT, and AWFT. The

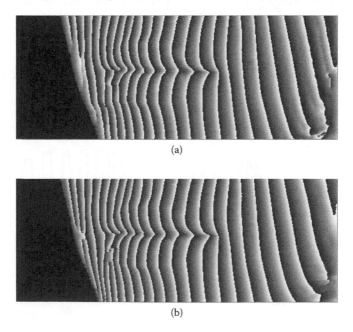

(a)

(b)

FIGURE 2.33   The wrapped phase using (a) FT, (b) WFT with $a = 30$, (c) GWT, and (d) AWFT methods. *(continued)*

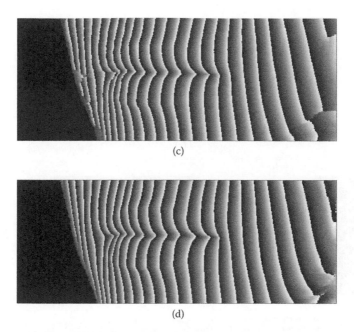

(c)

(d)

FIGURE 2.33 *(continued)* The wrapped phase using (a) FT, (b) WFT with *a* = 30, (c) GWT, and (d) AWFT methods.

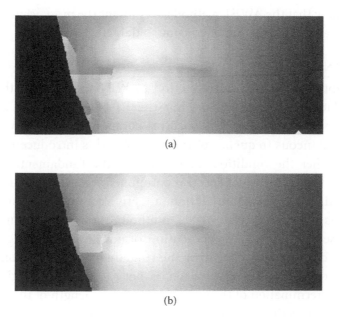

(a)

(b)

FIGURE 2.34 The unwrapped demodulated phase using (a) FT, (b) WFT with *a* = 30, (c) GWT, and (d) AWFT methods. *(continued)*

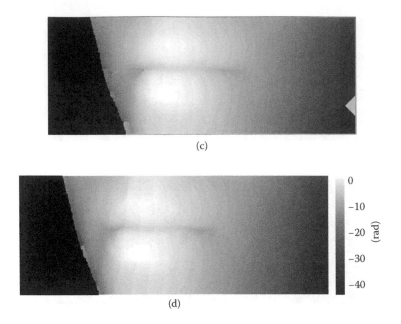

FIGURE 2.34 *(continued)* The unwrapped demodulated phase using (a) FT, (b) WFT with $a = 30$, (c) GWT, and (d) AWFT methods.

results show that the AWFT method can present the most accurate result for the phase demodulation of nonsinusoidal fringe patterns.

2.7 CONCLUSION

This chapter provides an introduction to WFT methods for the phase demodulation of fringe signals. For the FT-based methods, accurate extraction of the fundamental spectral component is the key process. The instantaneous frequency of the local signal is introduced to determine whether the condition for separating the fundamental spectral component from the other spectral components is satisfied. To balance the spatial and frequency resolution, adaptive control of the size of the window is introduced to the WFT method. The AWFT technique has been presented for phase demodulation of nonsinusoidal fringe signals when the condition for separating the fundamental spectral component from the other spectral components is not satisfied for the global signal. The determination of the window size by the length of the local stationary signal is discussed in theory. First, the local stationary signal is defined as the signal whose fundamental spectral component is separated from all the other spectral components within the local signal area. Then, the appropriate scale factor of the Gaussian window is determined by ensuring that the width of the Gaussian window is equal to the local stationary length. In addition, the minimum size of the employed window is discussed. Finally, the results of the comparison of phase demodulation by FT, WFT, AWFT, and GWT in numerical analysis and by an optical metrology experiment show the superiority of the AWFT localization analysis. AWFT can present the most accurate result for the phase demodulation of nonsinusoidal fringe patterns.

REFERENCES

1. Buytaert Jan and Dirckx Joris. Fringe projection profilometry. In: *Optical Methods for Solid Mechanics: A Full-Field Approach.* Edited by Rastogi Pramod and Hack Erwin. Weinheim, Germany: Wiley-VCH, 2012, 303–344.
2. Mitsuo Takeda and Kazuhiro Mutoh. Fourier transform profilometry for the automatic measurement of 3-D object shapes. *Applied Optics* **22**(24), 3977–3982 (1983).
3. Qican Zhang and Xianyu Su. High-speed optical measurement for the drumhead vibration. *Optics Express* **13**(8), 3110–3116 (2005).
4. Cesar A. Sciammarella and Taeeeui Kim. Determination of strains from fringe patterns using spatial-frequency representations. *Optical Engineering* **42**(11), 3182–3193 (2003).
5. Yves Surrel. Design of algorithms for phase measurements by the use of phase stepping. *Applied Optics* **35**(1), 51–60 (1996).
6. Fujun Yang and Xiaoyuan He. Two-step phase-shifting fringe projection profilometry: Intensity derivative approach. *Applied Optics* **46**(29), 7172–7178 (2007).
7. Bing Pan, Qian Kemao, Lei Huang, and Anand Asundi. Phase error analysis and compensation for nonsinusoidal waveforms in phase-shifting digital fringe projection profilometry. *Optics Letters* **34**(4), 416–418, 2009.
8. Mitsuo Takeda, Hideki Ina, and Seiji Kobayashi. Fourier-transform method of fringe-pattern analysis for computer-based topography and interferometry. *Journal of the Optical Society of America* **72**(1), 156–160 (1982).
9. Xianyu Su and Wenjing Chen. Fourier transform profilometry: a review. *Optics and Lasers in Engineering* **35**(5), 263–284 (2001).
10. Mitsuo Takeda. Measurements of extreme physical phenomena by Fourier fringe analysis. *International Conference on Advanced Phase Measurement Methods in Optics and Imaging, AIP Conference Proceedings* **1236**, 445–448 (2010).
11. Jingang Zhong and Jiawen Weng. Spatial carrier-fringe pattern analysis by means of wavelet transform: Wavelet transform profilometry. *Applied Optics* **43**(26), 4993–4998 (2004).
12. Jingang Zhong and Jiawen Weng. Phase retrieval of optical fringe patterns from the ridge of a wavelet transform. *Optics Letters* **30**(19), 2560–2562 (2005).

13. Munther A. Gdeisat, David R. Burton, and Michael J. Lalor. Spatial carrier fringe pattern demodulation by use of a two dimensional continuous wavelet transform. *Applied Optics* **45**(34), 8722–8732 (2006).
14. Jiawen Weng, Jingang Zhong, and Cuiying Hu. Digital reconstruction based on angular spectrum diffraction with the ridge of wavelet transform in holographic phase-contrast microscopy. *Optics Express* **16**(26), 21971–21981 (2008).
15. Zibang Zhang and Jingang Zhong. Applicability analysis of wavelet-transform profilometry. *Optics Express* **21**(16), 18777–18796 (2013).
16. Dennis Gabor. Theory of communication. Part 1: The analysis of information. *Journal of the Institution of Electrical Engineers–Part III: Radio and Communication Engineering* **93**(26),429–441 (1946).
17. Jiawen Weng and Jingang Zhong. Application of Gabor transform to 3-D shape analysis. *Acta Photonica Sinica* **32**(8), 993–996 (2003) (Chinese).
18. Kemao Qian, Seah Hock Soon, and Anand Asundi. Phase-shifting windowed Fourier ridges for determination of phase derivatives. *Optics Letters* **28**(18), 1657–1659 (2003).
19. Kemao Qian. Windowed Fourier transform for fringe pattern analysis. *Applied Optics* **43**(13), 2695–2702 (2004).
20. Kemao Qian. Windowed Fourier transform for fringe pattern analysis: addendum. *Applied Optics* **43**(17), 3472–3473 (2004).
21. Kemao Qian. Windowed Fourier transform method for demodulation of carrier fringes. *Optical Engineering* **43**(7), 1472–1473 (2004).
22. Kemao Qian, Haixia Wang, and Wenjing Gao. Windowed Fourier transform for fringe pattern analysis: Theoretical analyses. *Applied Optics* **47**(29), 5408–5419 (2008).
23. P. Hlubina, J. Luňáček, D. Ciprian, et al. Windowed Fourier transform applied in the wavelength domain to process the spectral interference signals. *Optics Communications* **281**(9), 2349–2354 (2008).
24. Kemao Qian and Seah Hock Soon. Two-dimensional windowed Fourier frames for noise reduction in fringe pattern analysis. *Optical Engineering* **44**(7), 075601 (2005).
25. Kemao Qian. Two-dimensional windowed Fourier transform for fringe pattern analysis: Principles, applications and implementations. *Optics and Lasers in Engineering* **45**(2), 304–317 (2007).
26. Kemao Qian. On window size selection in the windowed Fourier ridges algorithm. *Optics and Lasers in Engineering* **45**(12), 1186–1192 (2007).
27. Jingang Zhong and Jiawen Weng. Dilating Gabor transform for the fringe analysis of 3-D shape measurement. *Optical Engineering* **43**(4), 895–899 (2004).
28. Suzhen Zheng, Wenjing Chen, and Xianyu Su. Adaptive windowed Fourier transform in 3-D shape measurement. *Optical Engineering* **45**(6), 063601 (2006).
29. Jingang Zhong and Huiping Zeng. Multiscale windowed Fourier transform for phase extraction of optical fringe pattern. *Applied Optics* **46**(14), 2670–2675 (2007).
30. Jingang Zhong and Jiawen Weng. Generalized Fourier analysis for phase retrieval of fringe pattern. *Optics Express* **18**(26), 26806–26820 (2010).

31. Stephane Mallat. *A Wavelet Tour of Signal Processing.* New York: Academic Press, 1998.

32. Jingang Zhong and Yu Huang. Time-frequency representation based on an adaptive short-time Fourier transform. *IEEE Transactions on Signal Processing* **58**(10), 5118–5128 (2010).

33. Delprat Nathalie, Bernard Escudié, Philippe Guillemain, et al. Asymptotic wavelet and Gabor analysis: Extraction of instantaneous frequencies. *IEEE Transactions on Information Theory* **38**(2), 644–664 (1992).

CHAPTER 3

Continuous Wavelet Transforms

Lionel R. Watkins

Department of Physics
University of Auckland
Auckland, New Zealand

3.1 INTRODUCTION

Wavelets are simple, continuous functions that are scaled, translated, and possibly rotated in a systematic way such that the entire family of wavelets thus generated forms a suitable basis for representation of signals. The name *wavelet* was coined in the early 1980s by French physicists and mathematicians, who called these functions *ondelettes*, literally "small waves," because they are functions (with zero mean) that have a few oscillations that decay to zero. There are many types of "wavelet transform," but the problem of recovering phase distributions from interference patterns is almost exclusively the preserve of the continuous wavelet transform (CWT). As we show, the coefficients of this transform provide the information necessary to retrieve either the local instantaneous frequency or the phase of the signal.

The aim of this chapter is to describe both the one- (1-D) and two-dimensional (2-D) CWTs and to demonstrate, with the aid of numerical examples, the practical aspects of recovering the phase distribution from fringe patterns.

The chapter is organized as follows. We begin with the 1-D CWT and a simple numerical example to demonstrate how one can extract both the instantaneous frequency and the phase from a fringe pattern. Along the way, we define some essential properties of wavelets and establish a notation for the ensuing mathematics. The CWT produces a surface in the 2-D time-frequency plane called the *scalogram*, and because this surface is key to recovering either frequency or phase, we spend some time discussing its properties. This allows us to introduce the *wavelet ridge*, a path that follows the maximum modulus of the wavelet transform. The wavelet ridge, in turn, can be used with either the gradient or the phase method to recover, respectively, the instantaneous frequency or the phase. On a practical note, we explain how the CWT can be evaluated with the aid of the fast Fourier transform (FFT) because this considerably speeds up the computation and how the effect of discontinuities at the signal edges may be mitigated. We conclude this part of the chapter by listing many of the functions that are commonly used to construct 1-D wavelets.

The 2-D CWT is the subject of the rest of the chapter. Fortunately, many of the properties of the 1-D transform carry across into 2-D, but there are additional aspects that require further explanation. In 2-D, the wavelets may be either isotropic or anisotropic, and we give examples of both, outlining the trade-offs inherent in each. In addition, 2-D wavelets transform a 2-D signal into a four-dimensional (4-D) representation that is not easily visualized. For our purposes, the *position representation* approach is the most useful, and the way in which the wavelet ridge is extracted from this 4-D space is explained. We revisit a previous practical example and show that it can be advantageous to extract the phase from a 2-D fringe pattern using the 2-D wavelet transform.

3.2 THE ONE-DIMENSIONAL CONTINUOUS WAVELET TRANSFORM

A wavelet is a function $\psi(t)$, centered at $t = 0$ with a few oscillations and zero mean, a property that we can express mathematically as

$$\int_{-\infty}^{\infty} \psi(t)\,dt = 0. \tag{3.1}$$

This equation is known as the "admissibility condition." To give a good idea of the type of functions we have in mind here, some examples of typical wavelets are given in Example 3.1 and Section 3.10. Almost all the signals in which we are interested are of finite duration and therefore have finite energy. For the moment, we assume that our signals are continuous functions of time, although the extension to sampled or discrete signals is straightforward. A measure of a signal's energy is obtained by integrating the squared modulus of the signal over the signal duration. Because our signals are of finite duration, we can write this as

$$\int_{-\infty}^{\infty} |f(t)|^2 \, dt < \infty. \tag{3.2}$$

A signal $f(t)$ that satisfies Equation (3.2) has finite energy. For real valued signals, $|\cdot|^2$ simply means the amplitude squared, but for complex valued signals, it means the product of the signal and its complex conjugate.

It will help if we introduce some notation at this point. The inner product of two functions $f(t)$ and $g(t)$, both of which have finite energy, is defined as

$$\langle f(t), g(t) \rangle = \int_{-\infty}^{\infty} f(t) g^*(t) dt \tag{3.3}$$

where $g^*(t)$ denotes the complex conjugate of g. With this definition of inner product, it is easy to see that the energy of a signal $f(t)$ is

$$\|f\|^2 = \langle f(t), f(t) \rangle = \int_{-\infty}^{\infty} |f(t)|^2 \, dt \tag{3.4}$$

where $\|\cdot\|$ denotes the "norm" of a function and has a meaning analogous to the length of a vector or absolute value of a complex number.

In principle, there is no constraint on the amplitude of the wavelet, but it is usual practice to normalize $\psi(t)$ such that it has unit energy, that is, $\|\psi\|^2 = 1$. Next, we generate a family of functions, denoted $\psi_{a,b}(t)$, from this "mother wavelet" by translations and dilations according to

$$\psi_{a,b}(t) = \frac{1}{\sqrt{a}} \psi\left(\frac{t-b}{a}\right). \tag{3.5}$$

There are several points to note here. First, the literature is divided between those who scale the wavelet amplitudes by $1/a$ and those who choose $1/\sqrt{a}$. We note that a factor $1/\sqrt{a}$ ensures that $\psi_{a,b}$ maintains its norm with scaling factor a so that no scale is unduly weighted. Second, the scaling factor a is inversely proportional to frequency; wavelets with large values of a have low-frequency oscillations and vice versa. However, it is important to note that the function is scaled in such a way that each scaled wavelet always has exactly the same number of oscillations under its envelope, as can be seen in Example 3.1, and it is this property of wavelets that sets them apart from the windowed Fourier transform (WFT). The scale parameter a may take any value on the positive real line, obviously excluding zero. Finally, the translation parameter b may have any value on the real line, although with images and sampled fringe patterns, it is common practice to let b take on the integer values of the sample or pixel number.

The 1-D CWT of a function $f(t)$ is defined as

$$W_f(a,b) = \langle f, \psi_{a,b} \rangle = \int_{-\infty}^{\infty} f(t) \frac{1}{\sqrt{a}} \psi^* \left(\frac{t-b}{a} \right) dt \qquad (3.6)$$

This is nothing other than the inner product of our signal, $f(t)$, with the translations and dilations of the family of wavelets generated by Equation (3.5). At this point, we are ready to consider a simple numerical example that will illustrate some practical aspects of the CWT and provide motivation for the more detailed discussion to follow. A summary of the key definitions may be found in Box 3.1.

BOX 3.1   SUMMARY OF DEFINITIONS

- Admissibility condition:

$$\int_{-\infty}^{\infty} \psi(t) dt = 0 \Leftrightarrow \hat{\psi}(0) = 0.$$

- Inner product of two finite energy functions f and g:

$$\langle f(t), g(t) \rangle = \int_{-\infty}^{\infty} f(t) g^*(t) dt$$

- Energy of a signal f:

$$||f||^2 = \langle f(t), f(t) \rangle = \int_{-\infty}^{\infty} |f(t)|^2 dt$$

- Generation of wavelet family:

$$\psi_{a,b}(t) = \frac{1}{\sqrt{a}} \psi \left(\frac{t-b}{a} \right)$$

- The one-dimensional (1-D) continuous wavelet transform:

$$W_f(a,b) = \langle f, \psi_{a,b} \rangle = \int_{-\infty}^{\infty} f(t) \frac{1}{\sqrt{a}} \psi^* \left(\frac{t-b}{a} \right) dt$$

Example 3.1: Linear Chirp

The example chosen is a computer-generated fringe pattern with a quadratic phase and hence a linearly increasing spatial frequency, as can be seen in Figure 3.1. The fringe pattern is intended to replicate an image that might be captured by a typical CCD (charge-coupled device) camera, so the amplitude represents a grayscale discretization, and each row is 512 pixels long. For any row, the intensity is given by

$$I(x) = 32\left(1 + \cos\left[\left(\frac{3\pi x}{512}\right)^2 + \frac{2\pi x}{32}\right]\right), \quad x = 1,\ldots,512. \quad (3.7)$$

where x is the pixel position.

One of the attractions of the CWT is the wide variety of functions that can be used as the mother wavelet, but for this example we use the well-known Morlet wavelet (also known as the Gabor wavelet in the literature), which is given by

$$\psi_{\text{Morlet}}(t) = \frac{1}{\sqrt[4]{\sigma^2 \pi}} \exp(i\omega_0 t) \exp\left(\frac{-t^2}{2\sigma^2}\right). \quad (3.8)$$

Figure 3.2 shows a few representations of this function for various values of the parameters a, b, and σ. ω_0, which sets the frequency of oscillation of the mother wavelet, is bounded by the requirement for the function to satisfy the admissibility condition. The Fourier transform $\mathcal{F}\{\psi_{\text{Morlet}}(t)\} = \hat{\psi}(\omega)$ is a Gaussian function centered at ω_0, and in the Fourier domain, Equation (3.1) is equivalent to requiring $\hat{\psi}(0) = 0$, which is not strictly possible for a function with a Gaussian envelope. However, for all practical purposes, $\hat{\psi}(0)$ will be small enough provided

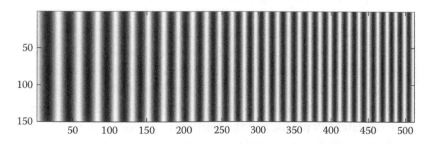

FIGURE 3.1 Fringe pattern with quadratic phase and a spatial carrier of $2\pi/32$ rad/pixel.

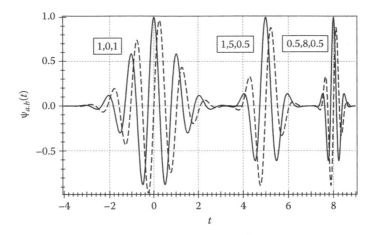

FIGURE 3.2 The Morlet wavelet of Equation (3.8). The values of a, b, and σ, respectively, are given by the annotations next to each function. Solid lines represent the real part of the function and dashed lines the imaginary part.

$\omega_0 \geq \pi \, (2/\ln 2)^{1/2} \approx 5.34$, a value suggested by Daubechies [1,2]. Many experimentalists choose $\omega_0 = 6$, which is the value that has been used in Figure 3.2. The Morlet wavelet has an additional parameter σ that enables us to control the width of the Gaussian envelope and hence the number of periods, a feature that is useful experimentally because this can be done independently of the oscillation frequency ω_0.

Until now, we have treated t as a continuous variable, but our fringe pattern is discretized with t (or x) taking integer values from 1 to 512. If we use the Morlet wavelet with $a = 1$, the result is a very sharp function indeed that spans approximately 8 pixels, much shorter than the highest instantaneous frequency in the fringe. Although a may, in principle, take any value on the positive real line (excluding zero), it is necessary to pick initial and final values of a and an appropriate step size a_{step} if we wish to implement Equation (3.6) on a computer. The function idisp.m in the Robotics Toolbox, which may be downloaded from http://www.petercorke.com/Robotics_Toolbox. html, can be used to estimate initial and final values of a [3]. This function shows that the maximum period of our fringe pattern is approximately 29 pixels, and the minimum period is around 12 pixels. These periods correspond to spatial frequencies of approximately 0.22 rad/pixel and 0.52 rad/pixel, respectively. As we show in further discussion [see Equation (3.23c)], the center frequency of our scaled

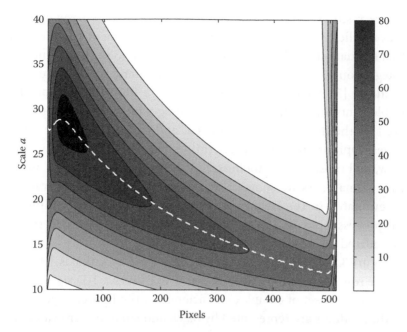

FIGURE 3.3 $|Wf(a,b)|$ of one row of Figure 3.1 using the Morlet wavelet with $\omega_0 = 6$ and $\sigma = 0.5$. The dashed white line shows the wavelet ridge.

wavelet is ω_0/a, so if we set $\omega_0 = 6$ rad/pixel for our Morlet wavelet, then these spatial frequencies correspond to values of a of 27.3 and 11.5, respectively. These values of a should be regarded as a guideline, although one can see from Figure 3.3 that most of the signal energy does indeed lie between these values of a, but some trial and error does not go amiss. An appropriate value for a_{step} depends on whether one is trying to recover the instantaneous frequency or the phase [4], with the former requiring much finer steps.

To analyze the fringe pattern in Figure 3.1, we pick any row, set $\omega_0 = 6.0$ and $\sigma = 0.5$, and generate a family of wavelets with $a \in \{10,40\}$, $a_{step} = 0.1$, and $b \in \{1,512\}$. Suppose that a and b are stored in vectors of appropriate length. Then, a MATLAB-like algorithm for calculating the CWT of our row might look something like:

```
for j = 1:length(a)
    for k = 1:length(b)
        c(j,k) = cwt(f,a(j),b(k));
    end
end
```

where f stores the row we wish to analyze, c is an array of complex coefficients, and cwt(f,a,b) is a subroutine or function call that evaluates the CWT according to Equation (3.6). The numerical integration can be done using one of the many quadrature rules available [5]; for example, MATLAB has a built-in function that implements Simpson's rule. The advantages of this direct approach are that the data are not periodized, and one can treat the signal edges appropriately. However, this approach is computationally slow, and most wavelet toolboxes do not evaluate the CWT in the time domain. As we show in further discussion, the integral can be evaluated in the frequency domain, considerably speeding up the process, but with the disadvantage that the data are necessarily periodized [6].

Figure 3.3 is a plot of the modulus of the 1-D CWT, that is, $|W_f(a,b)|$ of one row of this fringe pattern. The CWT takes a 1-D signal f and returns a 2-D set of complex coefficients. In the figure, large values of the modulus are represented by black and small or zero values by white. To find the instantaneous fringe frequency at a given pixel position, we look down the column corresponding to that pixel and seek the value of a that yields the maximum $|W_f(a,b)|$. Once we have done this for every pixel, we obtain the dashed white line shown in the figure, which is known as the *wavelet ridge*.

In principle, the instantaneous frequency can easily be calculated from values of $a(b)$ on the ridge, using Equation (3.23c), and when plotted against the exact frequencies as determined from Equation (3.7), yields the graph shown in Figure 3.4. In practice, we do use this equation, but there is a small correction factor that is discussed elsewhere in the chapter. Note that the recovered fringe frequency contains errors at the left- and right-hand edges that are caused by the abrupt truncation of the data here. We discuss ways to reduce this error further in the chapter. The approach we have just illustrated is frequently referred to in the literature as the *gradient method* for obvious reasons, but there is an alternative approach, called the *phase method*, that we describe fully further in the chapter. For the moment, we note that the recovered phase using this latter method is in excellent agreement with the expected values, as shown in Figure 3.4, and conclude this example by noting a few practical considerations, given in Box 3.2.

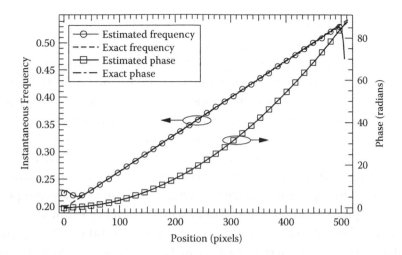

FIGURE 3.4   Comparison of recovered and exact instantaneous frequencies and phases for the fringe pattern of Figure 3.1 using the gradient and phase methods.

BOX 3.2   PRACTICAL CONSIDERATIONS

- We chose the Morlet wavelet in this example for convenience, but an important feature of the continuous wavelet transform (CWT) is one's ability to choose a wavelet that best suits the problem at hand.
- To recover the phase or local frequency content of a signal, it is *essential* that the wavelet be analytic [7]. A function $f(t)$ is said to be analytic if its Fourier transform $\hat{f}(\omega) = 0$ for all $\omega < 0$. If we choose a wavelet with an even (symmetric) real part and an odd (antisymmetric) imaginary part, then the properties of the Fourier transform guarantee that its transform $\hat{f}(\omega)$ will be real. If we further choose the modulation frequency ω_0 appropriately, we can ensure that our wavelet is analytic, at least for practical purposes.
- The fringe pattern *must* be superimposed on a temporal or spatial carrier. If the fringe pattern goes through zero spatial (or temporal) frequency, not only is there an ambiguity about the sign of the phase after the zero crossing but also, by definition, it is not analyzable by wavelets because $\hat{\psi}(0) = 0$.
- For computation efficiency, a must be discretized, but there is no point in reducing it below a certain minimum step size, determined by the resolution of the chosen wavelet.
- The most natural discretization for b is the pixel or sample number, which was used in the example.

3.3 WAVELET CENTERS AND BANDWIDTHS

The representation of the CWT shown in Figure 3.3 is called a *scalogram*, and because its properties are important to the wavelet transform and the process of phase recovery, it merits further discussion. However, before we can do this, we need to define some additional properties of wavelets. Each wavelet $\psi_{a,b}(t)$ has a certain center, or mean value, in both the time and the frequency domains. It also has a certain "spread" about these values in both domains that is intrinsic to its resolving ability. The way in which individual wavelets cover or "tile" the scalogram is indicative of their ability to resolve the frequency content of a signal at a given point in time, a property often referred to as "time-frequency estimation." We need, therefore, the mean value of a function in both time and frequency domains and the variances or root-mean-square (rms) bandwidths of the function about those mean values.

The mean position u of a function f is defined as

$$u = \frac{1}{\|f\|^2} \int_{-\infty}^{\infty} t \,|f(t)|^2 \, dt \tag{3.9}$$

and similarly, its average frequency is

$$\eta = \frac{1}{2\pi \|f\|^2} \int_{-\infty}^{\infty} \omega \,|\hat{f}(\omega)|^2 \, d\omega. \tag{3.10}$$

The variances about these values are, respectively,

$$\sigma_t^2 = \frac{1}{\|f\|^2} \int_{-\infty}^{\infty} (t-u)^2 \,|f(t)|^2 \, dt \tag{3.11}$$

and

$$\sigma_\omega^2 = \frac{1}{2\pi \|f\|^2} \int_{-\infty}^{\infty} (\omega-\eta)^2 \,|\hat{f}(\omega)|^2 \, d\omega. \tag{3.12}$$

We also need to make use of the Fourier transform, and conventions for this vary, but we define the Fourier transform $\hat{f}(\omega)$ of a function $f(t)$ as

$$\hat{f}(\omega) = \int_{-\infty}^{\infty} f(t)e^{-i\omega t} \, dt \tag{3.13}$$

and its inverse as

$$f(t) = \frac{1}{2\pi} \int_{-\infty}^{\infty} \hat{f}(\omega)e^{i\omega t}\, d\omega. \qquad (3.14)$$

More detailed information on the Fourier transform and its properties may be found in Brigham [6] and Mallat [7], but the essential properties that we need are listed in Box 3.3.

BOX 3.3   SELECTED PROPERTIES OF THE FOURIER TRANSFORM

- Convolution

$$f_1(t) * f_2(t) \Leftrightarrow \hat{f}_1(\omega)\hat{f}_2(\omega)$$

- Translation

$$f(t-u) \Leftrightarrow e^{-iu\omega}\hat{f}(\omega)$$

- Modulation

$$e^{i\xi t}f(t) \Leftrightarrow \hat{f}(\omega - \xi)$$

- Scaling

$$f(t/a) \Leftrightarrow |a|\hat{f}(a\omega)$$

- Complex conjugate

$$f^*(t) \Leftrightarrow \hat{f}^*(-\omega)$$

- Time reversal

$$f(-t) \Leftrightarrow \hat{f}(-\omega)$$

Example 3.2: Centers and Bandwidths of the Morlet Wavelet

As an illustrative example, we now find the centers and bandwidths of the Morlet wavelet given in Equation (3.8). For this wavelet, the corresponding family is given by

$$\psi_{a,b}(t) = \frac{1}{\sqrt[4]{\sigma^2 \pi}} \frac{1}{\sqrt{a}} \exp\left[i\omega_0\left(\frac{t-b}{a}\right)\right]\exp\left[-\frac{1}{2}\left(\frac{t-b}{\sigma a}\right)^2\right].$$

$$(3.15)$$

Several of our calculations will require $|\psi_{a,b}(t)|^2$. Recall that

$$|\psi_{a,b}(t)|^2 = \psi_{a,b}(t)\psi_{a,b}^*(t) \tag{3.16a}$$

which, substituting from Equation (3.15) gives

$$|\psi_{a,b}(t)|^2 = \frac{1}{\sigma a \sqrt{\pi}} \exp\left[-\left(\frac{t-b}{\sigma a}\right)^2\right] \tag{3.16b}$$

We should first check the norm of this wavelet. According to our definition in Equation (3.4),

$$\|\psi_{a,b}(t)\|^2 = \frac{1}{\sigma a \sqrt{\pi}} \int_{-\infty}^{\infty} \exp\left[-\left(\frac{t-b}{\sigma a}\right)^2\right] dt \tag{3.17a}$$

which can be simplified using the substitution $x = (t - b)/\sigma a$:

$$= \frac{1}{\sqrt{\pi}} \int_{-\infty}^{\infty} e^{-x^2}\, dx \tag{3.17b}$$

so

$$\|\psi_{a,b}(t)\|^2 = 1 \tag{3.17c}$$

because $\int_{-\infty}^{\infty} \exp(-x^2)dx$ evaluates to $\sqrt{\pi}$. So, our wavelet does indeed have unit energy.

Because $\psi(t)$ is centered at $t = 0$, we should expect $\psi_{a,b}(t)$ to be centered at $t = b$. Using the definition of Equation (3.9), we have

$$u = \frac{1}{\sigma a \sqrt{\pi}} \int_{-\infty}^{\infty} t \exp\left[-\left(\frac{t-b}{\sigma a}\right)^2\right] dt \tag{3.18a}$$

which may again be simplified using the substitution $x = (t - b)/\sigma a$:

$$= \frac{1}{\sqrt{\pi}} \int_{-\infty}^{\infty} (\sigma a x + b)e^{-x^2}\, dx \tag{3.18b}$$

and because the product of an odd function x, with an even function $\exp(-x^2)$, is odd, the first term in this integral evaluates to zero. Hence,

$$u = b \qquad (3.18c)$$

as expected. The rms time bandwidth of our wavelet is, according to Equation (3.11),

$$\sigma_t^2 = \frac{1}{\sigma a \sqrt{\pi}} \int_{-\infty}^{\infty} (t-b)^2 \exp\left[-\left(\frac{t-b}{\sigma a}\right)^2\right] dt \qquad (3.19a)$$

which, with the same substitution as before, yields

$$= \frac{\sigma^2 a^2}{\sqrt{\pi}} \int_{-\infty}^{\infty} x^2 e^{-x^2}\, dx \qquad (3.19b)$$

and thus

$$\sigma_t^2 = \frac{\sigma^2 a^2}{2}. \qquad (3.19c)$$

We must use an analytic wavelet if we wish to recover the local phase or frequency content of a signal [7]. As noted previously, a function $f(t)$ is said to be analytic if $\hat{f}(\omega) = 0$ for all $\omega < 0$. The Morlet wavelet has an even real part and odd imaginary part, as can been seen in Figure 3.2; therefore, its Fourier transform is real. Moreover, if we choose $\omega_0 \geq 5.34$, then $\hat{\psi}(\omega)$ will be approximately zero for all $\omega < 0$, and it therefore qualifies as an analytic wavelet. Given the Fourier transform pair

$$f(t) = e^{-t^2/2} \Leftrightarrow \hat{f}(\omega) = \sqrt{2\pi}e^{-\omega^2/2} \qquad (3.20)$$

we can use the properties of the Fourier transform to find $\hat{\psi}_{a,b}(\omega)$:

$$\hat{\psi}_{a,b}(\omega) = \frac{\sqrt{2\pi\sigma^2 a}}{\sqrt[4]{\sigma^2\pi}} \exp(-ib\omega)\exp\left[\frac{-(a\sigma\omega - \sigma\omega_0)^2}{2}\right]. \qquad (3.21)$$

We will need $|\hat{\psi}_{a,b}(\omega)|^2$, which is given by

$$|\hat{\psi}_{a,b}(\omega)|^2 = \hat{\psi}_{a,b}(\omega)\hat{\psi}^*_{a,b}(\omega) \tag{3.22a}$$

$$= \frac{2\pi a\sigma}{\sqrt{\pi}}\exp[-(a\sigma\omega - \sigma\omega_0)^2]. \tag{3.22b}$$

The mean frequency of the Morlet wavelet, by Equation (3.10), is therefore

$$\eta = \frac{a\sigma}{\sqrt{\pi}}\int_{-\infty}^{\infty}\omega\exp[-(a\sigma\omega - \sigma\omega_0)^2]d\omega \tag{3.23a}$$

which may be simplified by substituting $x = \sigma(a\omega - \omega_0)$:

$$= \frac{1}{a\sqrt{\pi}}\int_{-\infty}^{\infty}(\frac{x}{\sigma}+\omega_0)e^{-x^2}\,dx. \tag{3.23b}$$

The first term of this integral evaluates to zero; therefore,

$$\eta = \frac{\omega_0}{a}. \tag{3.23c}$$

Notice that the width of the Gaussian envelope σ does not affect the modulation frequency ω_0, as asserted previously. Finally, we calculate the rms frequency bandwidth of our wavelet. According to Equation (3.12), this is given by

$$\sigma^2_\omega = \frac{a\sigma}{\sqrt{\pi}}\int_{-\infty}^{\infty}\left(\omega - \frac{\omega_0}{a}\right)^2\exp[-(a\sigma\omega - \sigma\omega_0)^2]d\omega \tag{3.24a}$$

which, with the same substitution as previously, can be simplified to

$$= \frac{1}{\sqrt{\pi}}\int_{-\infty}^{\infty}\frac{x^2}{a^2\sigma^2}e^{-x^2}\,dx \tag{3.24b}$$

so that

$$\sigma^2_\omega = \frac{1}{2a^2\sigma^2}. \tag{3.24c}$$

In a similar manner, one can calculate the centers and bandwidths of any other wavelet. This has been done for a number of popular wavelets; the results are listed in Table 3.1. A summary of the key results is given in Box 3.4.

TABLE 3.1   Center Frequency η and Root Mean Square Time and Frequency
Bandwidths $\sigma_t^2, \sigma_\omega^2$ for Several Wavelet Functions

Name	η	σ_t^2	σ_ω^2	$\sigma_t^2 \sigma_\omega^2$
Morlet	ω_0	$\dfrac{1}{2}$	$\dfrac{1}{2}$	$\dfrac{1}{4} = 0.25000$
Paul, $n = 2$	$\dfrac{5}{2}$	$\dfrac{1}{3}$	$\dfrac{5}{4}$	$\dfrac{5}{12} = 0.41667$
Paul, $n = 4$	$\dfrac{9}{2}$	$\dfrac{1}{7}$	$\dfrac{9}{4}$	$\dfrac{9}{28} = 0.32142$
Mexican hat	$\dfrac{4}{3\sqrt{\pi}} = 0.75225$	$\dfrac{7}{6}$	0.40117	0.46804
Derivative of Gaussian (DoG), $n = 6$	$\dfrac{512}{231\sqrt{\pi}} = 1.25049$	$\dfrac{23}{22} = 1.0455$	0.90438	1.13092

BOX 3.4   WAVELET CENTERS AND BANDWIDTHS

If $\psi(t)$ is centered at $t = 0$ and has energy $\|\psi\|^2$, then $\psi_{a,b}(t)$ is centered at $t = b$ and has an rms time bandwidth given by

$$\sigma_t^2 = \frac{1}{\|\psi\|^2} \int_{-\infty}^{\infty} (t - b)^2 \, |\psi_{a,b}(t)|^2 \, dt.$$

If the mother wavelet has an rms bandwidth σ_t', then the bandwidth of the dilated wavelet $\sigma_t = a\sigma_t'$.

The center frequency η of the mother wavelet is

$$\eta = \frac{1}{2\pi \|\psi\|^2} \int_{-\infty}^{\infty} \omega \, |\hat{\psi}(\omega)|^2 \, d\omega.$$

The Fourier transform of $\psi_{a,b}(t)$ is

$$\hat{\psi}_{a,b}(\omega) = \sqrt{a}\,\hat{\psi}(a\omega)e^{-i\omega b}$$

and the center frequency of the dilated wavelet is therefore η/a.

The rms bandwidth of the dilated wavelet around this center frequency is

$$\sigma_\omega^2 = \frac{1}{2\pi \|\psi\|^2} \int_{-\infty}^{\infty} \left(\omega - \frac{\eta}{a}\right)^2 |\hat{\psi}_{a,b}(\omega)|^2 \, d\omega.$$

If the mother wavelet has rms bandwidth σ_ω', then the dilated wavelet has bandwidth $\sigma_\omega = \sigma_\omega'/a$.

3.3.1 Heisenberg Principle

The foregoing discussion about the rms bandwidth of a function leads naturally to an important principle in signal processing often called the Heisenberg uncertainty because it bears a resemblance to its quantum mechanical equivalent. In essence, this principle states that we cannot simultaneously have arbitrarily accurate information about the time and frequency content of a signal. This follows naturally from the properties of the Fourier transform and from the fact that time and frequency are conjugate variables. Suppose, for example, that we have some function $f(t)$ that is reasonably well localized in the time domain. We can reduce its support by creating a new function $f_a(t) = f(t/a)$ by scaling by a factor $a < 1$. However, the Fourier transform of the new function $\hat{f}_a(\omega) = \sqrt{a}\hat{f}(a\omega)$ is now dilated compared with the original function. Ultimately, we must make a trade-off between the time and frequency localization of our wavelets.

The Heisenberg uncertainty states that the temporal and frequency variance of some finite energy function f satisfies

$$\sigma_t^2 \sigma_\omega^2 \leq \frac{1}{4}. \tag{3.25}$$

and is an equality only for Gaussians. For the Morlet wavelet we have just analyzed,

$$\sigma_t^2 \sigma_\omega^2 = \frac{\sigma^2 a^2}{2} \cdot \frac{1}{2\sigma^2 a^2} = \frac{1}{4} \tag{3.26}$$

and it therefore has optimal time-frequency localization for any value of σ. This feature accounts for the popularity of the Morlet wavelet in time-frequency analysis. As a corollary, other types of wavelets that do not have Gaussian envelopes cannot be expected to improve on the time-frequency localization of the Morlet wavelet.

3.4 SCALOGRAMS

We are now in a position to explain the physical significance of the scalogram. From our example, we see that each wavelet is centered at $(b, \omega_0/a)$. About these center points, we draw Heisenberg boxes one standard deviation either side of the mean, so for the Morlet wavelet, these boxes have lengths $\sqrt{2}\sigma a$ along the time axis and $\sqrt{2}/(\sigma a)$ along the frequency axis.

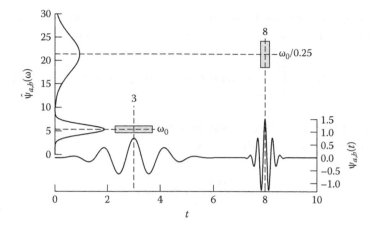

FIGURE 3.5   Heisenberg boxes for the Morlet wavelet for $(a,b,\sigma) = (1,3,1)$ and $(0.25,8,1)$, drawn to scale. $\omega_0 = 5.34$ rad/s and $\sigma = 1$.

For wavelets, the area of these rectangles remains constant at all scales, but the resolution in time and frequency depends on the scaling parameter a. Figure 3.5 shows Heisenberg boxes for the Morlet wavelet for $(a,b,\sigma) = (1,3,1)$ and $(0.25,8,1)$, drawn accurately to scale. As the wavelets are translated in time and scale, they are said to "tile" a 2-D surface usually referred to as the "time-frequency plane." The squared modulus of the CWT of a function $f(t)$ is therefore a measure of the local time-frequency energy density of f in the Heisenberg box associated with each wavelet, and the 2-D plot of this energy density is called a *scalogram*. $|W_f(a,b)|$ is the height of the CWT above the 2-D time-frequency plane, and a path that follows $\max|W_f(a,b)|$ is called the *wavelet ridge*. Finally, values of $a(b)$ along the wavelet ridge yield the phase or local instantaneous frequency of the function f, as demonstrated in the following section.

3.5 RIDGE OF THE CONTINUOUS WAVELET TRANSFORM

It is well known that the instantaneous frequency of an interference pattern is directly proportional to the phase gradient of the measured field; from the preceding discussion, it should be evident that wavelets are able to isolate the local frequency content of such nonstationary signals. We turn our attention now to the question of extracting the phase directly from the CWT and begin with an analogy from the Fourier world.

Suppose we have a fixed frequency signal $f(t)$ with some unknown phase ϕ, that is, let $f(t) = A \cos(\omega_0 t + \phi)$. To find the unknown phase with, say, a lock-in amplifier, we would multiply our signal with a pair of quadrature test signals and then low-pass filter. In effect, we would find

$$\langle f, \cos(\omega_0 t) \rangle = \frac{1}{T} \int_{t-T/2}^{t+T/2} A[\cos\omega_0 t \cos\phi - \sin\omega_0 t \sin\phi]\cos\omega_0 t \, dt \quad (3.27a)$$

where T represents the integration time of the lock-in. Because sin and cos are orthogonal functions, this leaves

$$= \frac{A\cos\phi}{T} \int_{t-T/2}^{t+T/2} \cos^2\omega_0 t \, dt. \quad (3.27b)$$

Substituting $\frac{1}{2}(1+\cos(2\omega_0 t))$ and noting that $\text{sinc}(\omega_0 T) \to 0$ for sufficiently large T, we have

$$\langle f, \cos(\omega_0 t) \rangle = \frac{A}{2}\cos\phi. \quad (3.27c)$$

Similarly, for the inner product with the sin $\omega_0 t$ test function,

$$\langle f, \sin(\omega_0 t) \rangle = \frac{1}{T} \int_{t-T/2}^{t+T/2} A\left[\cos\omega_0 t \cos\phi - \sin\omega_0 t \sin\phi\right]\sin\omega_0 t \, dt \quad (3.28a)$$

$$= \frac{A}{2}\sin\phi. \quad (3.28b)$$

That is, the phase can be obtained from the argument of $\langle f, \exp(i\omega_0 t) \rangle$. The CWT is similar in nature, with the obvious exception that we are using wavelets instead of sinusoids. Nonetheless, the ratio of the imaginary to real part of the CWT gives precisely the tangent of the phase.

Rather than use a particular wavelet, we create a generic analytic wavelet:

$$\psi(t) = g(t)\exp(i\omega_0 t). \quad (3.29)$$

Here, $g(t)$ is a symmetric window function with unit norm $||g|| = 1$, for example, a Gaussian envelope, although any symmetric window function will do, and ω_0 is the modulation frequency that we must choose so that $\hat{\psi}(\omega) \approx 0$ for all $\omega < 0$. Next, we generate a family of wavelets $\psi_{a,b}$ according to Equation (3.5) in the usual way.

Let $f(t) = V(t)\cos(\phi(t))$ represent the alternating current (AC) component of our fringe pattern. Recall that $\hat{\psi}(0) = 0$, so wavelets automatically filter out any direct current (DC) component of a signal. From Equation (3.6), the CWT of our signal is

$$W_f(a,b) = \int_{-\infty}^{\infty} V(t)\cos(\phi(t))\frac{1}{\sqrt{a}}g\left(\frac{t-b}{a}\right)\exp\left[-i\omega_0\left(\frac{t-b}{a}\right)\right]dt. \quad (3.30a)$$

Now, write $\cos\phi$ as

$$\cos\phi = \frac{1}{2}[e^{i\phi} + e^{-i\phi}]$$

and substitute this into the previous equation for $W_f(a,b)$ so that

$$W_f(a,b) = I(\phi) + I(-\phi) \quad (3.30b)$$

where

$$I(\phi) = \frac{1}{2}\int_{-\infty}^{\infty} V(t)\exp(i\phi(t))\frac{1}{\sqrt{a}}g\left(\frac{t-b}{a}\right)\exp\left[-i\omega_0\left(\frac{t-b}{a}\right)\right]dt \quad (3.30c)$$

and similarly for $I(-\phi)$. Changing variables and writing $t - b = x$ allows $I(\phi)$ to be written as

$$I(\phi) = \frac{1}{2}\int_{-\infty}^{\infty} V(x+b)\exp(i\phi(x+b))\frac{1}{\sqrt{a}}g\left(\frac{x}{a}\right)\exp\left(-i\omega_0\frac{x}{a}\right)dx. \quad (3.30d)$$

Next, over the support of the window function g, we expand the amplitude and phase terms in a Taylor's expansion to first order about b:

$$V(x+b) \approx V(b) + xV'(b) \quad (3.30e)$$

and

$$\phi(x+b) \approx \phi(b) + x\phi'(b). \quad (3.30f)$$

and substitute these into $I(\phi)$:

$$I(\phi) = \frac{1}{2} \int_{-\infty}^{\infty} [V(b) + xV'(b)] \exp(i\phi(b)) \frac{1}{\sqrt{a}} g\left(\frac{x}{a}\right)$$

$$\times \exp\left(-i\left[\omega_0 \frac{x}{a} - \frac{x}{a} a\phi'(b)\right]\right) dx. \tag{3.30g}$$

Because the product of an odd function $xV'(b)$ with the even window function is zero,

$$I(\phi) = \frac{V(b)}{2} \exp(i\phi(b)) \int_{-\infty}^{\infty} \frac{1}{\sqrt{a}} g\left(\frac{x}{a}\right) \exp\left(-i\left[\omega_0 \frac{x}{a} - \frac{x}{a} a\phi'(b)\right]\right) dx. \tag{3.30h}$$

With x/a as the variable of integration, and from the definition of the Fourier transform,

$$I(\phi) = \frac{\sqrt{a}}{2} V(b) \exp[i\phi(b)] \hat{g}(\omega_0 - a\phi'(b)). \tag{3.30i}$$

A similar analysis for $I(-\phi)$ yields

$$I(-\phi) = \frac{\sqrt{a}}{2} V(b) \exp[-i\phi(b)] \hat{g}(\omega_0 + a\phi'(b)) \tag{3.30j}$$

and provided $\phi'(b) \geq \omega_0/a, I(-\phi) \ll 1$. Hence, $W_f(a,b) \approx I(\phi)$, and

$$|W_f(a,b)|^2 = \frac{a}{4} V^2(b) |\hat{g}(\omega_0 - a\phi'(b))|^2 . \tag{3.30k}$$

$\hat{g}(\omega)$ has its maximum modulus when $\omega = 0$, that is, when

$$\frac{\omega_0}{a(b)} = \phi'(b). \tag{3.30l}$$

The wavelet ridge is the path that follows the maximum modulus of the CWT and from Equations (3.30l) and (3.30i) yields either the instantaneous fringe frequency or the phase, respectively. The instantaneous frequency is obtained from values of $a(b)$ on the ridge while the phase is returned, modulo 2π, from

$$\tan\phi = \frac{\Im\{W_f(a,b)\}}{\Re\{W_f(a,b)\}} \qquad (3.31)$$

where $\Re$ and $\Im$ denote the real and imaginary parts of the CWT, respectively, and are evaluated on the ridge. The exact form of the expression for $\phi'(b)$ depends on the normalization adopted, that is, whether the wavelets are scaled by $1/\sqrt{a}$ or by $1/a$. Li et al. [4] provided expressions for both cases, in addition to useful formulae for determining the minimum and maximum values of the scaling parameter a to use in the numerical analysis.

A Taylor's expansion of the amplitude $V(x)$ and phase $\phi(x)$ to second order shows that second-order terms are negligible if [7]

$$\frac{\omega_0^2}{|\phi'(b)|^2} \frac{|V''(b)|}{|V(b)|} \ll 1 \qquad (3.32a)$$

and

$$\omega_0^2 \frac{|\phi''(b)|}{|\phi'(b)|^2} \ll 1 \qquad (3.32b)$$

Essentially, V' and ϕ' must have slow variations if the instantaneous frequency ϕ' is small. Detailed and rigorous expansions for the case of moderate to strongly modulated signals can be found in Lilly and Olhede [8].

3.6 THE GRADIENT METHOD

We are now in a position to explain some features of Example 3.1 and to discuss more fully the relative advantages and disadvantages of the gradient method. It is clear from Equation (3.30l) that values of ω_0/a along the wavelet ridge give the instantaneous fringe frequency as a function of position b. This approach is usually termed the *gradient method* in the literature, and we have shown that phase distributions can be successfully recovered from a single interference pattern using this method [9,10] even in the presence of noise, although great care must be taken at the edges.

In our case, we found that linear predictive extrapolation (LPE) [5] was necessary to ensure the best results. The advantage of this approach lies in the fact that the phase map does not need to be unwrapped. However, it is the case that small errors in the recovered instantaneous frequency will tend to propagate through the recovered phase as a consequence of the integration [4,11]. In addition, the uncertainty in the instantaneous frequency depends on how finely the scale parameter a is discretized [11], with small errors requiring that a be finely discretized indeed with a concomitant increase in computation time. By contrast, the phase method can achieve small errors in the recovered phase with a relatively coarse discretization of a. Recent numerical simulations using the Morlet wavelet [4] found that the gradient method performed more poorly than the phase method, even in the absence of noise, despite truncating the left-hand edge of the retrieved phase gradient map to limit the propagation of errors during integration. If one is interested in recovering the phase, then it seems clear from the literature that the gradient method should not be used despite its intuitive appeal. There are, however, some applications in which the phase gradient or instantaneous frequency is precisely the quantity required, for example, the dispersion of bending waves in beams [12], birefringence dispersion of liquid crystals [13], and the extraction of strain [14].

3.6.1 Correcting the Instantaneous Frequency

We have shown that if the mother wavelet has a center frequency η, then scaled versions of it have center frequencies η/a; hence, in principle, we can calculate the instantaneous frequency of a signal from values of a along the wavelet ridge. In practice, these calculated frequencies may need correcting. As a concrete example, consider the Morlet wavelet. If $\omega_0 = 6$ and we have a local fringe frequency of $2\pi/32$ rad/pixel, we would expect that the value of a that maximizes the CWT would be $6/a = 2\pi/32 \Rightarrow a = 30.56$. In practice, it is a little larger than this value.

We can find the optimal value of a algebraically in the following way: The CWT of our constant-frequency cosinusoid using the Morlet wavelet is

$$W_f(a) \propto \frac{1}{\sqrt{a}} \int_{-\infty}^{\infty} \cos\left(\frac{6t}{a}\right) \exp\left[-\frac{1}{2}\left(\frac{t}{\sigma a}\right)^2\right] \cos(k_0 t)\,dt \qquad (3.33a)$$

where $k_0 = 2\pi/32$, and we have dropped the b dependence because we are dealing with a constant-frequency signal. Note also that the quadrature term involving $\sin(6t/a)$ evaluates to zero. Substituting $u = t/(a\sigma)$, we have

$$= \sqrt{a}\sigma \int_{-\infty}^{\infty} \exp\left(-\frac{1}{2}u^2\right)\cos(6\sigma u)\cos(k_0\sigma a u)\,du \qquad (3.33b)$$

$$= \frac{\sqrt{a}\sigma}{2} \int_{-\infty}^{\infty} \exp\left(-\frac{1}{2}u^2\right)[\cos k_1 u + \cos k_2 u]\,du \qquad (3.33c)$$

where $k_1 = (6\sigma - k_0\sigma a)$ and $k_2 = (6\sigma + k_0\sigma a)$. Evaluating the integral gives

$$W_f(a) = \frac{\sqrt{a\pi}\sigma}{\sqrt{2}}\left[\exp\left(-\frac{1}{2}k_1^2\right) + \exp\left(-\frac{1}{2}k_2^2\right)\right]. \qquad (3.33d)$$

To find the value of a that maximizes the CWT, we solve

$$\frac{dW_f(a)}{da} = 0 \qquad (3.33e)$$

which yields

$$0 = \exp\left(-\frac{1}{2}k_1^2\right) + \exp\left(-\frac{1}{2}k_2^2\right)$$
$$+ 2a\sigma k_0\left[k_1 \exp\left(-\frac{1}{2}k_1^2\right) - k_2 \exp\left(-\frac{1}{2}k_2^2\right)\right]. \qquad (3.33f)$$

Solving this equation numerically gives $a_{opt} = 32.17$. All of the simulations in this chapter have been performed with a free MATLAB toolbox, the "yet another wavelet toolbox" (yawtb) [15]. As a check, the same constant-frequency signal was analyzed using this toolbox, which returned an optimal value for a of 32.20 with the difference entirely caused by setting $a_{step} = 0.2$. If the gradient method is to be used, it is prudent to use a known constant-frequency signal to check for any additional correction factors.

3.7 THE PHASE METHOD

Alternatively, along the ridge, we observe from Equation (3.30i) that the argument of the complex CWT yields the phase directly:

$$\arg(W_f(a,b)) = \phi(b). \qquad (3.34)$$

Methods that use this approach to recover the phase are generally termed *phase methods* in the literature. A significant advantage of this approach is that any errors in the recovered phase, which have as their limit the approximation implicit in Equation (3.30f), are locally isolated and do not propagate. On the other hand, it is now necessary to unwrap the phase map because the phase will be returned modulo 2π. To demonstrate how this works in practice, we return to Example 3.1.

Example 3.3: Linear Chirp Revisited

We already have the wavelet ridge from our previous analysis, and this step remains the same irrespective of whether one intends to use the gradient method or the phase method. In Figure 3.6, we have plotted the argument of the complex CWT for one row of our

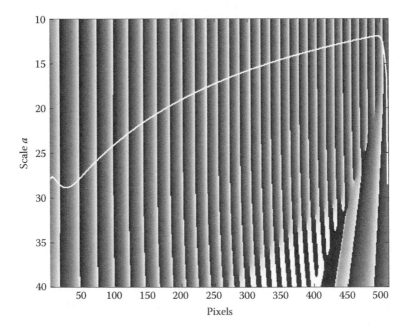

FIGURE 3.6 The arg *Wf(a,b)* of one row of Figure 3.1 using the Morlet wavelet with $\omega_0 = 6$ and $\sigma = 0.5$. The solid white line is the wavelet ridge.

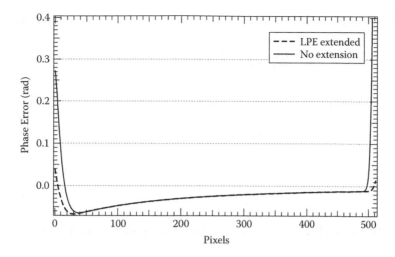

FIGURE 3.7 Phase error for the fringe pattern of Figure 3.1 using the phase method, with and without edge extension using linear predictive extrapolation (LPE).

fringe pattern, with the wavelet ridge shown superimposed as a solid white line.

The arg $W_f(a,b)$ at each point $a(b)$ along the ridge is saved and subsequently unwrapped. The spatial carrier must also be removed, which amounts to subtracting $2\pi t/32$, $t \in \{1,512\}$ from the phase. The recovered phase is plotted in Figure 3.4, along with the values expected from Equation (3.7). The phase error is shown in Figure 3.7 and is modest except at the left- and right-hand edges. However, if the edges of the signal are extended using LPE as discussed further in the chapter, then the phase errors near the edges can be substantially reduced, as can be seen in the figure.

3.8 FOURIER APPROACH TO CWT

In practice, the CWT is not usually evaluated in the time domain. For reasons of efficiency and speed, most wavelet toolboxes make use of the FFT, despite the fact that the CWT is clearly *not* a convolution integral. The final step, then, is to show how this is done.

Let

$$\tilde{\psi}_a(t) = \frac{1}{\sqrt{a}} \psi^*\left(-\frac{t}{a}\right) \tag{3.35a}$$

so that the CWT of Equation (3.6) *can* be written as a convolution integral:

$$W_f(a,b) = \int_{-\infty}^{\infty} f(t)\tilde{\psi}_a(b-t)\,dt. \tag{3.35b}$$

Hence, by the properties of the Fourier transform,

$$\mathcal{F}\{W_f(a,b)\} = \hat{f}(\omega)\hat{\tilde{\psi}}_a(\omega). \tag{3.35c}$$

Because

$$\hat{\tilde{\psi}}_a(\omega) = \sqrt{a}\,\hat{\psi}^*(a\omega), \tag{3.35d}$$

the CWT is readily calculated from the inverse FFT according to

$$W_f(a,b) = \mathcal{F}^{-1}\{\sqrt{a}\,\hat{f}(\omega)\hat{\psi}^*(a\omega)\} \tag{3.35e}$$

$$= \frac{\sqrt{a}}{2\pi}\int_{-\infty}^{\infty} \hat{f}(\omega)\hat{\psi}^*(a\omega)e^{i\omega b}\,d\omega. \tag{3.35f}$$

If we intend to implement the CWT using a FFT, then the data will obviously be sampled, and we should write Equation (3.35f) as [16]

$$W_f(a,b) = \sqrt{a}\sum_{k=0}^{N-1} \hat{f}(\omega_k)\hat{\psi}^*(a\omega_k)\exp(i\omega_k b) \tag{3.36a}$$

where

$$w_k = \begin{cases} \dfrac{2\pi k}{N} & k \le \dfrac{N}{2} \\[2ex] -\dfrac{2\pi k}{N} & k > \dfrac{N}{2} \end{cases} \tag{3.36b}$$

and we have assumed unit sampling interval. This could be implemented reasonably efficiently as follows:

```
F = fft(f)
for k = 1:length(a)
    c(k,:) = sqrt(a(k)) * ifft(F*conj(morlet1d(a(k)*k)));
end
```

The vectors f, a, and k contain, respectively, the signal to be analyzed, a list of scaling factors, and a list of frequencies as per Equation (3.36b). Here, morlet1d is a function that evaluates the Morlet wavelet in the *frequency* domain. MATLAB enables this computation to be vectorized so that the entire calculation may be done without looping, a feature that yawtb exploits to further reduce computation time. Free wavelet toolboxes may be downloaded from the Web, and a list is maintained that is helpful in this regard (http://www.wavelet.org).

3.9 EFFECT OF DISCONTINUITIES AT THE SIGNAL EDGE

The edges of signals can be problematic if some care is not taken because wavelets are particularly good at locating discontinuities. In particular, these edges not only can produce significant errors in the instantaneous frequency (as can be seen in Figure 3.4), but also will cause phase errors. We give some qualitative reasoning for these errors with the aid of a simple example.

To simulate the effect of an edge, we generated a cosinusoidal signal $y(t) = 0.5 \cos(2\pi t/32)$ (which corresponds to the spatial carrier of our first example) and found the wavelet ridge for two cases: (1) with no step discontinuity and (2) with $y(t) = 0$ for $t = 0, \ldots, 96$. Figure 3.8 shows the case for

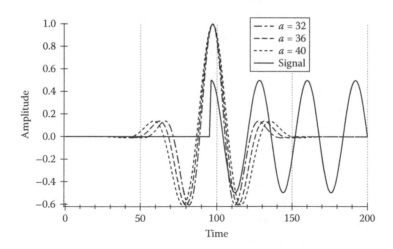

FIGURE 3.8 Step discontinuity corresponding to a simulated edge and Morlet wavelets of various scales.

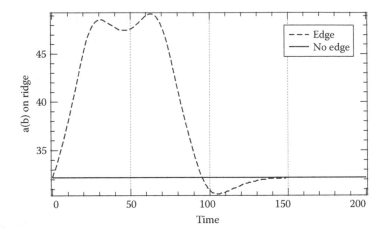

FIGURE 3.9   Optimal values of $a(b)$ along the ridge, with and without the step discontinuity.

which the signal has a step discontinuity, together with the Morlet wavelet for three different values of a and $\sigma = 0.5$. For $a = 32$ (which is close to the optimal value of 32.2), we can see from the figure that the scaled wavelet is a good match to our signal. We would therefore expect that the CWT should evaluate to a large number, indicative of the good match between wavelet and signal. Unfortunately, exactly half the wavelet does not overlap the signal. A common strategy, either implicitly or explicitly, is to "zero pad" the signal at each end, so in this particular instance, the CWT only has half its true value. If we increase a to, say, $a = 36$, then the value of the CWT actually *increases* as the slightly wider wavelet offers more overlap with the continuous part of the signal, pushing the wavelet ridge to higher values of a than it should. One would then conclude that the instantaneous frequency at the signal edge is rather smaller than its true value, and a significant error would ensue. Figure 3.9 shows the values of $a(b)$ on the wavelet ridge for both cases. We see that when there is an edge, the optimal value of a increases (so therefore there is a lower estimated instantaneous frequency) as the wavelet approaches the edge, decreasing slightly as it passes through the edge, before reaching the correct value of 32.2.

Although an obvious solution is to acquire more data so that the left- and right-hand edges are arbitrarily extended, this is not possible for fixed-format devices such as CCD cameras. One possible solution is simply to truncate those parts of the CWT that are deemed to be unreliable because of edge effects. If the mother wavelet has support $[-t_s, t_s]$, then the scaled wavelet has support $[-at_s, at_s]$, so the amount by which the CWT should be truncated depends on the choice of mother wavelet as well as the optimal value of a at each edge.

An alternative approach is to *smoothly* extend the data at each end, such that the function and at least its first derivative are continuous across the edge.* LPE [5] provides a straightforward way to achieve this. For the noise-free example given, LPE with a history of 64 (two periods) and an order of 2 extends the oscillations beyond the edge perfectly, so that it is indistinguishable from the original signal. This approach still works well, even in the presence of noise [9,10].

3.10 ONE-DIMENSIONAL WAVELET FUNCTIONS

One of the major attractions of the CWT is the multiplicity of functions that can be used for the mother wavelet. Listed next is a selection of commonly used functions, each of which has been scaled so that it has unit energy.

$$\psi_{Paul}(t) = k_n(1 - it)^{-(n+1)} \tag{3.37a}$$

$$\psi_{Morlet}(t) = \frac{1}{\sqrt[4]{\sigma^2 \pi}} \exp(i\omega_0 t) \exp\left(\frac{-t^2}{2\sigma^2}\right) \tag{3.37b}$$

$$\psi_{Mexican}(t) = \frac{2}{\sqrt[4]{9\pi}} (1 - t^2) \exp\left(\frac{-t^2}{2}\right) \tag{3.37c}$$

$$\psi_{Hermitian}(t) = \frac{2}{\sqrt[4]{25\pi}} (1 + it - t^2) \exp\left(\frac{-t^2}{2}\right) \tag{3.37d}$$

$$\psi_{DoG}(t) = \frac{(-1)^{n+1}}{\sqrt{\Gamma\left(n + \frac{1}{2}\right)}} \frac{d^n}{dt^n} \exp\left(\frac{-t^2}{2}\right) \tag{3.37e}$$

$$\psi_{GMW}(t) = a_{\beta,\gamma} \omega^\beta \exp(-\omega^\gamma) H(\omega). \tag{3.37f}$$

The Paul wavelet of order n is complex and is generally used with $n = 2$ or 4 [17,18]. For these values of n, the corresponding scaling factors k_n needed to ensure that the wavelet has unit energy are $k_2 = \sqrt{8/3\pi}$ and $k_4 = \sqrt{128/35\pi}$.

---

* See the technical note on vanishing moments in Box 3.5.

BOX 3.5   TECHNICAL NOTE VANISHING MOMENTS

A wavelet is said to have n vanishing moments if

$$\int_{-\infty}^{\infty} t^k \, \psi(t) \, dt = 0 \quad \text{for} \quad 0 \le k < n.$$

A function f that is m times differentiable in the interval $[v - h, v + h]$ can be approximated by a Taylor polynomial p_v such that

$$f(t) = p_v(t) + \epsilon_v(t) \quad \text{with} \quad |\epsilon_v(t)| \le K |t - v|^\alpha$$

where K is a constant and

$$p_v(t) = \sum_{k=0}^{m-1} \frac{f^{(k)}(v)}{k!}(t - v)^k.$$

If $n > \alpha$, one can show [7] that $W_{p_v}(a,b) = 0$, so that

$$W_f(a,b) = W_{\epsilon_v}(a,b).$$

This is sometimes referred to as the "zoom-in" ability of wavelets, and the decay of the wavelet coefficients across the scales can be used to measure the regularity α of f.

The Morlet wavelet has already been discussed in some detail, but we note that in the literature it is also known as the Gabor wavelet. This wavelet is often presented in its simplest form, with $\sigma = 1$, but its performance can be considerably improved by setting $\sigma = 0.5$ [19]. The Mexican hat function is sometimes called the Laplacian-of-Gaussian (LoG) wavelet because it is the second derivative of a Gaussian, and because it is real, the fringes must be converted to complex form, which can easily be done via the FFT [20]. Interestingly, the Mexican hat wavelet does have a complex extension called the Hermitian hat wavelet [21,22], the real and imaginary parts of which correspond to the second and first derivatives of a Gaussian, respectively. The derivative-of-Gaussian (DoG) wavelet is the nth derivative of a Gaussian and generates the Mexican hat for $n = 2$ (sometimes called the Marr wavelet [16]).

The last wavelet in our list is the generalized Morse wavelet (GMW) [8,23]. These are designed to maximize the eigenvalues of a particular joint time-frequency localization operator [24] and are controlled by two parameters, β and γ. For a valid wavelet, we must have $\beta > 0$ and $\gamma > 0$. Here, $\alpha_{\beta,\gamma}$ is a normalization constant, and $H(\omega)$ is the Heaviside or unit step function. Varying these two parameters enables these wavelets to have a broad range of characteristics while remaining exactly analytic. The GMWs are in fact a "superfamily" of wavelets and subsume two other families of wavelets: the Morse or Paul wavelets for $\gamma = 1$ and the DoG wavelets for $\gamma = 2$ [25]. Examples of GMWs for a range of values of (β, γ) may be found in Lilly and Olhede [25].

There is another family of functions that is commonly used to create wavelets and that has so far been ignored. These are the cardinal B-splines. The B-spline wavelet of order $n + 1$ is defined recursively by $B_n(t) = B_{n-1} *$ $\chi_{[0,1)}$ where χ is the indicator function on the unit interval and $*$ denotes convolution [26]. By the convolution theorem, therefore, its Fourier transform is the product of sinc functions. Conversely, the Shannon wavelet is generated from a sinc function in the time domain and has, as its Fourier transform, the indicator function in the frequency domain. Because both of these wavelets produce large phase errors when used for phase recovery, their use in this context is not recommended [18], and we do not consider them any further.

The time and frequency domain responses of some of the more commonly used wavelets, carefully scaled so that they have unit energy, are shown in Figure 3.10. We also provide, in Table 3.1, the center frequency η and rms time and frequency bandwidths for some of the more popular wavelets.

Example 3.4: The "Peaks" Function

To complete this section on 1-D wavelets, we show how they can be used to recover a 2-D phase map. We need a suitable phase object; one that is often used in the literature is the "peaks" function, a built-in MATLAB function that generates a phase distribution in the xy-plane from Gaussian distributions according to

$$\phi(x,y) = 3(1-x)^2 \exp(-x^2 - (y+1)^2) - 10(x/5 - x^3 - y^5)\exp(-x^2 - y^2)$$
$$-1/3\exp(-(x+1)^2 - y^2)$$

$$(3.38)$$

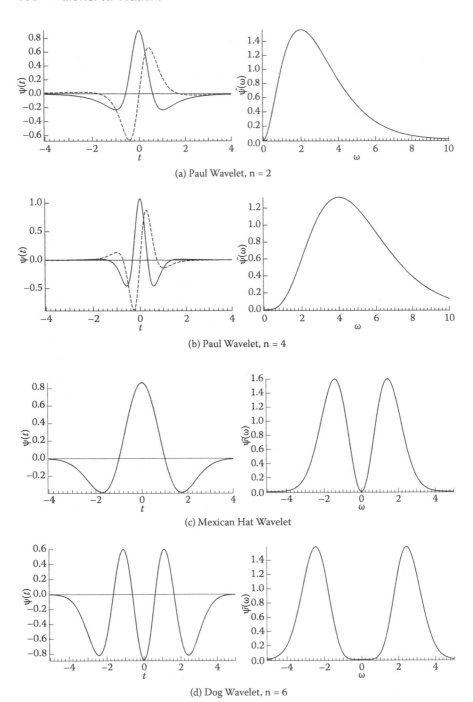

FIGURE 3.10 Mother wavelets $\psi(t)$ and their Fourier transforms $\hat{\psi}(\omega)$. The real and imaginary parts of the functions are shown as solid and dashed lines, respectively.

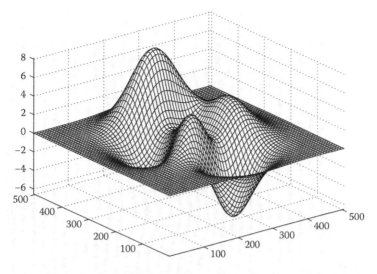

FIGURE 3.11   The 2-D phase map $\phi(x,y)$ generated by the MATLAB peaks function.

and that is shown in Figure 3.11. With a spatial carrier in the x-direction only, the interference pattern $I(x,y)$ shown in Figure 3.12 was produced from

$$I(x,y) = 32\left(1 + \cos\left[\frac{2\pi x}{16} + \phi(x,y)\right]\right), \quad x,y = 1,\ldots,512 \quad (3.39)$$

where x,y are the pixel positions, and $\phi(x,y)$ is the phase distribution given by the peaks function. Recovering the phase map then proceeds in the following way: For each row of the interference pattern (because this is the direction of the spatial carrier), we extend the left- and right-hand edges by 100 pixels using LPE, then use our chosen wavelet (the Morlet wavelet) and the phase method to extract the phase along that row. The recovered phase is unwrapped and the linear term caused by the carrier subtracted. This is repeated for each row of the interference pattern, after which the entire phase map may be reassembled. Subtracting the recovered phase from the known, expected values yields the 2-D error map shown in Figure 3.13. It is almost certainly the case that a different choice of wavelet will result in a different error map, but the objective of this example is to demonstrate the steps that must be followed to recover the phase. A detailed study regarding the relative merits of different wavelets may be found in Gdeisat et al. [18].

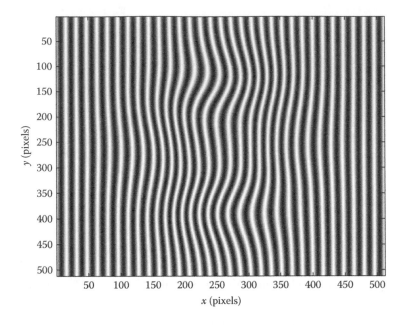

FIGURE 3.12 The 2-D interference pattern with the peaks function and a spatial carrier of $2\pi/16$ rad/pixel.

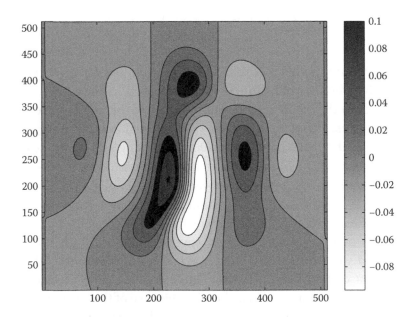

FIGURE 3.13 Phase error map (in radians) for the peaks function using the Morlet wavelet with $\omega_0 = 6$, $\sigma = 0.5$.

3.11 TWO-DIMENSIONAL CONTINUOUS WAVELET TRANSFORM

Although our previous example has just shown that a 2-D phase map can successfully be recovered using 1-D wavelets, smaller phase errors, especially when the fringes are contaminated by noise, are achieved if 2-D wavelets are used [27–29]. We therefore turn our attention to the 2-D CWT. Fortunately, many of the concepts already discussed for the 1-D case carry over into two dimensions.

A 2-D wavelet is a function $\psi(\mathbf{x})$ with zero mean and unit energy satisfying the same admissibility condition as its 1-D counterpart, namely,

$$\iint \psi(\mathbf{x})d^2\mathbf{x} = 0 \Leftrightarrow \hat{\psi}(\mathbf{0}) = 0. \tag{3.40}$$

Because we are now working in two dimensions, it is convenient to introduce vector notation so that $\mathbf{x} = (x, y)$, $\mathbf{b} = (b_x, b_y)$, and similarly for other quantities. $\hat{\psi}(\mathbf{k})$ is the 2-D Fourier transform of $\psi(\mathbf{x})$ where $\mathbf{k}$ is the spatial frequency with $|\mathbf{k}|^2 = \mathbf{k} \cdot \mathbf{k} = k_x^2 + k_y^2$.

In 2-D, we can define three unitary operators on a finite energy signal s, namely,

$$\text{translations: } (T_{\mathbf{b}}s)(\mathbf{x}) = s(\mathbf{x} - \mathbf{b}), \mathbf{b} \in \mathbb{R}^2 \tag{3.41a}$$

$$\text{dilations: } (D_a s)(\mathbf{x}) = a^{-1}s(a^{-1}\mathbf{x}), a > 0 \tag{3.41b}$$

$$\text{rotations: } (R_\theta s)(\mathbf{x}) = s(r_{-\theta}, (\mathbf{x})), \theta \in [0, 2\pi) \tag{3.41c}$$

where r_θ is the usual 2×2 rotation matrix:

$$r_\theta = \begin{bmatrix} \cos(\theta) & -\sin(\theta) \\ \sin(\theta) & \cos(\theta) \end{bmatrix}.$$

For the machinery of the CWT to work correctly in 2-D, our family of wavelets needs to be generated by translations, dilations, *and* rotations [30]:

$$\psi_{\mathbf{b},a,\theta} = a^{-1}\psi(a^{-1}r_{-\theta}(\mathbf{x} - \mathbf{b})). \tag{3.42}$$

The 2-D CWT of a signal f is then readily defined in an analogous manner to the 1-D case:

$$W_f(a,\mathbf{b},\theta)=\langle f,\psi_{\mathbf{b},a,\theta}\rangle=a^{-1}\iint f(\mathbf{x})\psi^*(a^{-1}r_{-\theta}(\mathbf{x}-\mathbf{b}))d^2\mathbf{x} \quad (3.43)$$

Before we move on to consider some practical examples, there are several aspects of the 2-D CWT that need to be discussed. First, a note about the representation of this transform. The 1-D CWT takes a 1-D signal and produces a 2-D representation (scale and translation) that is easily plotted and visualized. Here, we have taken a 2-D signal and produced a 4-D representation that cannot be readily plotted or visualized. Two approaches to this visualization problem are discussed in detail elsewhere [30, 31]; in essence, either one suppresses one of the variables and considers a subsection of the entire parameter space or one can integrate out one of the variables, thereby producing various partial energy densities. The first approach is the more appropriate for our phase recovery problem, but even so, there exist several possible representations [30]. The most useful for our purposes is the *position representation*, in which a and θ are fixed and the 2-D CWT is considered a function of $\mathbf{b}$ alone. New values of (a,θ) are then considered so that a set of snapshots is produced, labeled by the (a,θ) pair as shown schematically in Figure 3.14 for an original fringe pattern that was 512 pixels square. Each of the 512×512 matrices in the position representation

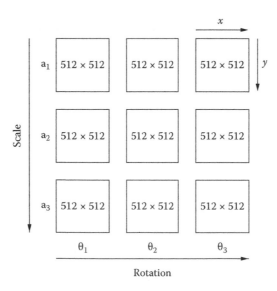

FIGURE 3.14   Position representation of the 2-D CWT for a 512×512 pixel signal.

may be considered the scalogram for its particular (a,θ) pair. For a given pixel position, we scan through the set of snapshots, looking for the (a,θ) values that maximize the CWT for that particular pixel, repeating the process for each pixel in our image. Our wavelet ridge is a $512 \times 512 \times 2$ array that contains $a(\mathbf{b}), \theta(\mathbf{b})$. We show in the Appendix that the 2-D CWT is maximized when the scale and orientation of the wavelet match that of the local fringe pattern; hence, we can find either the instantaneous frequency or the phase from the 2-D CWT exactly as we did for the 1-D CWT.

The second important difference is that wavelets in 2-D are either isotropic or directional. Because isotropic wavelets are rotationally invariant, we can drop the θ variable in Equation (3.43), which both simplifies the 2-D CWT and reduces the parameter space in which we must search. Listed next are some typical isotropic wavelets, of which the 2-D Mexican hat function, Equation (3.44a), is the archetypical:

$$\psi_{\text{Mexican}}(\mathbf{x}) = (1 - |\mathbf{x}|^2)\exp\left(-\frac{1}{2}|\mathbf{x}|^2\right) \qquad (3.44\text{a})$$

$$\psi_{\text{Paul}}(\mathbf{x}) = i^n (1 - i|\mathbf{x}|)^{-(n+1)} \qquad (3.44\text{b})$$

$$\psi_H^n(\mathbf{x}) = (-\nabla^2)^n \exp\left(-\frac{1}{2}|\mathbf{x}|^2\right) \qquad (3.44\text{c})$$

$$\hat{\psi}_{\text{Bessel}}(\mathbf{k}) = \begin{cases} 1 & R/\gamma \leq |\mathbf{k}| \leq R \\ 0 & \text{otherwise} \end{cases} \qquad (3.44\text{d})$$

$$\hat{\psi}_{\text{Pet}}(\mathbf{k}) = \begin{cases} \cos^2\left(\dfrac{\pi}{2}\log_2\dfrac{|\mathbf{k}|}{2\pi}\right) & \pi < |\mathbf{k}| < 4\pi \\ 0 & |\mathbf{k}| < \pi \text{ and } |\mathbf{k}| > 4\pi \end{cases} \qquad (3.44\text{e})$$

$$\hat{\psi}_{\text{Halo}}(\mathbf{k}) = \exp\left(-\frac{1}{2}(|\mathbf{k}| - |\mathbf{k}_0|)^2\right) \qquad (3.44\text{f})$$

$$\psi_D(\mathbf{x}) = \alpha^{-2}\phi(\alpha^{-1}\mathbf{x}) - \phi(\mathbf{x}) \quad 0 < \alpha < 1. \qquad (3.44\text{g})$$

The 2-D Fourier transform of a function $h(x, y)$ is defined as [6]

$$H(k_x, k_y) = \int_{-\infty}^{\infty} h(x, y)e^{-i(k_x x + k_y y)}\, dx\, dy. \qquad (3.45)$$

The 2-D Fourier transform can be viewed as two successive 1-D transforms. To see this, we rewrite Equation (3.45) as

$$H(k_x, k_y) = \int_{-\infty}^{\infty} e^{-ik_y y} \left[\int_{-\infty}^{\infty} h(x, y) e^{-ik_x x} \, dx \right] dy \qquad (3.46a)$$

The term in square brackets is simply the 1-D Fourier transform of $h(x, y)$ with respect to x:

$$Z(k_x, y) = \int_{-\infty}^{\infty} h(x, y) e^{-ik_x x} \, dx \qquad (3.46b)$$

so that

$$H(k_x, k_y) = \int_{-\infty}^{\infty} Z(k_x, y) e^{-ik_y y} \, dy. \qquad (3.46c)$$

Thus, the properties that we outlined previously for the 1-D Fourier transform carry over into the 2-D version. For example, the Mexican hat mother wavelet is obtained from $-\nabla^2 \exp(-x^2/2)$, where $\nabla^2 = \partial^2/\partial x^2 + \partial^2/\partial y^2$ is the Laplacian operator. Because it is the second derivative of a Gaussian, we can easily use the properties of the Fourier transform to find $\hat{\psi}$:

$$\hat{\psi}(\mathbf{k}) = (k_x^2 + k_y^2) \exp(-(k_x^2 + k_y^2)/2). \qquad (3.47)$$

Applying higher-order Laplacians to the Gaussian function generates the third wavelet listed in an analogous manner to the 1-D DoG wavelet we saw previously. The Bessel wavelet, so called because its Fourier transform is a Bessel function [30],

$$\psi_{\text{Bessel}}(\mathbf{x}) \approx \frac{J_1(r)}{r} - \frac{1}{\gamma^2} \frac{J_1(r/\gamma)}{r/\gamma} \qquad r = |\mathbf{x}| \qquad (3.48)$$

has inner and outer radii of R/γ and R, respectively. The Pet wavelet, which has better resolving power than the Mexican hat, was originally designed for efficient sorting of objects in astrophysical images [32]. The Halo wavelet [33] is real and selects an annular region $|\mathbf{k}| \approx |\mathbf{k}_0|$. The final type of isotropic wavelet is the difference wavelet ψ_D. If ϕ is a smooth non-negative function with all moments of order 1 vanishing at the origin, then the function ψ_D given by Equation (3.44g) is a wavelet that satisfies the admissibility condition [30]. A typical example is the difference of Gaussian wavelets for which an isotropic Gaussian function is used for ϕ. Some of these isotropic wavelets are shown in Figure 3.15. In particular, note that in the Fourier domain, each function has its energy distributed

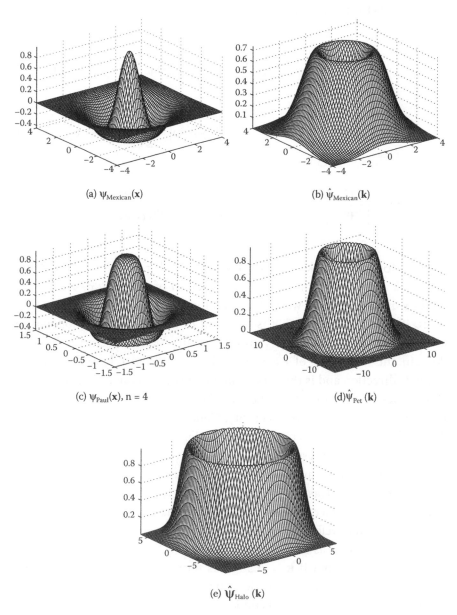

(a) $\psi_{Mexican}(\mathbf{x})$

(b) $\hat{\psi}_{Mexican}(\mathbf{k})$

(c) $\psi_{Paul}(\mathbf{x})$, $n = 4$

(d) $\hat{\psi}_{Pet}(\mathbf{k})$

(e) $\hat{\psi}_{Halo}(\mathbf{k})$

FIGURE 3.15　A selection of 2-D isotropic wavelets.

symmetrically in the k_x, k_y plane and that $\hat{\psi}(0) = 0$, as required by the admissibility condition.

In contrast to the isotropic wavelets, a directional wavelet is one whose support in $\mathbf{k}$-space is contained in a convex cone, the apex of which is at the origin [30]. We list next some common directional wavelets, of which the Morlet is the archetypical:

$$\psi_{\text{Morlet}} = \exp(i\mathbf{k}_0 \cdot \mathbf{x}) \exp\left(\frac{1}{2}|A\mathbf{x}|^2\right) \tag{3.49a}$$

$$\psi_{\text{MM}} = \left(2 - \frac{1}{\varepsilon}x^2 - (y - ik_0)^2\right) \exp(ik_0 y) \exp\left[-\frac{1}{2}\left(\frac{1}{\varepsilon}x^2 + y^2\right)\right] \tag{3.49b}$$

$$\hat{\psi}_m^{(\alpha)}(\mathbf{k}) = \begin{cases} (\mathbf{k} \cdot \mathbf{e}_{\tilde{\alpha}})^m (\mathbf{k} \cdot \mathbf{e}_{-\tilde{\alpha}})^m \exp(-\mathbf{k} \cdot \eta), & \mathbf{k} \in C(-\alpha, \alpha) \\ 0 & \text{otherwise} \end{cases} \tag{3.49c}$$

For the Morlet wavelet, $\mathbf{k}_0$ is the wave vector and $A = \text{diag}[\varepsilon^{-1/2}, 1]$, and $\varepsilon \geq 1$ is a 2×2 anisotropy matrix that has the effect of mapping $x \to x/\sqrt{\varepsilon}$ but leaves y unchanged. The wavelet is therefore elongated in the x-direction (if $\varepsilon > 1$) and has a phase that is constant in the direction orthogonal to $\mathbf{k}_0$ and linear in $\mathbf{x}$, $\text{mod}(2\pi/|\mathbf{k}_0|)$ along the direction of $\mathbf{k}_0$. In the Fourier domain, the "footprint" of $\hat{\psi}_{\text{Morlet}}$ is an ellipse centered at $\mathbf{k}_0$ and elongated in the k_y direction and is thus contained in a convex cone. The ratio of the ellipse axes is equal to $\sqrt{\varepsilon}$, so the cone becomes narrower as ε increases, which increases its angular selectivity. Generally, this wavelet is used with $\mathbf{k}_0 = (0, k_0)$, with $k_0 = 6$ a common choice. Figure 3.16 shows the time and frequency domain representations of this wavelet. We note that in the frequency domain, this wavelet does indeed have its energy concentrated in a cone, in clear contrast to the isotropic wavelets. Finally, as a practical note, we can introduce another parameter σ that controls the rms width of the Gaussian envelope in exactly the same way as its 1-D counterpart. For the same reason as its 1-D counterpart, the 2-D Morlet wavelet is not strictly admissible, but if $\mathbf{k}_0 > 5.34$, it is numerically acceptable. On the other hand, the modulated Mexican hat wavelet ψ_{MM} is admissible as it stands, without any restrictions on $\mathbf{k}_0$.

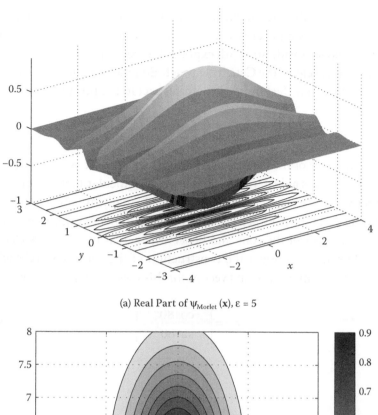

(a) Real Part of $\psi_{\text{Morlet}}(\mathbf{x})$, $\varepsilon = 5$

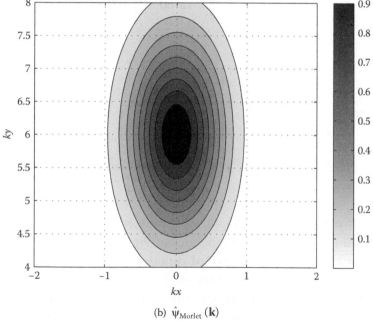

(b) $\hat{\psi}_{\text{Morlet}}(\mathbf{k})$

FIGURE 3.16   The 2-D Morlet wavelet with $\mathbf{k}_0 = (0,6)$ and $\varepsilon = 5$.

To create a truly directional wavelet, all one requires is a smooth function $\hat{\psi}^{(C)}(k)$ with support in a convex cone C in frequency space, with polynomial behavior inside and with exponential decay [30, 34]. An example of this type of wavelet is the Cauchy wavelet, Equation (3.49c). It is supported by the convex cone C, symmetric with respect to the positive k_x axis where

$$C \equiv C(-\alpha, \alpha), \quad -\alpha \le \arg k \le \alpha, \quad \alpha < \pi/2. \tag{3.50}$$

That is, the sides of the cone are determined by the unit vectors $\mathbf{e}_{-\alpha}, \mathbf{e}_\alpha, \alpha < \pi/2$. This cone has a dual, whose sides are perpendicular to it, determined by the unit vectors $\mathbf{e}_{-\tilde{\alpha}}, \mathbf{e}_{\tilde{\alpha}}$, where $\tilde{\alpha} = -\alpha + \pi/2$ Finally, $\eta = (\eta, 0), \eta > 0$, is a fixed vector, and the parameter m controls the number of vanishing moments of $\hat{\psi}$ on the edges of the cone and hence the regularity of the wavelet. As a practical example, consider the construction of the Cauchy wavelet $\hat{\psi}_4^{(10)}$ in the frequency domain. The unit vectors for this wavelet are nothing other than

$$\mathbf{e}_{\tilde{\alpha}} = \begin{bmatrix} \cos 80° \\ \sin 80° \end{bmatrix} \tag{3.51a}$$

and

$$\mathbf{e}_{-\tilde{\alpha}} = \begin{bmatrix} \cos 100° \\ \sin 100° \end{bmatrix} \tag{3.51b}$$

so that, if the arg($\mathbf{k}$) < 10°,

$$\hat{\psi}_4^{(10)} = (\cos(80°)k_x + \sin(80°)k_y)^4 (\cos(100°)k_x + \sin(100°)k_y)^4 \exp(-\eta k_x). \tag{3.51c}$$

The value of η controls how rapidly this wavelet decays, with larger values of η obviously causing more rapid decay. Figure 3.17 shows this wavelet with $\eta = 2$. Notice that it is indeed contained within a cone ±10° either side of the k_x axis. The advantage of these wavelets is that the vanishing moments and opening angle of the cone are completely controllable. Finally, we note that this wavelet has an analytic representation in the spatial domain, given by

$$\psi_m^{(\alpha)}(\mathbf{x}) = \frac{(-1)^{m+1}}{2\pi}(m)^2 \frac{(\sin 2\alpha)^{2m+1}}{[\mathbf{z} \cdot \sigma(\alpha)\mathbf{z}]^{m+1}} \tag{3.52a}$$

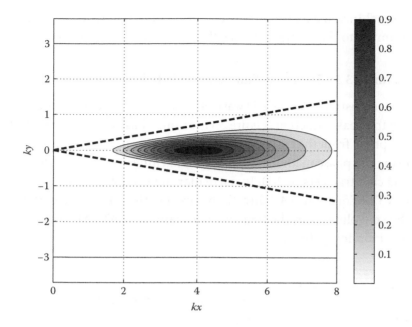

FIGURE 3.17 Cauchy wavelet $\hat{\psi}_4^{(10)}$ with $\eta = (2,0)$. The dashed lines show the cone $C(-10°,10°)$.

where $\mathbf{z} = x + i\eta$ and

$$\sigma(\alpha) = \begin{bmatrix} \cos^2 \alpha & 0 \\ 0 & \sin^2 \alpha \end{bmatrix}. \qquad (3.52b)$$

To conclude this discussion on anisotropic wavelets, we note that given a directional wavelet ψ, one can construct *multidirectional* wavelets with n-fold symmetry by summing n rotated copies of ψ according to [30]

$$\psi_n(\mathbf{x}) = \frac{1}{n} \sum_{k=0}^{n-1} \psi(r_{-\theta_k}(\mathbf{x})), \quad \theta_k = k\frac{2\pi}{n}, \, k = 0,1,\ldots,n-1. \qquad (3.53)$$

which corresponds to moving the wavelet along a circular locus in **k**-space. Summing 2-D Morlet wavelets over 2π, for example, will produce a quasi-isotropic wavelet, whereas summing over π with carefully controlled azimuthal increments will produce a directional fan wavelet with geometry similar to the arc wavelet of Dallard and Spedding [33, 35].

Example 3.5: The Peaks Function Revisited

To demonstrate how the 2-D wavelet transform is used in practice, we use the peaks function (Figure 3.11) as our phase object and extract the phase from the corresponding interference pattern (Figure 3.12) using the phase method. Much of the machinery established for the 1-D case transfers to 2-D. In particular, we do not evaluate the 2-D CWT of Equation (3.43) directly, but exploit the 2-D FFT to perform the convolutions in the frequency domain. A MATLAB code fragment for doing this is

```
F = fft2(f)
for j = 1:length(a)
    c(:,:,j) = a(j) * ifft2(F*conj(mexican2d(kx,ky,sigma)));
end
```

The 2-D fringe pattern is stored in the matrix f while a contains the vector of scaling parameters, as before. One must calculate the scaled (and possibly rotated) **k**-vectors, which are stored in matrices kx and ky, so that the wavelet of choice can be evaluated in the frequency domain. The complex array c stores the corresponding 2-D CWT coefficients. The yawtb contains MATLAB functions for the continuous 2-D case and is also used here.

For this example, we use an isotropic wavelet as this eliminates the θ dependence in the 2-D CWT, reducing the parameter space that must be searched, and facilitates the explanation to follow. Our choice of wavelet, then, is the 2-D Mexican hat wavelet. We again select a suitable range for the scale parameter a and choose a step size a_{step}. In position representation, the 2-D CWT produces 512 × 512 arrays of coefficients, a few of which are shown in Figure 3.18.

We would emphasize that this is just a subset of the coefficient arrays generated; in practice, we use $a_{\text{step}} \ll 1$. To find the wavelet "ridge," now a 512 × 512 array, we imagine that the arrays are stacked vertically, and for a given pixel, we scan through the snapshots

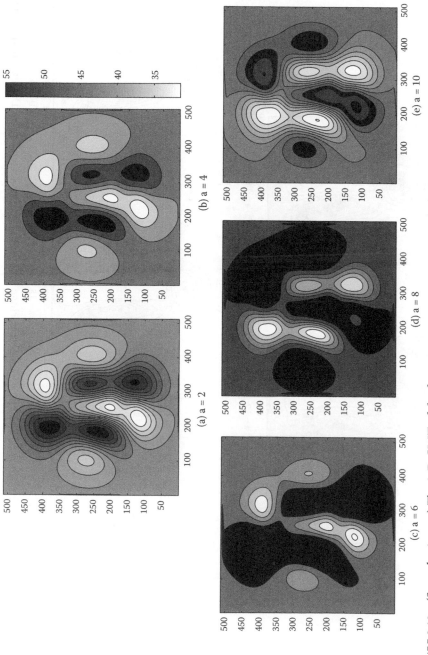

FIGURE 3.18 (See color insert.) The 2-D CWT of the fringe pattern in Figure 3.12 with the 2-D Mexican hat with σ = 0.5. (continued)

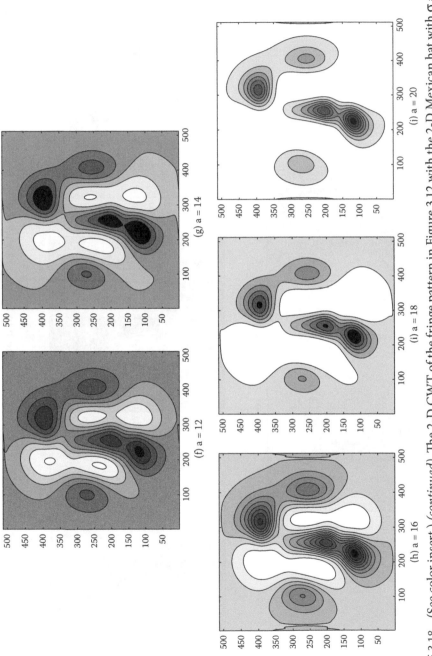

FIGURE 3.18 *(See color insert.) (continued)* The 2-D CWT of the fringe pattern in Figure 3.12 with the 2-D Mexican hat with σ = 0.5.

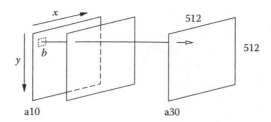

FIGURE 3.19   Extracting the wavelet ridge for the 512 × 512 image.

vertically, looking for the value of *a* that maximizes the CWT for that particular pixel value **b**. A selection of these snapshots are displayed in Figure 3.18 with the process illustrated schematically in Figure 3.19. As a concrete example of this, consider the top leftmost pixel in the snapshots of Figure 3.18, that is, the pixel with coordinates (1,1). Figure 3.20 shows that $a = 8$ maximizes the CWT for this particular pixel, so this value would occupy position (1,1) of our wavelet ridge array. Once we have done this for every pixel, we will have a 512 × 512 array containing a_{opt} for every pixel in the interference pattern. The phase is obtained, modulo 2π, from the argument of the CWT as previously and the linear phase term caused by the spatial carrier subtracted. The resulting error map is shown in Figure 3.21. The first and last 12 rows of this have been removed as these contained large errors because of the edge effects discussed

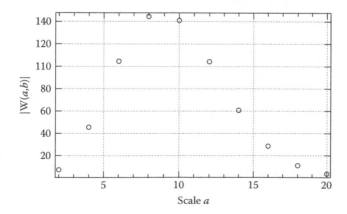

FIGURE 3.20   Maximum $|Wf(a,b)|$ as a function of scale parameter a for the pixel with coordinate (1,1) in Figure 3.18.

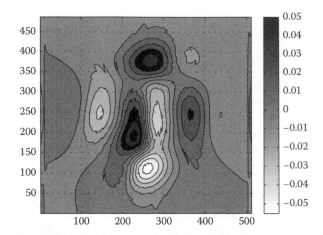

FIGURE 3.21 Error map for the peaks function using the 2-D CWT with a Mexican hat wavelet.

previously. The left- and right-hand edges of the original interference pattern were not extended for this example. A comparison with Figure 3.13 shows that the overall errors are somewhat smaller than the case when the interference pattern was processed row by row.

3.12 CONCLUSIONS

Wavelets are so called because they are functions with zero mean that have a few oscillations that decay to zero. Literally, they are small waves. When such a function, usually called the mother wavelet, is scaled and translated in a systematic way, an entire family of nonorthogonal functions is created that spans the time-frequency plane. Each wavelet in this family occupies a certain finite region of this plane, determined by the Heisenberg box associated with the wavelet. The CWT, which is the inner product of our signal with each of the wavelets in the family, is a measure of how much of the signal's energy resides in each of these Heisenberg boxes; that is, the CWT is a measure of the local time-frequency energy density. A path that follows the maximum modulus of the CWT, termed the wavelet ridge, therefore tells us how the instantaneous frequency of the signal evolves as a function of time. Because values of the scaling parameter on the ridge yield the instantaneous frequency directly, recovering the phase is simply a matter of integration. Although this is intuitively attractive, it is error prone. A more robust approach is to calculate the phase directly from the argument of the complex CWT along the ridge because errors in the phase will then remain locally isolated.

Although 2-D fringe patterns may be processed as a series of 1-D signals, there are advantages to processing these using 2-D wavelets. The same general principles that applied in 1-D carry over into 2-D, with two important additions. First, we find that our wavelets may be either isotropic or anisotropic, with the latter requiring an additional rotation parameter to create acceptable wavelet families. Second, the 2-D CWT takes a 2-D signal and produces a 4-D representation in transform space that is not easily visualized. This problem, and incidentally a method for finding the wavelet ridge, is solved by regarding the 2-D CWT as a series of snapshots. Each snapshot is a function of translation only, with a fixed value of scale and rotation, and can be regarded as the scalogram for that particular scale-rotation pair. We then build a series of snapshots, one for each scale-rotation pair in our transform space. For each pixel or coordinate pair, we search through the stacked snapshots, looking for the pair of scale-rotation parameters that maximizes the modulus of the CWT. If our original image was $m \times n$, then the wavelet ridge is now an $m \times n \times 2$ array. The phase is recovered, modulo 2π, from the argument of the complex CWT along this ridge, as previously done. The 2-D wavelets are recommended for phase recovery from 2-D fringe patterns because several studies have shown that they outperform their 1-D counterparts.

The other popular methods of phase recovery are Takeda's Fourier transform method and the WFT, so the question naturally arises: Why use wavelets? Takeda's method is a global one, and because fringe patterns are best described as nonstationary signals, it seems more appropriate to use local time-frequency estimators such as wavelets or the WFT. And, indeed, a careful comparison of these three methods by Huang et al. [27] showed that the WFT and the 2-D wavelet transform both outperform the Fourier transform method. A common criticism of the WFT is the need to select the window width a priori, and this width, once selected, cannot be adjusted dynamically. Wavelets, as we have seen, do not suffer from this restriction. Finally, in contrast to Fourier methods, many functions can be used as the mother wavelet, yielding wavelet families with a rich and diverse set of properties that can be tailored to the particular problem at hand.

APPENDIX A

RIDGE OF THE TWO-DIMENSIONAL CWT

Following the approach of Wang and Ma [19] and Ma et al. [36], we show that the 2-D CWT with a directional wavelet is maximized when its scale and orientation match that of the local fringe pattern. Let the local fringe period and orientation at an arbitrary point $\mathbf{b}$ in an interferogram be A and Θ ($0 \le \Theta < 2\pi$), respectively. The interference pattern at $\mathbf{x}$ around $\mathbf{b}$ can be expressed as

$$
\begin{aligned}
I(\mathbf{x}) &= I_b + I_a \cos[2\pi A^{-1}\mathbf{n}\cdot(\mathbf{x}-\mathbf{b}) + \phi(\mathbf{b})] \\
&= I_b + I_a \cos\left[2\pi \frac{(x-b_x)\cos\Theta + (y-b_y)\sin\Theta}{A} + \phi(\mathbf{b})\right] \quad (3A.1)
\end{aligned}
$$

where $\mathbf{n} = (\cos\Theta, \sin\Theta)$ is the local normal to the fringes. For the sake of convenience, we use the 2-D Morlet wavelet:

$$
\psi_{\text{Morlet}} = \exp(i\mathbf{k}_0 \cdot \mathbf{x})\exp(-m|\mathbf{x}|^2) \quad (3A.2)
$$

setting the wave vector $\mathbf{k}_0 = (2\pi, 0)$. Substituting Equations (3A.1) and (3A.2) into Equation (3.43), one eventually finds that

$$
\begin{aligned}
W_f(a,\mathbf{b},\theta) = &\frac{\pi}{m} I_b \exp\left(-\frac{\pi^2}{m}\right) \\
&+ \frac{\pi}{2m} I_a \exp\left(-\frac{\pi^2}{m}\left\{\left(\frac{a}{A}-1\right)^2 + 2\frac{a}{A}[1-\cos(\Theta-\theta)]\right\}\right)\exp[i\phi(\mathbf{b})] \\
&+ \frac{\pi}{2m} I_a \exp\left(-\frac{\pi^2}{m}\left\{\left(\frac{a}{A}+1\right)^2 - 2\frac{a}{A}[1-\cos(\Theta-\theta)]\right\}\right)\exp[-i\phi(\mathbf{b})].
\end{aligned}
$$

$$(3A.3)$$

from which it is clear that the modulus of the 2-D CWT reaches a maximum when $a = A$ and $\theta = \Theta$.

REFERENCES

1. I. Daubechies. *Ten Lectures on Wavelets*. Philadelphia: SIAM 1992.
2. I. Daubechies. The wavelet transform, time-frequency localization and signal analysis. *IEEE Transactions* **IT-36**(5):961–1005, 1990.

3. Peter I. Corke. *Robotics, Vision and Control: Fundamental Algorithms in MATLAB.* New York: Springer, 2011.

4. Sikun Li, Xianyu Su, and Wenjing Chen. Wavelet ridge techniques in optical fringe pattern analysis. *Journal of the Optical Society of America A–Optics Image Science and Vision* **27**(6):1245–1254, 2010.

5. W. H. Press, B. P. Flannery, S. A. Teukolsky, and W. T. Vetterling. *Numerical Recipes in FORTRAN: The Art of Scientific Computing.* 2nd ed. New York: Cambridge University Press, 1992.

6. E. O. Brigham. *The Fast Fourier Transform and Its Applications.* Englewood Cliffs, NJ: Prentice Hall, 1988.

7. S. Mallat. *A Wavelet Tour of Signal Processing.* 2nd ed. San Diego, CA: Academic Press, 2001.

8. Jonathan M. Lilly and Sofia C. Olhede. On the analytic wavelet transform. *IEEE Transactions On Information Theory* **57**(8):4135–4156, 2010.

9. L. R. Watkins, S. M. Tan, and T. H. Barnes. Determination of interferometer phase distributions by use of wavelets. *Optics Letters* **24**(13):905–907, 1999.

10. L. R. Watkins. Phase recovery from fringe patterns using the continuous wavelet transform. *Optics and Lasers in Engineering* **45**(2):298–303, 2007.

11. A. Dursun, S. Ozder, and F. N. Ecevit. Continuous wavelet transform analysis of projected fringe patterns. *Measurement Science and Technology* **15**(9):1768–1772, 2004.

12. J. C. Hong. Determination of the optimal Gabor wavelet shape for the best time-frequency localization using the entropy concept. *Experimental Mechanics* **44**(4):387–395, August 2004.

13. O. Koysal, S. E. San, S. Ozder, and F. N. Ecevit. A novel approach for the determination of birefringence dispersion in nematic liquid crystals by using the continuous wavelet transform. *Measurement Science and Technology* **14**(6):790–795, 2003.

14. Q. Kemao, S. H. Soon, and A. Asundi. Instantaneous frequency and its application to strain extraction in moire interferometry. *Applied Optics* **42**(32):6504–6513, 2003.

15. Yet Another Wavelet Toolbox homepage. rhea.tele.ucl.ac.be/yawtb/.

16. C. Torrence and G. P. Compo. A practical guide to wavelet analysis. *Bulletin of the American Meteorological Society* **79**(1):61–78, January 1998.

17. M. Afifi, A. Fassi-Fihri, M. Marjane, K. Nassim, M. Sidki, and S. Rachafi. Paul wavelet-based algorithm for optical phase distribution evaluation. *Optics Communications* **211**(1–6):47–51, 2002.

18. Munther A. Gdeisat, Abdulbasit Abid, David R. Burton, Michael J. Lalor, Francis Lilley, Chris Moore, and Mohammed Qudeisat. Spatial and temporal carrier fringe pattern demodulation using the one-dimensional continuous wavelet transform. Recent progress, challenges, and suggested developments. *Optics and Lasers in Engineering* **47**(12):1348–1361, September 2009.

19. Zhaoyang Wang and Huanfeng Ma. Advanced continuous wavelet transform algorithm for digital interferogram analysis and processing. *Optical Engineering* **45**(4):045601, 2006.

20. S. L. Marple. Computing the discrete-time "analytic" signal via FFT. *IEEE Transactions on Signal Processing* **47**(9):2600–2603, 1999.

21. H. Szu, C. Hsu, L. D. Sa, and W. Li. Hermitian hat wavelet design for singularity-detection in the PARAGUAY river level data analyses. *Proceedings of SPIE* **3078**:96–115, 1997.

22. Hui Li, Yuping Zhang, and Haiqi Zheng. Application of Hermitian wavelet to crack fault detection in gearbox. *Mechanical Systems and Signal Processing* **25**(4):1353–1363, January 2011.

23. S. C. Olhede and A. T. Walden. Generalized Morse wavelets. *IEEE Transactions on Signal Processing* **50**(11):2661–2670, 2002.

24. Jonathan M. Lilly and Sofia C. Olhede. Higher-order properties of analytic wavelets. *IEEE Transactions on Signal Processing* **57**(1):146–160, 2009.

25. Jonathan M. Lilly and Sofia C. Olhede. Generalized Morse wavelets as a superfamily of analytic wavelets. *IEEE Transactions on Signal Processing* **60**(11):6036–6041, 2012.

26. C. Chui. *An Introduction to Wavelets*, volume 1 of *Wavelet Analysis and Its Applications*. San Diego, CA: Academic Press, 1992.

27. Lei Huang, Qian Kemao, Bing Pan, and Anand Krishna Asundi. Comparison of Fourier transform, windowed Fourier transform, and wavelet transform methods for phase extraction from a single fringe pattern in fringe projection profilometry. *Optics and Lasers in Engineering* **48**(2):141–148, November 2009.

28. Sikun Li, Xianyu Su, and Wenjing Chen. Spatial carrier fringe pattern phase demodulation by use of a two-dimensional real wavelet. *Applied Optics* **48**(36):6893–6906, 2009.

29. Munther A. Gdeisat, David R. Burton, and Michael J. Lalor. Spatial carrier fringe pattern demodulation by use of a two-dimensional continuous wavelet transform. *Applied Optics* **45**(34):8722–8732, 2006.

30. J.-P. Antoine, R. Murenzi, P. Vandergheynst, and S. T. Ali. *Two-Dimensional Wavelets and Their Relatives*. Cambridge, UK: Cambridge University Press, 2004.

31. J. P. Antoine and R. Murenzi. Two-dimensional directional wavelets and the scale-angle representation. *Signal Processing* **52**(3):259–281, 1996.

32. P. Frick, R. Beck, E. M. Berkhuijsen, and I. Patrickeyev. Scaling and correlation analysis of galactic images. *Monthly Notices of the Royal Astronomical Society* **327**(4):1145–1157, 2001.

33. T. Dallard and G. R. Spedding. 2-D wavelet transforms—generalization of the hardy space and application to experimental studies. *European Journal of Mechanics B-Fluids* **12**(1):107–134, 1993.

34. J.-P. Antoine, R. Murenzi, and P. Vandergheynst. Directional wavelets revisited: Cauchy wavelets and symmetry detection in patterns. *Applied and Computational Harmonic Analysis* **6**(3):314–345, April 1999.

35. J. F. Kirby. Which wavelet best reproduces the Fourier power spectrum? *Computers and Geosciences* **31**(7):846–864, 2005.

36. Jun Ma, Zhaoyang Wang, Bing Pan, Thang Hoang, Minh Vo, and Long Luu. Two-dimensional continuous wavelet transform for phase determination of complex interferograms. *Applied Optics* **50**(16):2425–2430, 2011.

CHAPTER 4

The Spiral Phase Transform

Kieran G. Larkin

Nontrivialzeros Research
Sydney, Australia

4.1 INTRODUCTION

The spiral phase transform (SPT) is one of the simplest algorithms for demodulation of two-dimensional (2-D) fringe patterns. In its most compact form, the transform requires only two multiplication operations (one in the spatial domain, one in the Fourier domain) and one addition. Essential theory and practical details for implementation are presented.

4.2 THEORY

4.2.1 Demodulation in One and Two Dimensions

Given the two-dimensional (2-D) image represented by the following equation, an intensity fringe pattern (viz. *interferogram* or *hologram*) formed from the interference of two coherent wave-fields w_1 and w_2,

$$g(x,y) = a(x,y) + b(x,y)\cos\psi(x,y) = |w_1(x,y) + w_2(x,y)|^2 \quad (4.1)$$

we would like to solve for the three unknowns a, b, and ψ. Each of these unknowns corresponds to a 2-D distribution or "image" with spatial coordinates (x,y). It is often just one unknown, the phase image $\psi(x,y)$, that is of interest for interferometric applications, although the resurgent area of digital holography [1] and fingerprint demodulation [2] require all three.

The ideal solution to Equation (4.1) is perhaps best illustrated by the concept of the *analytic signal*:

$$\text{AnalyticSignal}\{g(x,y) - a(x,y)\} = b(x,y)\exp[i\psi(x,y)] \quad (4.2)$$

The sought-after phase angle is simply the argument of this complex image and can be computed in practice using the arctangent of the real and imaginary parts. The concept only applies to oscillating (direct-current- [DC-] free) signals, so the offset term $a(x,y)$ has been removed by some form of high-pass filtering before generating the analytic signal. In 1-D, the corresponding problem was solved in various ways during the nineteenth and twentieth centuries in radio communications theory, the most well known being the phase locked loop demodulator in FM (frequency modulation) radio receivers. Takeda first proposed a 1-D solution for fringe patterns, now known as the Fourier transform method (FTM) [3]. It is essentially a variation of direct Hilbert transform demodulation sometimes used in radio and TV communications, especially single-sideband systems. The method was formally extended for 2-D fringe patterns by Bone [4]. However, these solutions are restricted to open fringe patterns with strictly monotonic phase. Normally, this is ensured by adding a strong-enough carrier (tilt fringes) to the fringe pattern. The more general case for closed fringe patterns cannot be solved by the FTM. Various other approaches, such as regularized phase tracking [5], can provide a solution at the cost of increased computation because of the iterative optimization. The failure to find a direct solution for the closed fringe pattern demodulation problem can be attributed to the lack of a meaningful extension of the Hilbert transform (and the corresponding analytic signal) in 2-D.

The key to finding a solution for 2-D demodulation lies in a suitable definition of the signum function. In 1-D, the Hilbert transform can be simply defined in terms of its Fourier transform.

First, define the Hilbert transform operator:

$$H\{g(x)\} = \tilde{g}(x) = \int_{-\infty}^{+\infty} g(x') \frac{1}{\pi(x-x')} dx' \qquad (4.3)$$

The right-hand side is a convolution integral and should be interpreted as the Cauchy principal value because the Hilbert kernel is singular and the integral diverges. Our approach is to work as much as possible in the Fourier domain, where divergences are avoided and the Hilbert kernel becomes a bounded multiplier via the convolution theorem. The Fourier frequency coordinate here is denoted by u.

The forward Fourier transform operator is defined

$$F\{g(x)\} = G(u) = \int_{-\infty}^{+\infty} g(x)\exp[-2\pi iux]\,dx \qquad (4.4a)$$

and the inverse Fourier transform operator

$$F^{-1}\{G(u)\} = g(x) = \int_{-\infty}^{+\infty} G(u)\exp[+2\pi iux]\,du \qquad (4.4b)$$

The Fourier transform of the Hilbert transform is simply

$$F\{\tilde{g}(x)\} = \tilde{G}(u) = -i\,\mathrm{sgn}(u)G(u) \qquad (4.5)$$

In 1-D, the signum is defined thus:

$$\mathrm{sgn}(u) = \frac{u}{|u|} = \begin{cases} -1, & u < 0 \\ 0, & u = 0 \\ +1, & u > 0 \end{cases} \qquad (4.6a)$$

The two key properties of the signum are its unit magnitude and odd symmetry. Extending this idea into 2-D seems to have perplexed researchers over the last 40 years. The usual approach is to generate a 2-D function that looks like a step function (see Figure 1 in [6]). But, another approach is possible. It transpires that spiral symmetry can satisfy the two key properties and has no preferred direction, as follows:

$$\mathrm{sgn}(u,v) = \frac{u+iv}{|u+iv|} = \exp(i\phi), \begin{cases} u = q\cos\phi \\ v = q\sin\phi \end{cases} \qquad (4.6b)$$

The 2-D frequency coordinates are (u,v), and their polar representation is (q,ϕ). The definition is a pure spiral phase function in spatial frequency space. This central insight is the basis of the spiral phase implementation of the Riesz transform (see Box 4.1). Signums are not well defined at the origin,

BOX 4.1 RIESZ TRANSFORM

Readers interested in learning more about the Riesz transform and its applications can read the introductory papers and patents by Felsberg [7] and Larkin [6,8–10] and the references therein. Riesz transforms arise naturally in the elegant mathematical framework of Clifford algebra. Unfortunately, most engineers and physicists are not trained in the use of Clifford algebra, so we have instead concentrated on a particular embedding of the two-dimensional (2-D) Riesz transform within the much commoner and widespread algebra of 2-D complex variables as shown in patents [11,12] and more recent academic papers [13,14]. Some generality and versatility is lost, but potential users can quickly grasp the mathematical essentials and implement the algorithm in conventional image-processing software.

and the convention is to set the magnitude to zero at this singular point. For most (namely DC-free) signals, the singularity is irrelevant anyway.

4.2.2 Quadrature Signals

4.2.2.1 The 1-D Case

Closely related to the idea of Hilbert transformation is the idea of a quadrature signal or function. Quadrature here means the concept of an oscillating signal being 90° (or a quarter period) out of phase with another signal. The idea is best illustrated by an amplitude- and frequency-modulated (i.e., bandpass) signal, with phase offset α, written as follows:

$$s(x) = b(x)\cos(2\pi ux + \psi(x) + \alpha)$$

$$\equiv b(x)\frac{\exp(2i\pi ux + i\psi(x) + i\alpha) + \exp(-2i\pi ux - i\psi(x) - i\alpha)}{2} \quad (4.7)$$

In communication theory, a quadrature filter generates the following output, given the signal in Equation (4.7) as input. Essentially, all frequencies in the signal are delayed by 90° ($\pi/2$ radians):

$$\tilde{s}(x) = b(x)\cos(2\pi ux + \psi(x) + \alpha - \pi/2) = b(x)\sin(2\pi ux + \psi(x) + \alpha)$$

$$= b(x)\frac{\exp(2i\pi ux + i\psi(x) + i\alpha) - \exp(-2i\pi ux - i\psi(x) - i\alpha)}{2i} \quad (4.8)$$

It turns out that in most cases of practical interest (namely, bandpass-limited signals), the Hilbert transform is synonymous with quadrature (for details, see [15]). The reason is simply that the Hilbert transform flips the symmetry of the conjugate Fourier side lobes in a bandpass signal. The Fourier spectrum of a bandpass signal, centered on frequency u_0, has two (nonoverlapping) lobes:

$$F\{s(x)\} = S(u)$$

$$= F\left\{b(x)\frac{\exp(+i2\pi u_0 x + i\psi(x) + i\alpha) + \exp(-i2\pi u_0 x - i\psi(x) - i\alpha)}{2}\right\}$$

$$= \frac{B(+u - u_0)\exp(+i\alpha) + B(-u - u_0)\exp(-i\alpha)}{2}$$

$$(4.9A)$$

where the (reflected) side-lobe Fourier transforms are defined by

$$F\{b(x)\exp(+i\psi(x))\} = B(u)$$

$$F\{b(x)\exp(-i\psi(x))\} = B(-u)$$

$$(4.9B)$$

whereas the Hilbert transform spectrum has flipped (odd) symmetry:

$$F\{\tilde{s}(x)\} = -i\,\text{sgn}(u)S(u)$$

$$= \frac{B(-u - u_0)\exp(-i\alpha) - B(+u + u_0)\exp(+i\alpha)}{2i}$$

$$(4.10)$$

It is clear [comparing Equations (4.7) and (4.8) with (4.9) and (4.10)] that cosines flip to sines, and vice versa, under the influence of the Hilbert transform signum multiplier.

4.2.3 Intrinsically 1-D Structure of 2-D Fringe Patterns

A familiar concept in fringe analysis is the local AM-FM (amplitude modulation–frequency modulation) structure; the intensity is changing in one direction and more or less constant in the orthogonal direction: The structure is intrinsically 1-D. It is the central idea behind the windowed Fourier transform (WFT), for example. Here, we concentrate on the orientation, rather that the amplitude and frequency properties. For those wishing to know more, Granlund and Knutsson delved more deeply into intrinsic dimensionality [16].

Assuming that the local fringe structure induced by phase changes faster than the amplitude, then a small region of the fringe pattern can be expanded as a Taylor series of the phase Ψ:

$$\psi \approx \psi_{00} + \psi_{10}x + \psi_{01}y$$

$$g(x,y)-a(x,y)=\frac{b}{2}\{\exp[+i(\psi_{00}+\psi_{10}x+\psi_{01}y)]+\exp[-i(\psi_{00}+\psi_{10}x+\psi_{01}y)]\}$$

(4.11)

The forward and inverse Fourier transforms in 2-D are defined by

$$G(u,v)=\mathrm{F}\{g(x,y)\}=\int_{-\infty}^{+\infty}\int_{-\infty}^{+\infty}g(x,y)\exp[-2\pi i(ux+vy)]dxdy \quad (4.12a)$$

and

$$g(x,y)=\mathrm{F}^{-1}\{G(u,v)\}=\int_{-\infty}^{+\infty}\int_{-\infty}^{+\infty}G(u,v)\exp[+2\pi i(ux+vy)]dudv \quad (4.12b)$$

The (local) 2-D Fourier transform of these equations is simply two Dirac delta side lobes:

$$\mathrm{F}\{g-a\}=G-A$$

$$=\frac{b}{2}\{\delta(u-u_0,v-v_0)\exp[+i\psi_{00}]+\delta(u+u_0,v+v_0)\exp[-i\psi_{00}]\}$$

(4.13)

where the local phase ψ_{00} and local spatial frequencies are defined as

$$u_0=\psi_{10}/(2\pi), \ v_0=\psi_{01}/(2\pi), \ \nabla\psi=\mathbf{i}\psi_{10}+\mathbf{j}\psi_{01},$$

$$\tan(\beta)=\left(\frac{\psi_{01}}{\psi_{10}}\right) \qquad (4.14)$$

The local spatial and Fourier properties are illustrated in Figures 4.1 and 4.2. Especially important is the correspondence of fringe normal angle and spectral lobe polar angle β. Note that a pure phasor function shown in Figure 4.1a has a well-defined direction angle β, whereas a real fringe pattern, shown in Figure 4.1b has an orientation angle indistinguishable

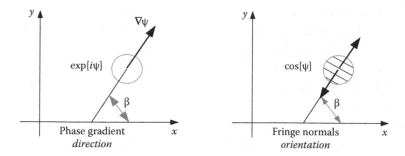

FIGURE 4.1   Phase gradient direction and fringe orientation.

from the opposite direction β + π. The ambiguity is related to the twin lobes of Figure 4.2b compared to the single lobe of Figure 4.2a.

4.2.4 SIGNUM Returns

In 2-D, the spiral phase signum multiplier [Equation (4.6)] takes the following symmetric spectral lobes of the cosine fringe pattern (note the plus sign):

$$g(x,y) - a(x,y) = \frac{b(x,y)}{2}\{\exp[+i\psi(x,y)] + \exp[-i\psi(x,y)]\} \quad (4.15)$$

and turns them into the antisymmetric spectral lobes of the sine fringe pattern (note the minus sign between the exponentials):

$$\${g(x,y) - a(x,y)\} \cong \frac{b(x,y)}{2}\exp[i\beta(x,y)]\{\exp[+i\psi(x,y)]$$

$$- \exp[-i\psi(x,y)]\} \quad (4.16)$$

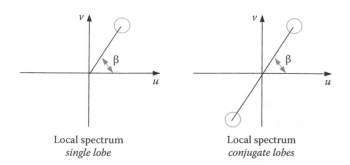

FIGURE 4.2   Spectra corresponding to complex phase pattern and to real cosine pattern.

The spiral phase multiplies the positive frequency lobe by the direction phasor and the negative frequency lobe by the negation of the direction phase. To understand this, it may help to think in terms of the WFT discussed in Chapter 2. Locally, the fringe pattern has a particular fringe orientation and spacing. Each spatial patch maps onto two lobes in the Fourier spectrum, shown in Figure 4.2b. The polar coordinates of the two lobes are at angles β and $\beta + \pi$. So, the spectral spiral phasor at these location is $\exp[i\beta]$ and $\exp[i\beta + i\pi]$, and the lobes flip signs because of Euler's identity:

$$\exp[+i\pi] \equiv -1 \qquad (4.17)$$

The SPT equation, Equation (4.16), can now be written as follows:

$$\${g(x,y) - a(x,y)\} \cong ib(x,y)\exp[+i\beta(x,y)]\sin[\psi(x,y)] \qquad (4.18)$$

This important equation is the basis for many variations on the demodulation theme, and the approximation accuracy has been precisely quantified [6].

4.3 IMPLEMENTATION

4.3.1 Vortex Operator Algorithm

The simplest implementation of the spiral phase demodulator requires forward and inverse discrete Fourier transforms (DFTs; see Box 4.2), a fringe orientation algorithm, and an orientation phase unwrapper. In our experience, one of the most reliable and accurate orientation estimators is the Riesz-based energy operator of Larkin [19]. Further discussion of the importance of orientation and direction is also contained in Larkin [9].

The "vortex operator" combines the spiral phasor spectral multiplier with the direction phasor spatial multiplier as follows: First, the spiral phase generates a complex function with orientation and quadrature intertwined. The process entails just one complex multiplication in the spectral domain. In operator notation, where the box represents the input image:

$$\$\{\square\} \equiv F^{-1}\left\{\exp[i\phi(u,v)].F\{\square\}\right\} \qquad (4.19)$$

Then, the "vortex" operator disentangles direction from the quadrature term, which ends up in the imaginary part of the complex output. The process entails just one complex multiplication in the spatial domain. Again, in operator notation,

$$V\{\square\} \equiv \exp[-i\beta(x,y)]\$\{\square\} \qquad (4.20)$$

BOX 4.2 DISCRETE FOURIER TRANSFORM IMPLEMENTATION

Although most of the mathematical analysis of the spiral phase transform (SPT) is in the continuous domain, the method transfers flawlessly to the discrete domain of digital images and discrete Fourier transforms (DFTs). Fast Fourier transforms (FFTs) should be used as these are no longer limited to powers-of-two image dimensions (see [17]).

Windowing is not required at any stage of the processing, and the fringe patterns are only prone to Fourier artifacts close to edges and boundaries.

In the algorithms presented, all Fourier operations occur in forward/inverse pairs, so the usual normalization factors can be implemented at the end to reduce overall multiplicative operations.

A Linux-based implementation of the SPT algorithm is freely available online [18].

DFT CENTERING

The DFT/FFT algorithms utilized should have the zero frequency in the center of the data set, not at the edges. This makes computation of the spiral phasor straightforward. If a noncentered DFT is used, the spiral computation must be compensated accordingly. In software such as MATLAB, centering is achieved with the fftshift operation.

LOW-PASS FILTERING

The SPT assumes that the background or direct current– (DC-) like component has been removed. Many methods for this are known in fringe analysis. The simplest is just to block out a suitable size disk region around the Fourier origin. The most effective is to subtract two phase-shifted fringes, but this is clearly not an option for single-fringe patterns.

Full demodulation occurs by disregarding the real part (if any) and combining the imaginary output with the original real input. The process entails just one imaginary addition.

$$b(x,y)\exp[i\psi(x,y)] \cong \Box + \Im\left\{V\left\{\Box\right\}\right\} \qquad (4.21)$$

This sequence is executed by the algorithm shown in Figure 4.3, noting that the low-frequency background component $a(x,y)$ must be removed as a preprocessing step.

The output of the vortex operator is nominally imaginary, but various systematic and random errors can introduce a residual real part. Error leakage can be reduced by keeping the imaginary part and discarding the real component. The DFTs can utilize all the latest improvements in fast

$f(x, y) = a(x, y) + b(x, y)\cos\psi(x, y)$

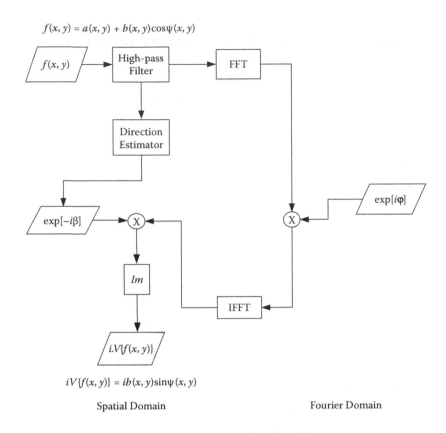

$iV\{f(x, y)\} = ib(x, y)\sin\psi(x, y)$

Spatial Domain   Fourier Domain

FIGURE 4.3 The vortex (quadrature) algorithm flowchart.

Fourier transforms (FFTs), and even prime length image dimensions attain the $N\log N$ speed of the power-of-two FFT (where N represents the total number of pixels in the fringe image). We recommend the use of FFTW [17] for researchers wishing to have the fastest software implementations.

The vortex operator mentioned (see Box 4.3) can be interpreted as a spectrally and spatially adaptive (complex) filter with unit magnitude response. In one sense, it occupies the intermediate world between the purely 2-D FTM method (see Chapter 1) and the computationally challenging four-dimensional (4-D) WFT method (see Chapter 2). Essentially, 4-D of adaptive performance is achieved with just 2-D of computational load. Furthermore, the guaranteed applicability of FFT algorithms ensures a final computational load of just $N\log N$, which translates into a fraction of a second for current computer processors and megapixel fringe patterns.

Figure 4.3 shows the basic vortex algorithm. The algorithm requires just two multiplication operations to obtain the quadrature image.

BOX 4.3   SOME TERMINOLOGY

Spiral phase transform (SPT). The SPT strictly means the transform defined by (two-dimensional, 2-D) Fourier transformation, spiral phasor multiplication, then inverse Fourier transformation. This is equivalent to the Riesz transform embedded in complex algebra. It is now used synecdochically to mean the full 2-D demodulation process, which includes the complex Riesz transform mentioned, the formation of the analytic image, and the extraction of the phase using the arctangent operation.

Vortex operator. This combines spectral spiral phasor multiplication with spatial direction phasor multiplication to yield the quadrature fringe pattern from an input fringe pattern.

Fourier transform method (FTM). The FTM is essentially a one-dimensional (1-D) Hilbert transform demodulation method.

4.3.2 Orientation and Direction Estimation

The simplest implementation of the SPT does not require estimation of the fringe orientation. Without orientation or direction, it is still possible to estimate the absolute value of the quadrature sine wave using Equation (4.19).

To disentangle the underlying structure of fringe phase, it is necessary to know, or at least estimate, the fringe direction. Returning to the interferometric fringe pattern in Equation (4.1), there is an assumption that there is an underlying phase pattern $\psi(x,y)$. The direction vector associated with the gradient of this 2-D scalar function is $\beta(x,y)$, shown in Figure 4.1a.

$$\nabla\psi(x,y)=\mathbf{i}\frac{\partial\psi}{\partial x}+\mathbf{j}\frac{\partial\psi}{\partial y}=|\nabla\psi|\{\mathbf{i}\cos\beta+\mathbf{j}\sin\beta\},$$

$$\tan\beta(x,y)=\left(\frac{\partial\psi}{\partial y}\right)\bigg/\left(\frac{\partial\psi}{\partial x}\right) \tag{4.22}$$

The problem, of course, is that we do not know ψ, only $\cos\psi$. There is a sign ambiguity related to

$$\cos[+\psi]=\cos[-\psi] \tag{4.23}$$

Much has been written in the research literature about how to resolve the ambiguity. To implement the full SPT, you can either use your own fringe direction estimator or use the robust orientation estimator in

Larkin [19] and follow the unwrapping instructions in Larkin [9]. The end result is one of the following orientation or direction phasors:

$$\exp[i\beta_{dir}] = \pm\sqrt{\exp[2i\beta_{orient}]} = \begin{cases} \exp[i\beta_{orient}], \text{ or} \\ \exp[i\beta_{orient} + i\pi] \end{cases} \tag{4.24}$$

Unwrapping involves a procedure for choosing the plus/minus sign in this equation to remove the π discontinuities. An alternative approach is to stay with the doubled orientation angle (2β) and then apply any conventional phase-unwrapping method into the range -2π to $+2\pi$. The result is then simply halved to reveal the direction β. Figure 4.4 shows an example of steps in the process of orientation and direction estimation.

(a)

(b)

(c)

(d)

FIGURE 4.4 Fringe pattern (a) and its orientation phase (b) shown between -π (black) and +π (white). Half orientation phase (c) and final unwrapped direction phase (d).

4.4 WHEN TO USE THE SPIRAL PHASE TRANSFORM

4.4.1 Single Frame: Open or Closed Fringes

Everyone seems to have a favorite fringe demodulation algorithm, but it is not clear which algorithm is best in each particular case. Unfortunately, there is little published on the systematic intercomparison of single shot fringe demodulation algorithms. The SPT is based on minimal assumptions: that the fringe pattern has a visible, single fringe orientation that is smoothly spatially varying and that the background component is also slowly varying. This means that the algorithm is unlikely to work (directly) with speckle patterns and other noisy fringe patterns. Conversely, the algorithm works well with any fringe pattern that satisfies the minimal constraints. One obvious area where the algorithm fails is in regions where the underlying phase has a stationary point, for example, where the fringe direction changes sign and the frequency is zero. Algorithms that work in this situation must impose strong constraints on phase continuity and differentiability. These constraints have unwanted side effects when fringe patterns have real discontinuities.

It is fair to say that the SPT will work in virtually all situations for which the 2-D FTM works, with better, or at least equivalent, accuracy. The required accuracy of the fringe direction estimator is relaxed because the error propagation is essentially second order (see Larkin [6]).

The SPT can be used as a good starting point or first approximation for iterative phase estimation algorithms.

4.4.2 Amplitude Demodulation and Fringe Normalization

An obvious, but nevertheless useful, property of the simpler SPT is that it gives the magnitude of the quadrature function directly, that is, without the need for fringe direction estimation.

$$\left|\$\{b(x,y)\cos[\psi(x,y)]\}\right| \cong \left|b(x,y)\sin[\psi(x,y)]\right| \tag{4.25}$$

$$\{b(x,y)\cos[\psi(x,y)]\}^2 + \left|\$\{b(x,y)\cos[\psi(x,y)]\}\right|^2 \cong b^2(x,y) \tag{4.26}$$

Thus, the amplitude $b(x,y)$ can be estimated without the orientation estimation step. This provides an efficient method for the many other algorithms that need to normalize the fringe pattern as a first step (see Quiroga and Servin [20], for example). The intrinsic accuracy of this estimate is good (see error analysis in Larkin, Bone, et al. [8]).

4.4.3 Multiframe Sequences with Arbitrary (and Unknown) Phase Shifts

The SPT can be used advantageously for sequences of two or more phase-shifted fringe patterns. For three or more frames, the SPT allows good estimation of the unknown shifts, and then the Lai and Yatagai generalized phase-shifting algorithm [21] can be used to compute the phase pattern. Orientation estimation is required, but the direction phase-unwrapping step can be omitted, and just the positive square root is utilized (see [10]) to compute a "contingent" phase. Estimation of the unknown phase shift is unaffected by the arbitrary choice of the positive root.

$$\exp[i\beta_c] = +\sqrt{\exp[2i\beta_{orient}]} \qquad (4.27)$$

4.4.4 Other Fringe-like Patterns

The SPT can be applied to other patterns not formed by the interference of light beams. For example, the ridges in human fingerprints have a structure similar to optical fringe patterns. In addition, fingerprints have structures crucial to identification, known as minutiae, that correspond to spiral phase singularities. SPT analysis of fingerprints allows easy detection of singularities and extreme compression based on phase demodulation of closed ridge patterns [2,22].

4.5 PRACTICAL DEMODULATION EXAMPLE

A comparison of the SPT and the FTM is instructive. Consider the simple heart-shaped fringe pattern in Figure 4.5. This fringe pattern exhibits some crucial spectral properties that make the FTM approach doubly problematic.

The problem for the half-plane Hilbert method (FTM), even in this simplified case, is to recover the spectrum of Figure 4.6a from the spectrum of Figure 4.6b with its overlapping and intertwined side-lobe structure. In general, it is just not possible and inevitably leads to substantial phase estimation errors. For the SPT, as we shall see, the intertwining is intrinsically resolved and undone by fringe direction estimation. Figure 4.7 shows details of the the quadrature estimation.

FIGURE 4.5   Synthetic 256 × 256 heart-shaped fringe pattern with unit amplitude.

The demodulation errors for the SPT are concentrated on the fringe orientation discontinuities at the origin and along the horizontal axis. The FTM errors are widely dispersed and mostly related to the half-plane discontinuity and the corresponding horizontal fringes (at 1 o'clock and 5 o'clock in Figure 4.8c). Zero error is denoted by midgray. In this example,

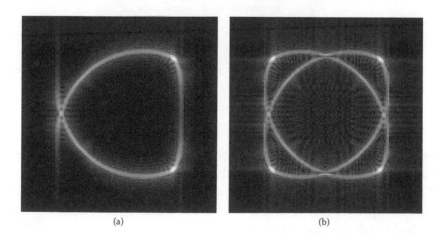

(a)   (b)

FIGURE 4.6   Fourier spectral magnitude for (a) ideal (complex) fringe and (b) cosine (real) fringe.

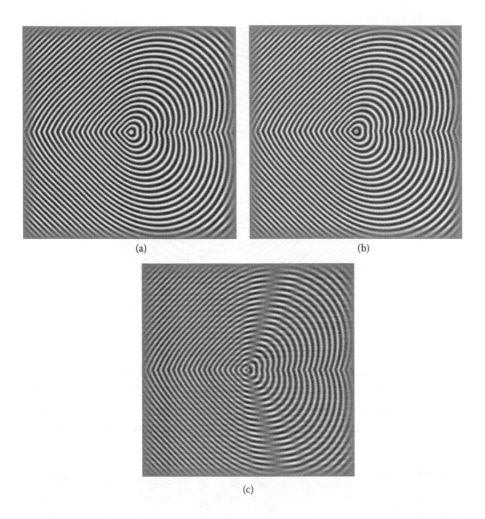

(a)   (b)

(c)

FIGURE 4.7 Demodulated quadrature fringe patterns compared to idealized sinusoid: (a) ideal; (b) vortex; (c) Hilbert.

the root mean square (rms) errors for the Hilbert demodulator are more than 10 times those of the SPT demodulator, as shown in Table 4.1. Spectra of the FTM and vortex methods are compared to the ideal in Figure 4.9. The FTM spectral discrepancy is clear in this instance, whereas the vortex method is indistinguishable from the ideal.

(a) (b)

(c)

FIGURE 4.8   Error map for the estimated quadrature fringe pattern compared to a unit amplitude ideal: (a) ideal |sine|; (b) SPT |sine| error; (c) Hilbert |sine| error.

TABLE 4.1   Comparison of Quadrature Demodulation Errors for the SPT and FTM Shown in Figure 4.8

Modulus of Sine Errors	SPT	Hilbert FTM
Max	0.230	0.250
Min	−0.430	−1.000
Mean	0.000	0.280
Standard deviation (rms)	0.014 (1.4% of fringe peak)	0.190 (19% of fringe peak)

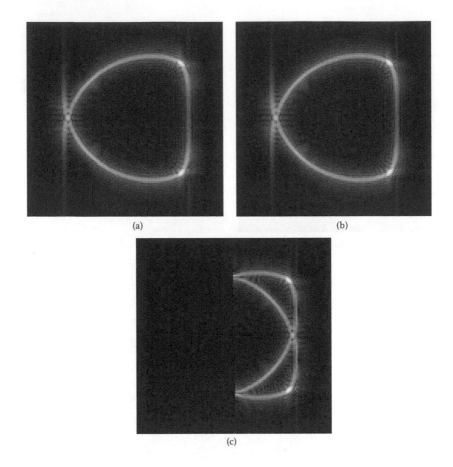

FIGURE 4.9   Demodulated fringe spectra compared to idealized spectrum: (a) ideal; (b) vortex; (c) Hilbert.

4.6 SUMMARY

The SPT is a fast ($N\log N$) and accurate 2-D demodulation algorithm suitable for a variety of fringe analysis problems. When applied to fringe normalization, the algorithm is particularly trivial to implement, yet accurate.

REFERENCES

1. Schnars, U., and W. Jüptner (2005). *Digital Holography*. New York: Springer.
2. Larkin, K. G., and P. A. Fletcher (2007). A coherent framework for fingerprint analysis: Are fingerprints holograms? *Optics Express* **15**(14): 8667–8677.
3. Takeda, M., H. Ina, et al. (1982). Fourier-transform method of fringe-pattern analysis for computer-based topography and interferometry. *Journal of the Optical Society of America* **72**(1): 156–160.

4. Bone, D. J., H.-A. Bachor, et al. (1986). Fringe-pattern analysis using a 2-D Fourier transform. *Applied Optics* **25**(10): 1653–1660.

5. Servin, M., J. L. Marroquin, et al. (1997). Demodulation of a single interferogram by use of a two-dimensional regularized phase-tracking technique. *Applied Optics* **36**(19): 4540–4548.

6. Larkin, K. G. (2001). Natural demodulation of two-dimensional fringe patterns: II. Stationary phase analysis of the spiral phase quadrature transform. *Journal of the Optical Society of America A* **18**(8): 1871–1881.

7. Felsberg, M., and G. Sommer (2001). The monogenic signal. *IEEE Transactions on Signal Processing* **49**(12): 3136–3144.

8. Larkin, K. G., D. Bone, et al. (2001). Natural demodulation of two-dimensional fringe patterns: I. General background to the spiral phase quadrature transform. *Journal of the Optical Society of America A* **18**(8): 1862–1870.

9. Larkin, K. G. (2001). *Natural Demodulation of 2D Fringe Patterns*. Fringe'01—The Fourth International Workshop on Automatic Processing of Fringe Patterns. Bremen, Germany: Elsevier.

10. Larkin, K. G. (2001). A self-calibrating phase-shifting algorithm based on the natural demodulation of two-dimensional fringe patterns. *Optics Express* **9**(5): 236–253.

11. Larkin, K. G., M. A. Oldfield, et al, Demodulation and phase estimation of two-dimensional patterns. US Patent 7043082 B2, January 6, 2000 (orientation estimation).

12. Larkin, K. G, M. A. Oldfield, et al. Demodulation and phase estimation of two-dimensional patterns. US Patent 7444032 B2, April 14, 2000 (spiral demodulation).

13. Reisert, M., and H. Burkhardt (2008). Complex derivative filters. *IEEE Transactions on Image Processing* **17**(12): 2265–2274.

14. Unser, M., and N. Chenouard (2013). A unifying parametric framework for 2D steerable wavelet transforms. *SIAM Journal on Imaging Science* **6**(1): 102–135.

15. Nuttall, A. H. (1966). On the quadrature approximation to the Hilbert transform of modulated signals. *IEEE Proceedings* **54**: 1458–1459.

16. Granlund, G. H., and H. Knutsson (1995). *Signal Processing for Computer Vision*. Dordrecht, Netherlands: Kluwer.

17. FFTW: The fastest Fourier transform in the West. http://www.fftw.org.

18. Fourier and spiral phase transform software application in Linux. https://github.com/magister-ludi/eigenbroetler/#readme.

19. Larkin, K. G. (2005). Uniform estimation of orientation using local and nonlocal 2-D energy operators. *Optics Express* **13**(20): 8097–8121.

20. Quiroga, J. A., and M. Servin (2003). Isotropic n-dimensional fringe pattern normalization. *Optics Communications* **224**(4–6): 221–227.

21. Lai, G., and T. Yatagai (1991). Generalized phase-shifting interferometry. *Journal of the Optical Society of America A* **8**(5): 822–827.

22. Larkin, K. G. Method and apparatus for highly efficient representation and compression of images. US Patent 6,571,014 B1, May 7, 1998 (fingerprint image compression).

CHAPTER 5

Regularized Phase Estimation Methods in Interferometry

Moises Padilla and Manuel Servin

Centro de Investigaciones en Optica A. C.
Leon Guanajuato, Mexico.

5.1 INTRODUCTION

In digital image processing and digital interferometry, one finds two broad classes of problems: direct problems and inverse problems. A direct problem is one that, acting on an input signal $I(x,y)$, produces an output signal $f(x,y)$ by means of a well-defined operation. In other words, a direct problem transforms the input continuously and in a unique way. That is,

$$f(x,y) = T[I(x,y)], \tag{5.1}$$

where $T[\cdot]$ describes a general direct problem. As the name suggests, inverse problems $T^{-1}[\cdot]$ work in the opposite direction with respect to direct problems by analyzing the output $f(x,y)$ to estimate the input $I(x,y)$:

$$I(x,y) = T^{-1}[f(x,y)]. \tag{5.2}$$

In turn, inverse problems can also be classified as well-posed or ill-posed. A well-posed inverse problem is one in which a solution exists, is unique, and depends univocally and continuously on the data; otherwise, it is said to be an ill-posed inverse problem. Experience shows that most inverse problems are ill posed. Usually, one needs to add prior information about the functional set of the expected solution to obtain a stable estimate of the solution; this is called regularization of an ill-posed inverse problem.

The prototypical example of a direct problem in digital image processing is the moving average convolution low-pass filter (LPF), given by

$$f(x,y) = h_{Avg}(x,y) ** I(x,y),$$
(5.3)

where ** represents the two-dimensional (2-D) convolution product and

$$h_{Avg}(x,y) = \frac{1}{9}\begin{pmatrix} 1 & 1 & 1 \\ 1 & 1 & 1 \\ 1 & 1 & 1 \end{pmatrix}$$
(5.4)

is the 3×3 averaging mask. This low-pass filtering operator uniquely and continuously produces an output $f(x,y)$ to any input image function $I(x,y)$. For the inverse problem, if we want to estimate the original image $I(x,y)$ from its low-pass-filtered version $f(x,y)$, we need to perform high-pass filtering using the inverse LPF $[h_{Avg}(x,y)]^{-1}$ as follows (from this point, we label the estimated functions using a "hat" $\wedge$ over them):

$$\hat{I}(x,y) = [h_{Avg}(x,y)]^{-1} ** f(x,y) = [h_{Avg}(x,y)]^{-1} ** [h_{Avg}(x,y) ** I(x,y)].$$
(5.5)

Or, in the frequency domain, applying the standard convention for the Fourier transform FT[·], $H(u,v) = FT[h(x,y)]$, we have

$$\hat{I}(u,v) = H_{Avg}^{-1}(u,v)F(u,v) = H_{Avg}^{-1}(u,v)[H_{Avg}(u,v)I(u,v)].$$
(5.6)

We believe it is important to distinguish between the original signal $I(x,y)$ and its estimated value $\hat{I}(x,y)$ because, in general, the solution to an inverse problem is not unique, and these two may differ significantly.

In this particular case, it should be noted that the frequency transfer function (FTF) $H_{Avg}(u,v)$ has spectral zeros in $\{u_0,v_0\}=\pm 2\pi/3$. Thus, the information at those frequencies is lost during the direct low-pass filtering, $F(u_0,v_0)=0$, and the original image $I(u,v)$ cannot be exactly recovered from its filtered version $F(u,v)$. As $[h_{Avg}(x,y)]^{-1}$ turns out to be unfeasible, in practice one implements $[h_{Avg}(x,y)+\varepsilon]^{-1}$, where ε is an ideally small free parameter that provides an upper bound ε^{-1} for $FT\{[h_{Avg}(x,y)+\varepsilon]^{-1}\}$. A well-behaved (invertible) counterexample is found if we replace the averaging kernel with a Gaussian one, given respectively in the spatial and Fourier domains by

$$h_{Gauss}(x,y)=\frac{1}{2\pi\sigma^2}\exp\left(-\frac{x^2+y^2}{2\sigma^2}\right),$$

$$H_{Gauss}(u,v)=\frac{1}{2\pi}\exp\left(-\frac{\sigma^2(u^2+v^2)}{2}\right).$$

(5.7)

Here, σ determines the width of the Gaussian kernel. Unlike the averaging kernel, $H_{Gauss}(u,v)\neq 0$ for all $(u,v)\in\mathbb{R}^2$; thus, $H_{Gauss}^{-1}(u,v)$ is bounded, and $[h_{Gauss}(x,y)]^{-1}$ is perfectly defined. That is, $I(x,y)$ can be univocally recovered from its Gaussian low-pass filtered version $f(x,y)=h_{Gauss}(x,y)**I(x,y)$ using the high-pass filter $[h_{Gauss}(x,y)]^{-1}$. This different behavior between average and Gaussian low-pass filtering is illustrated in Figure 5.1 using one-dimensional (1-D) plots for ease of observation.

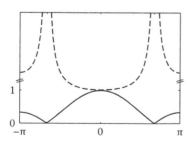 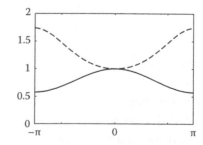

FIGURE 5.1   Frequency transfer function plot (continuous line) and its inverse (dashed line) for the three-step averaging linear filter (left) and the normalized five-step Gaussian low-pass filter with $\sigma = 0.5$ (right).

In interferometry, the synthesis of a fringe pattern by some physical system (interferometer, Moirè system, etc.) represents a direct problem; its analysis, so-called phase demodulation, is an inverse problem. Specifically, consider the following mathematical model of an ideal fringe pattern $I(x,y)$ given by

$$I(x,y) = a(x,y) + b(x,y)\cos[\varphi(x,y)]. \tag{5.8}$$

Here, $a(x,y)$ represents the background illumination, $b(x,y)$ is the fringe contrast term, and $\varphi(x,y)$ is the searched phase map. Clearly, the solution for the inverse problem $\hat{\varphi}(x,y) = T^{-1}[I(x,y)]$ is not unique, even for this ideal model. In fact, there are infinitely many valid solutions merely because the cosine function is even and periodic. This means that, in general, phase demodulation is an ill-posed inverse problem unless we modify the experiment or provide some prior information about the searched phase (such as spatial-temporal carrier).

In this chapter, we use classical regularization to translate ill-posed inverse interferometry problems into well-posed ones by adding restrictions to the functional space of the searched phase: In Section 5.2, we show that smoothness restrictions enable us to LPF noisy signals while preserving the frontiers of valid and invalid data. In Sections 5.3–5.6, we discuss several efficient ways to apply spatial regularization in linear phase carrier techniques. In Section 5.7, we apply temporal regularization methods to analyze dynamic signals. In Section 5.8, we review the regularized phase tracker method to analyze single-image interferograms with closed fringes. In Section 5.9, we test spatial regularization for phase and fringes extrapolation or interpolation. Finally, in Section 5.10 we analyze a regularization method for wavefront reconstruction in lateral shearing interferometry (LSI) with large displacement.

5.2 REGULARIZED LOW-PASS LINEAR FILTERING

Linear filtering is the most basic operation in signal processing. The LPF obtained by the convolution of the data with a small averaging window (also called a kernel or mask) is by far the most commonly used linear filter in fringe analysis. Now, the fundamental question to answer is why anyone should be interested in using regularized linear filtering instead of the standard convolution low-pass filtering. The short answer is that, unlike convolution filters, regularized linear filters prevent mixing of valid data and invalid background. More specifically, interferometric signals are bounded in space and time by a spatial or a temporal pupil. In the space domain, the fringe pattern is bounded by the optical pupil size of the testing interferometer. In the time domain, when several phase-shifted interferograms are taken, the fringe data are also bounded by this temporal interferometric sequence. Right at the edge of these pupil boundaries, convolution filters mix up well-defined interferometric data with invalid background outside the pupil where no fringe data (or phase information) is available or defined. This linear mixing of valid and invalid data translates into degradation of the demodulated phase near the boundaries of the interferogram pupil, in space, time, or space-time. This is the most serious drawback of convolution filtering in fringe pattern phase demodulation.

In spatial regularization filtering, the input image $I(x,y)$ and the filtered image $f(x,y)$ are both assumed to be 2-D surfaces in a Euclidian three-dimensional (3-D) space. In addition, the interferogram image will be assumed to be bounded by a pupil $P = P(x,y)$, where valid fringe data are located. To filter out some noise from the image, we assume that the filtered field has a free-field mathematical model of a rubber membrane or thin metallic plate. This interpretation of the filtered surface is the prior knowledge that one introduces as a regularizer, and this greatly enriches the capabilities of the linear filtering process.

According to Marroquin et al. [1], the regularized low-pass filtering problem may be stated as estimating a smooth and continuous function $f(x,y)$ from a noisy measured image $I(x,y)$ as follows:

$$I(x,y) = f(x,y) + n(x,y), \quad \forall (x,y) \in P. \tag{5.9}$$

Here, $n(x,y)$ is a noisy random field, usually white and Gaussian. A common mathematical formulation for this regularized low-pass filtering may be stated as the 2-D estimation field $f(x,y)$, which interacts with the data $I(x,y)$, minimizing a specific energy functional $U[f(x,y)]$. For instance, consider the following functional:

$$U[f(x,y)] = \iint_{(x,y) \in P} \left\{ [f(x,y) - I(x,y)]^2 + \eta \left[\frac{\partial f(x,y)}{\partial x} \right]^2 + \eta \left[\frac{\partial f(x,y)}{\partial y} \right]^2 \right\}.$$

$$\tag{5.10}$$

From this equation, the first term on the right $[f(x,y)-I(x,y)]^2$ is a quadratic fidelity metric between the smoothed filtered-estimated field $f(x,y)$ and the observed data $I(x,y)$ in a least-squares sense; this is called the interaction term. The remaining quadratic terms tend to keep the filtered-estimated image $f(x,y)$ within the space of continuous functions up to the first derivative, the C^1 functional space. This first-order regularizer is known as a rubber membrane regularizer because this functional corresponds to the mechanical energy of a 2-D rubber membrane. This membrane model is attached by linear springs to the observations; the spring energy is formalized by $[f(x,y)-I(x,y)]^2$. The parameter η is a weight factor that measures the stiffness of our membrane filtering model. For data with low noise, the parameter η must be set to a low value (wide bandwidth). On the other hand, for noisy images, the parameter η must be set to a high value to filter out most of the degrading image noise (narrow bandwidth).

By the same token, using second derivatives of $f(x,y)$ as a regularizer, one may have metallic thin-plate behavior for the filtered image field $f(x,y)$. The potential energy of a metallic thin plate is a quadratic form of second-order derivatives. In this way, the energy functional now reads

$$U[f(x,y)]= \iint_{(x,y)\in P} \left\{ [f(x,y)-I(x,y)]^2 \right.$$

$$\left. +\eta\left[\frac{\partial^2 f(x,y)}{\partial x^2}\right]^2 +\eta\left[\frac{\partial^2 f(x,y)}{\partial y^2}\right]^2 +\eta\left[\frac{\partial^2 f(x,y)}{\partial x\partial y}\right]^2 \right\}.$$

$$(5.11)$$

As can be seen from this equation, everything remains the same except for the change from first-order derivatives to second-order derivatives for the filtered-estimated field $f(x,y)$. In this case, the continuity of the solution remains up to its first-order derivatives. That is, the estimated field $f(x,y)$ that minimizes $U[f(x,y)]$ is searched within a more restricted C^2 functional space.

To this point, we have only defined two functionals (the most used ones), Equation (5.10) and (5.11), whose minimizer $f(x,y)$ is the searched filtered estimated field. Now, we need to search such a minimum. It is well known that minimization of quadratic functionals produces linear algebraic systems (or linear filters in our case). This is done by taking the derivative of $U[f(x,y)]$ with respect to $f(x,y)$ and equating it to zero,

$$\frac{\partial U[f(x,y)]}{\partial f(x,y)} = 0. \tag{5.12}$$

This gives rise to a set of simultaneous linear equations that can be solved using numerical methods such as gradient descent, Gauss-Seidel, conjugate gradient, and so on. Particularly, gradient descent is one of the easiest recursive methods to implement, and it is given by

$$f^{k+1}(x,y) = f^{k}(x,y) - \tau \frac{\partial U[f^{k}(x,y)]}{\partial f^{k}(x,y)}, \tag{5.13}$$

where k is the iteration number. We must use a small value for τ to ensure convergence, for example, $\tau < 0.1$. We stop this recursive process until the error signal $|f^{k+1}(x,y) - f^{k}(x,y)| < \varepsilon$ does change below a predefined small threshold value. Note that for large images, it may be necessary to implement more efficient methods, such as the conjugate gradient.

Although we have displayed continuous energy functionals, in practice the ones we actually implement in a digital computer are their discrete versions. The discrete functional for the first-order (rubber-band membrane) regularizer filter is given by [from Equation (5.10)]

$$U[f(x,y)] = \sum_{(x,y)\in P} \{[f(x,y) - I(x,y)]^2 + \eta[f(x,y)$$
$$- f(x-1,y)]^2 + \eta[f(x,y) - f(x,y-1)]^2\}, \tag{5.14}$$

And, the second-order (metallic thin-plate) regularizer functional is given by [from Equation (5.11)]

$$U[f(x,y)] = \sum_{(x,y)\in P} \left\{ \begin{array}{c} [f(x,y) - I(x,y)]^2 + \eta[f(x+1,y) - 2f(x,y) + f(x-1,y)]^2 \\ + \eta[f(x,y+1) - 2f(x,y) + f(x,y-1)]^2 \\ + \eta[f(x,y) + f(x,y-1) - f(x-1,y) - f(x-1,y-1)]^2 \end{array} \right\}. \tag{5.15}$$

Another practical consideration is the finite spatial pupil where well-defined data exist; this indicator function is $P = P(x, y)$, which equals 1 for valid fringes inside $P(x, y)$ and equals 0 otherwise. For instance, taking into consideration this indicator pupil function, the first-order regularizer may be rewritten as

$$U[f(x, y)]$$

$$= \sum_{(x, y) \in \mathbb{Z}^2} \left\{ \begin{array}{c} [f(x, y) - I(x, y)]^2 P(x, y) + \eta [f(x, y) - f(x-1, y)]^2 P(x, y) P(x-1, y) \\ + \eta [f(x, y) - f(x, y-1)]^2 P(x, y) P(x, y-1) \end{array} \right\}.$$

(5.16)

In this way, one automatically deletes the quadratic potentials that fall outside the indicator pupil $P(x, y)$. To summarize and emphasize, the first-order (rubber membrane) regularizer restricts the estimated field $f(x, y)$ to be within the C^1 functional space. On the other hand, a second-order (metallic thin-plate) regularizer further restricts the estimated field $f(x, y)$ within the C^2 functional space. This is shown in Figure 5.2 for a slice in the x direction for ease of observation.

5.2.1 Frequency Response of Low-Pass Regularizing Filters

Here, we show how the stiffness parameter η is related to the spectral behavior of the first- and second-order regularized low-pass filtering fields $f(x, y)$ in Equations (5.10) and (5.11). Note that for this spectral analysis we consider the data field $I(x, y)$ and the estimated field $f(x, y)$ to be spatially

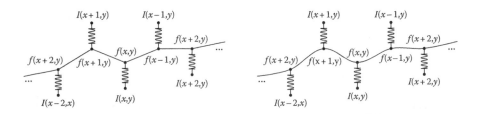

FIGURE 5.2 Schematic comparison between the estimated filtered field $f(x, y)$ using the first-order rubber membrane regularization (left) and second-order metallic thin-plate regularization (right).

unbounded. In other words, the bounding pupil equals 1, $P(x,y) = 1$, for the full 2-D Euclidian space $\mathbb{R}^2$.

For the discrete version of the first-order (rubber membrane) regularizer [Equation (5.14)], we have

$$U[f(x,y)]$$

$$= \sum_{(x,y)\in\mathbb{R}^2} \{[f(x,y)-I(x,y)]^2 + \eta[f(x,y)-f(x-1,y)]^2 + \eta[f(x,y)-f(x,y-1)]^2\}.$$

$$(5.17)$$

Taking the derivative of Equation (5.17) with respect to $f(x,y)$ and equating it to 0, we obtain

$$\frac{\partial U[f(x,y)]}{\partial f(x,y)} = f(x,y)-I(x,y)-\eta\left[\begin{array}{c} f(x+1,y)-2f(x,y)+f(x-1,y) \\ +f(x,y+1)-2f(x,y)+f(x,y-1) \end{array}\right]=0.$$

$$(5.18)$$

Taking the Fourier transform FT[·] of Equation (5.18) (i.e., FT[$f(x + 1, y+1)] = e^{i(u+v)}F(u,v)$) we obtain

$$F(u,v)[1+2\eta(2-\cos u-\cos v)]-I(u,v)=0. \qquad (5.19)$$

Finally, solving for the FTF $H_m(u,v)= F(u,v)/I(u,v)$, the FTF of the first-order (rubber membrane) regularizer filter is given by

$$H_m(u,v)=\frac{F(u,v)}{I(u,v)}=\frac{1}{1+2\eta(2-\cos u-\cos v)}. \qquad (5.20)$$

On the other hand, following identical steps for the second-order (metallic thin-plate) regularizer filter, we find that its FTF $H_p(u,v)= F(u,v)/I(u,v)$ is now given by

$$H_p(u,v)=\frac{1}{1+2\eta[8-6(\cos u+\cos v)+\cos 2u+\cos 2v+2\cos u\cos v]}. \qquad (5.21)$$

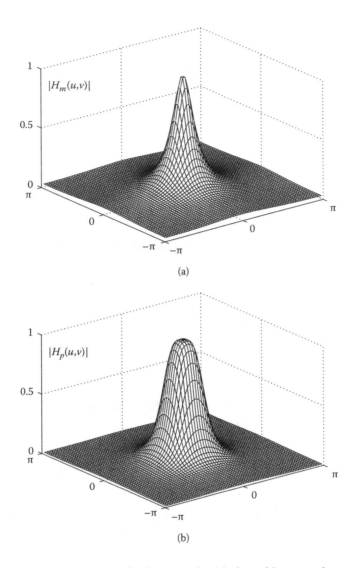

(a)

(b)

FIGURE 5.3 Frequency transfer function for (a) the rubber membrane and (b) the metallic thin-plate regularizer filters using the same value for the stiffness parameter, $\eta = 5$.

Note from Equations (5.20) and (5.21) that the FTFs for the first- and second-order regularizer filters behave somewhat like 2-D Lorentzian functions (see Figure 5.3), for which the bandwidth is controlled by the stiffness parameter η: Increasing the stiffness parameter η, one narrows the bandwidth of the FTF. For a quick reference and summary of some of the more important concepts in this section, see Box 5.1.

BOX 5.1 CLASSICAL REGULARIZATION FOR LOW-PASS FILTERING

- A noisy measured image $I(x,y)$ may be mathematically modeled as

$$I(x,y) = f(x,y) + n(x,y), \quad \forall (x,y) \in P.$$

- We can estimate $f(x,y)$ by minimizing quadratic regularizer functionals (using, for instance, gradient descend). The first-order regularizer functional is given by

$$U[f(x,y)] = \sum_{(x,y) \in P} \{[f(x,y) - I(x,y)]^2 + \eta[f(x,y) - f(x-1,y)]^2 + \eta[f(x,y) - f(x,y-1)]^2\}.$$

The second-order (metallic thin-plate) regularizer functional is given by

$$U[f(x,y)] = \sum_{(x,y) \in P} \left\{ \begin{array}{c} [f(x,y) - I(x,y)]^2 + \eta[f(x+1,y) - 2f(x,y) + f(x-1,y)]^2 \\ + \eta[f(x,y+1) - 2f(x,y) + f(x,y-1)]^2 \\ + \eta[f(x,y) + f(x,y-1) - f(x-1,y) - f(x-1,y-1)]^2 \end{array} \right\}.$$

- The main advantage of regularized filtering is that it prevents the mixture of valid data with background signals (outside the bounding pupil where no interferometric data are available).
- The frequency transfer function (FTF) of the first-order regularizer filter is given by

$$H_m(u,v) = \frac{1}{1 + 2\eta(2 - \cos u - \cos v)}.$$

The FTF of the second-order regularizer filter is given by

$$H_p(u,v) = \frac{1}{1 + 2\eta[8 - 6(\cos u + \cos v) + \cos 2u + \cos 2v + 2\cos u \cos v]},$$

where the bandwidth is controlled by the parameter η.

Example 5.1: Comparison between Regularized and Convolution Low-Pass Filtering of Noisy Interferograms

For this example, we have simulated three phase-shifted samples of a temporal interferogram with a phase step $\omega_0 = 2\pi/3$ and corrupted them by additive white Gaussian noise (AWGN). In this case, our bounding pupil $P(x, y)$ is a disk. Two sets of filtered fringe patterns were produced using, respectively, a rubber membrane regularizer filter and a running average LPF. Finally, both were phase demodulated using the well-known three-step temporal least-squares phase-shifting algorithm (LS-PSA) of Bruning et al. [2] to see their phase-estimating behavior. The results are shown in Figures 5.4 and 5.5. For a meaningful comparison, necessary care was taken to ensure that both filters had similar spectral bandwidth responses.

As illustrated in Figures 5.4 and 5.5, regularized filtering prevents the mixing of valid data and invalid data, preserving the estimated-phase behavior near the edges of $P(x, y)$. In contrast, the averaging (convolution) low-pass filtering mixes valid fringe data with the background outside $P(x, y)$, producing noticeable phase distortion at the region near the boundaries of the pupil. This is one of the advantages of using spatial regularized filtering.

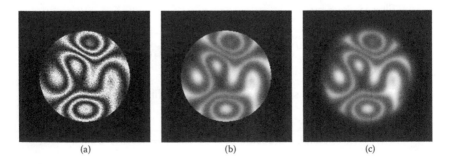

(a) (b) (c)

FIGURE 5.4 (a) A sample of the simulated three temporal phase-shifted interferograms with AWGN. (b) Smooth interferogram resulting from the regularized low-pass filtering. (c) Convolution low-pass filtered interferogram. Note the different behavior near and at the edge of the pupil of the interferogram (in both cases, the cutoff frequency was about 0.1 rad/pixel).

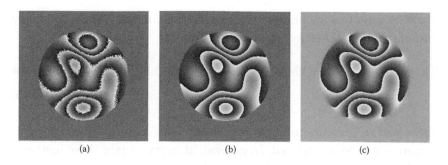

(a)   (b)   (c)

FIGURE 5.5 Wrapped phase as obtained from the three-step LS-PSA using (a) the raw data, (b) regularized low-pass filtering, and (c) convolution low-pass filtering, respectively. Note how the estimated phase in (c) is distorted near the edges of the pupil $P(x, y)$ of the interferogram.

5.3 CONVOLUTION-BASED TEMPORAL PHASE-SHIFTING INTERFEROMETRY

In this section we will review the standard theory of phase-shifting algorithms (PSAs) to highlight its intrinsic relation with convolution quadrature linear filters [3–6] and to make better use of the regularized techniques in the following sections. Consider the standard mathematical model of a digital interferogram defined in a regular discrete lattice $(x, y) \in \mathbb{Z}^2$:

$$I(x, y) = a(x, y) + b(x, y)\cos[\varphi(x, y)], \quad (x, y) \in \mathbb{Z}^2, \quad \forall(x, y) \in P(x, y).$$

$$(5.22)$$

Here, $a(x, y)$ and $b(x, y)$ represent, respectively, the local background and fringe contrast functions, and $\varphi(x, y)$ represents the searched modulating phase. As discussed, the region with valid interferometric data is spatially bounded by the optical pupil of the interferometer, the pupil of the instrument, or the object under test, with all of them represented by $P(x, y)$. Although a variety of techniques exists that may demodulate a single-image closed-fringes pattern [7–9], in interferometry we are always looking for experimental setups that allow the introduction of a linear carrier, whether spatial, temporal, or spatial-temporal. That is because carrier frequency (either spatial or temporal) phase demodulation is by far the easiest process to implement and analyze. The first carrier used in interferometry was the linear carrier. In particular, consider the following temporal phase-shifted interferogram sequence:

$$I(x,y,t) = a(x,y) + b(x,y)\cos[\varphi(x,y) + \omega_0 t], \quad (x,y) \in P, \quad t \in \mathbb{Z}, \quad \omega_0 \in (0,\pi).$$

$$(5.23)$$

where $\omega_0 t$ represents the temporal linear carrier and all other terms remain as previously defined. In general, at every instant these temporal interferograms contain closed spatial fringes (such as those presented in Figure 5.4a). Taking the Fourier transform of Equation (5.23), the spectrum of a long-run (infinitely many samples in theory) phase-shifted interferometric signal is given by

$$I(x,y,\omega) = a(x,y)\delta(\omega) + (1/2)b(x,y)\exp[i\varphi(x,y)]\delta(\omega - \omega_0)$$
$$+ (1/2)b(x,y)\exp[-i\varphi(x,y)]\delta(\omega + \omega_0).$$

$$(5.24)$$

Note that the searched modulating phase $\varphi(x,y)$ can be straightforwardly estimated from the analytic signal $(1/2)\, b(x,y)\, \exp[i\varphi(x,y)]$. As illustrated in Figure 5.6, this analytic signal can be isolated by means of a convolution temporal quadrature filter $h(t)$ with a spectral response or FTF given by $H(\omega) = F[h(t)]$. The spectrum of $H(\omega)$ must have at least two zeroes: one at $\omega = 0$ and the other at $\omega = -\omega_0$. That is,

$$H(-\omega_0) = 0, \quad H(0) = 0, \quad H(\omega_0) \neq 0, \tag{5.25}$$

A band-pass filter that fulfills this spectral criterion is called a quadrature linear filter.

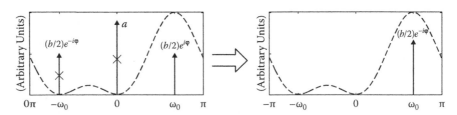

FIGURE 5.6 Representation in the Fourier domain of a quadrature linear filtering process. The magnitude of the frequency transfer function (FTF) of the temporal filter $|H(\omega)|$ is illustrated by the dashed line. We can see that the magnitude $|H(\omega)|$ must have at least two zeros: one at $|H(0)| = 0$ and the other at $|H(-\omega_0)| = 0$. Finally, the analytic signal that we are looking for is located at $|H(\omega_0)|$.

In general, an N-step quadrature linear filter [3–6] may be formally written as

$$h(t) = \sum_{n=0}^{N-1} c_n \delta(t-n), \qquad H(\omega) = F[h(t)] = \sum_{n=0}^{N-1} c_n \exp(-in\omega), \qquad (5.26)$$

where $\{c_n\} = \{a_n + i b_n\} \in \mathbb{C}$. In the temporal domain, the analytic signal $(1/2)$ $b(x,y) \exp[i\varphi(x,y)]$ is obtained by applying this quadrature linear filter to the temporal phase-shifted interferogram:

$$I(x,y,t) * h(t) = \left[\sum_{n=0}^{N-1} I(x,y,n)\delta(t-n) \right] * \left[\sum_{n=0}^{N-1} c_n \delta(t-n) \right]$$

$$= (1/2)H(\omega_0)b(x,y)\exp[i\varphi(x,y) + i\omega_0 t] \qquad (5.27)$$

where $\sum_{n=0}^{N-1} I(x,y,n)\delta(t-n)$ is the interferometric data composed of N phase-shifted samples, and $H(\omega_0) = F\{h(t)\}|_{\omega=\omega_0}$. The convolution product presented has full temporal support of $2N - 1$ samples, but we are only interested in the middle term at $t = N - 1$, where both the support of the filter and the data coincide and the estimated analytic signal is most reliable [5]. That is,

$$[I(x,y,t) * h(t)]\big|_{t=N-1} = \sum_{n=0}^{N-1} c_n I(x,y,N-n-1) = A_0 \exp[i\hat{\varphi}(x,y)], \quad \forall (x,y) \in P.$$

$$(5.28)$$

where $A_0 = (1/2)H(\omega_0)b(x,y)\exp[i\omega_0(N-1)]$. It should be noted that in the literature this is usually reported as $\sum_{n=0}^{N-1} c_n I(x,y,n)$ instead of $\sum_{n=0}^{N-1} c_n I(x,y,N-n-1)$; the first case corresponds to the middle term of a cross-correlation product, and the second one is the middle term of a convolution product. The only difference between both approaches is (at most) a global sign change, obtaining $A_0 \exp[-i\hat{\varphi}(x,y)]$ instead of $A_0 \exp[i\hat{\varphi}(x,y)]$. Furthermore, most popular PSAs have symmetrical real-valued coefficients; in those cases, the convolution and cross-correlation products coincide.

We call the mathematical form shown in Equation (5.20) the analytic formulation of a temporal PSA; this is not the standard form for presenting PSAs in the literature. Instead, it is usually presented in what we call the "arc-tangent formulation," obtained when solving for the estimated phase $\varphi(x, y)$ as follows:

$$[\varphi(x, y)]\bmod 2\pi = \arctan\left\{\frac{\displaystyle\sum_{n=0}^{N-1} b_n I(x, y, n)}{\displaystyle\sum_{n=0}^{N-1} a_n I(x, y, n)}\right\}. \tag{5.29}$$

In summary, linear PSAs are simply a convolution (or cross-correlation) product between N temporal samples of the interferometric real-valued data $\sum I(x, y, t)\delta(t-n)$ and a (complex-valued) quadrature linear filter $h(t) = \sum c_n \delta(t-n)$. For an encyclopedic list of the most popular PSAs (presented in the arc-tangent form) see, for instance, the two books of Malacara et al. [10, 11].

Understanding the linear PSAs as quadrature linear filters and working with its analytic signal formulation [Equation (5.28)], instead of the arc-tangent version, has many advantages when applying regularized phase estimation methods, as shown in the following sections.

Example 5.2: Analytic Formulation of Schwider-Hariharan's Five-Step PSA

Consider the popular five-step PSA by Schwider et al. [12] and Hariharan et al. [13]. This linear PSA uses a phase step of $\omega_0 = \pi/2$, and in the arc-tangent formulation it is given by

$$\varphi(x, y) = \arctan\left\{\frac{2[I(x, y, 3) - I(x, y, 1)]}{I(x, y, 0) - 2I(x, y, 2) + I(x, y, 4)}\right\}. \tag{5.30}$$

By direct comparison with Equations (5.28) and (5.29), from the denominator and numerator in the argument of the arctan(.) function, we have

$$[I(x, y, t) * h_1(t)]_{t=4} = \sum_{n=0}^{4} a_n I(x, y, 4-n) = I(x, y, 0) - 2I(x, y, 2) + I(x, y, 4),$$

$$[I(x, y, t) * h_2(t)]_{t=4} = \sum_{n=0}^{4} b_n I(x, y, 4-n) = I(x, y, 3) - I(x, y, 1).$$

$$\tag{5.31}$$

This interpretation of a linear PSA as the convolution between the temporal phase-shifted interferogram and two linear filters is from the work of Freischlad and Koliopoulos [3]. Furthermore, as pointed out by Surrel [4], we know that these two linear filters $h_1(t)$ and $h_2(t)$ are actually the imaginary and real part, respectively, of a (complex-valued) quadrature linear filter, given by

$$h(t) = h_1(t) + i\, h_2(t) = \sum_{n=0}^{N-1} (a_n + ib_n)\delta(t-n),$$

(5.32)

where $c_n = a_n + ib_n = \{1, -2i, -2, 2i, 1\}$. Thus, using the compact notation $I_n = I(x, y, n)$, the analytic formulation for Schwider-Hariharan's PSA is given by

$$A_0 \exp[i\varphi(x, y)] = \sum_{n=0}^{4} c_n I(x, y, 4-n) = I_0 - 2i\, I_1 - 2I_2 + 2i\, I_3 + I_4.$$

(5.33)

In summary, the analytic formulation [right side of Equation (5.33)] may be written directly from the arc-tangent formulation [Equation (5.30)], so once the theory is understood all intermediate steps can be omitted. We also want to highlight that the Fourier transform of $h(t)$, the FTF of the filter $H(\omega) = F[h(t)]$, is the natural tool (since at least the 1940s) for assessment and gauge convolution linear filters. In this case, the FTF of the Schwider-Hariharan five-step PSA is given by

$$H(\omega) = F\{h(t)\} = 1 - 2ie^{-i\omega} - 2e^{-i2\omega} + 2ie^{-i3\omega} + 2e^{-i4\omega},$$

$$= A_0[1 - e^{i\omega}][1 - e^{i(\omega+\pi/2)}]^2[1 - e^{i(\omega+\pi)}],$$

(5.34)

where $A_0 \in \mathbb{C}$ is an irrelevant amplitude factor (a global piston phase). That is, the FTF $H(\omega) = F[h(t)]$ of Schwider-Hariharan's PSA has a first-order spectral zero at $\omega = 0$, another single zero at $\omega = \pi$, and finally a second-order zero at $\omega = -\pi/2$. This second-order spectral zero provides its characteristic robust tolerance to detuning error. For more details, see Freischlad and Koliopoulos [3], Surrel [4], Servin et al. [5], and Gonzalez et al. [6].

5.4 SPATIALLY REGULARIZED TEMPORAL LINEAR CARRIER INTERFEROMETRY

In this section, we show how a spatial regularized filter may be used in temporal phase-shifting interferometry when the searched modulating phase $\varphi(x,y)$ does not change in time. That is, the only change in the measuring optical wavefront is caused by the (piston-like) phase shift introduced by the interferometer as a temporal carrier $(\omega_0 t)$. A temporal phase-shifted interferogram can be modeled as

$$I(x,y,t) = a(x,y) + b(x,y)\cos[\varphi(x,y) + \omega_0 t],$$

$$(x,y) \in P, \; t \in \{0,\ldots,N-1\}, \; \omega_0 \in [0,\pi]. \tag{5.35}$$

As seen in the previous section, temporal PSAs are given by the convolution product between the interferometric fringe data $I(x,y,t)$ and a (complex-valued) quadrature linear filter $h(t)$. Thus, in the analytic PSA formulation, the searched signal (for symmetrical PSAs) is given by

$$A_0 \exp[i\varphi(x,y)] = [I(x,y,t) * h(t)]\big|_{t=N-1} = \sum_{n=0}^{N-1} c_n I(x,y,n), \quad \forall (x,y) \in P, \tag{5.36}$$

where the amplitude factor is given by $A_0 = (1/2)b(x,y)H(\omega_0)\exp[i\omega_0(N-1)]$. This amplitude factor cancels out when solving for the searched modulating phase. With a minor variation in the energy functional $U[f(x,y)]$ analyzed in Section 5.2, we can use spatially regularized low-pass filtering directly on the analytic signal given by the PSA. In other words, instead of individually low-pass filtering the N interferograms in separate processes, we do it with a single spatially regularized low-pass filtering process. This may be achieved by minimizing the following energy functional:

$$U = \sum_{(x,y) \in P} \left\{ \left| f(x,y) - \sum_{n=0}^{N-1} c_n I(x,y,n) \right|^2 + \eta |f(x,y) - f(x-1,y)|^2 + \eta |f(x,y) - f(x,y-1)|^2 \right\}, N \geq 3 \tag{5.37}$$

for the first-order (rubber membrane) regularizer. The complex weighting factors in $\sum_{n=0}^{N-1} c_n I(x,y,n)$ must be taken from any desired PSA. Alternatively, from Equation (5.36), we have

$$U[f(x,y)] = \sum_{(x,y) \in P} \{|f(x,y) - A_0 e^{i\varphi(x,y)}|^2 + \eta |f(x,y) - f(x-1,y)|^2 + \eta |f(x,y) - f(x,y-1)|^2\}. \tag{5.38}$$

Note the change from real-valued differences to absolute differences, [·] $\Rightarrow$ |·| because now the noisy data input $A_0 \exp[i\varphi(x,y)] = \sum_{n=0}^{N-1} c_n I(x,y,n)$ and the filtered field $f(x,y)$ are both complex-valued fields. For example, consider the analytic formulation deduced in the previous example for the detuning-tolerant five-step PSA by Schwider et al. [12] and Hariharan et al. [13], replicated here for convenience:

$$A_0 \exp[i\varphi(x,y)] = \sum_{n=0}^{4} c_n I(x,y,n) = I_0 - 2iI_1 - 2I_2 + 2iI_3 + I_4. \quad (5.39)$$

Introducing this (complex-valued) analytic signal into the membrane LPF one obtains smooth phase estimations, preventing the mixture of valid and invalid fringe data near the optical pupil boundary. In this case, we see that the spatially regularized low-pass filtering is applied only once instead of the (five times) more time-consuming approach of filtering each temporal sample $\{I_0, I_1, I_2, I_3, I_4\}$ individually and later combining them with the convolution linear PSA.

It should be noted that in some interpreted computer languages (such as MATLAB), the minimization of $U[f(x,y)]$ might be much slower for complex fields. In that case, the more efficient approach is first computing the analytic signal $A_0 \exp[i\varphi(x,y)]$, then minimizing its imaginary and real parts separately, and finally computing the searched phase using the atan2(.,.) function. Next, in Box 5.2, we summarize the most important conclusions obtained so far for spatially regularized temporal linear carrier interferometry.

BOX 5.2 SPATIAL REGULARIZATION FOR TEMPORAL PHASE-SHIFTING INTERFEROMETRY

- A temporal phase-shifted interferogram can be mathematically modeled as

$$I(x,y,t) = a(x,y) + b(x,y)\cos[\varphi(x,y) + \omega_0 t], \quad (x,y) \in P, \quad \omega_0 \in (0,\pi), \quad t \in \mathbb{Z},$$

  where $\{a, b, \varphi\}$ are unknown, and $\varphi(x,y)$ is the searched modulating phase.

- In (complex) analytic formulation, temporal phase-shifting algorithms (PSAs) are given by

$$A_0(x,y)\exp[i\varphi(x,y)] = \sum_n c_n I(x,y,n),$$

where $\{c_n\}$ depend on the PSA, and $\varphi(x,y)$ is computed as the angle of this analytic signal.

- There are several advantages in working with the analytic signal formulation for spatial regularized filtering. For instance, using first-order regularized low-pass filtering,

$$U[f] = \sum_{(x,y)\in P} \{|f(x,y) - A_0 e^{i\,\varphi(x,y)}|^2 + \eta|f(x,y) - f(x-1,y)|^2 + \eta|f(x,y) - f(x,y-1)|^2\},$$

it is N times more efficient for an N-step PSA than preprocessing each temporal phase-shifted interferogram separately.

Example 5.3: Estimation of a Smooth Phase Map by Spatially Regularized Low-Pass Filtering of the Analytic Signal

For this illustrative example, we have simulated five samples of a temporal phase-shifted interferogram with a phase step $\omega_0 = \pi/2$ and corrupted by AWGN. Again, our bounding pupil $P(x,y)$ is given by a disk. According to the theory discussed in this section, we first computed an analytic signal by means of the five-step Schwider-Hariharan PSA, and then we applied (first-order) spatially regularized low-pass filtering to this analytic signal. The results are shown in Figures 5.7 and 5.8.

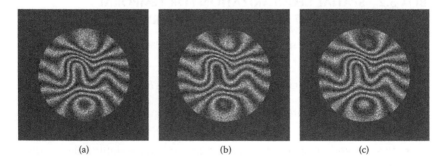

(a)   (b)   (c)

FIGURE 5.7 Three of five samples for the simulated temporal phase-shifted interferogram with AWGN. The relative phase step among them is $\omega_0 = \pi/2$. These five fringes-patterns are combined according to the PSA in Equation (5.39) to produce the complex-valued analytic signal.

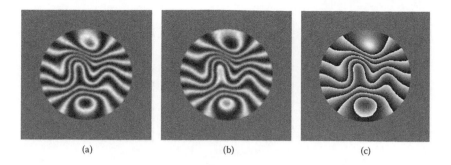

(a)   (b)   (c)

FIGURE 5.8   (a) and (b) The real and imaginary parts, respectively, of the smooth analytic signal obtained with the spatially regularized low-pass filtering obtained in Equation (5.30). (c) Resulting estimated phase. Note there is no boundary distortion because of the regularized low-pass filtering.

In Figure 5.8, we show the real and imaginary parts of the complex-valued membrane filtered field $f(x,y)$ in Equation 5.30. We can see how the demodulated phase $\varphi(x,y)$ is almost free of distortion because of the boundary $P(x,y)$ of the interferogram.

5.5 CONVOLUTION-BASED SPATIAL-CARRIER INTERFEROMETRY

Spatial linear carrier interferometry (also called the Fourier method) is credited to Ichioka and Inuiya [14] and Takeda et al. [15]. In this method, a linear spatial carrier is introduced to a single-image interferogram. Thus, the interferometric data may be written as

$$I(x,y) = a(x,y) + b(x,y)\cos[\varphi(x,y) + u_0 x + v_0 y],$$

$$(u_0, v_0) \in [0.\pi] \times [0,\pi], \forall(x,y) \in P. \tag{5.40}$$

If the spatial carrier values (u_0, v_0) are larger than the maximum variation of $\varphi(x,y)$ in the x and y directions, then Equation (5.40) represents an open-fringes interferogram such as the one illustrated in Figure 5.9.

Taking the Fourier transform of the spatial linear carrier interferogram [Equation (5.40)], we have

$$I(u,v) = A(u,v) + C(u - u_0, v - v_0) + C^*(u + u_0, v + v_0),$$

$$C(u,v) = F\{(1/2)b(x,y)\exp[i\varphi(x,y)]\}. \tag{5.41}$$

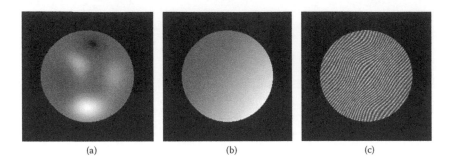

(b)   (c)

FIGURE 5.9   (a) Computer-simulated phase under study. (b) Tilted reference wavefront producing the spatial linear carrier. (c) Resulting open-fringes pattern corrupted with AWGN. Because of the spatial linear carrier, the modulating phase can be estimated using the single-image Fourier interferometric method.

Considering that the spectral lobes $A(u,v)$ and $C(u,v)$ are band-limited signals, Equation (5.41) means that the spatial linear carrier $u_0 x + v_0 y$ produces a spectral separation between the analytic components of the interferogram if the following conditions are fulfilled:

$$\max\left|\frac{\partial\varphi(x,y)}{\partial x}\right| < u_0, \quad \text{and} \quad \max\left|\frac{\partial\varphi(x,y)}{\partial y}\right| < v_0. \tag{5.42}$$

Note that these conditions are applied to $\varphi(x,y)$ because $a(x,y)$ and $b(x,y)$ are usually very-low-frequency functions in comparison with the spatial carrier. Because of the separation of the spectral lobes, we can isolate one of the analytic signals in Equation (5.41) by means of a complex quadrature linear filter, with an FTF given by

$$H(u-u_0, v-v_0) = F\{h(x,y)\exp[-i(u_0 x + v_0 y)]\}, \tag{5.43}$$

where $h(x,y)$ is a real-valued LPF. In this way, assuming that Equation (5.42) is fulfilled, $I(u,v)H(u-u_0, v-v_0)$ rejects the unwanted signals $A(u,v)$ and $C(u+u_0, v+v_0)$, centered at the baseband $(0,0)$ and at $(-u_0, -v_0)$, respectively. But, frequently a more convenient approach is to spectrally displace the information of interest toward the origin by multiplying the data by the carrier; formally,

$$F\{I(x,y)\exp[-i(u_0 x + v_0 y)]\} = I(u-u_0, v-v_0). \tag{5.44}$$

FIGURE 5.10 Alternative implementations for the Fourier demodulation method for the open-fringes pattern in Figure 5.9. At the left, the analytic signal is isolated with quadrature linear filtering, $I(u,v)H(u-u_0, v-v_0)$. At the right, the interferometric data are spectrally displaced, and then the analytic signal is isolated with ordinary low-pass filtering, $I(u-u_0, v-v_0)\,H(u,v)$.

Then, we can isolate the now baseband analytic signal $c(x,y) = (1/2) b(x,y) \exp[i\varphi(x,y)]$ by means of a spatial LPF, usually given by the convolution with an averaging mask $h(x,y)$. That is,

$$(1/2)b(x,y)\exp[i\varphi(x,y)] = \{I(x,y)\exp[-i(u_0 x + v_0 y)]\}**h(x,y), \quad (5.45)$$

where ** stands for the 2-D convolution operation. Both alternative approaches for the Fourier demodulation method [14,15] are illustrated in Figure 5.10.

5.6 REGULARIZATION IN GENERAL SPATIAL CARRIER INTERFEROMETRY

The so-called Fourier method analyzed in the previous section can be considered a particular case of the synchronous demodulation method using a generalized spatial carrier $c(x,y)$. In general, any spatial carrier interferogram may be stated as

$$I(x,y) = a(x,y) + b(x,y) \cos[\varphi(x,y) + c(x,y)], \quad \forall (x,y) \in P. \quad (5.46)$$

As before, $a(x,y)$ and $b(x,y)$ are the background and fringe contrast functions, respectively, $\varphi(x,y)$ is the searched modulating phase, and $P(x,y)$ is the bounding pupil where we have valid interferometric data. The function $c(x,y)$ describes a general spatial carrier, which in order to introduce enough spectral separation between the spectral lobes of the data, must fulfill the following condition:

$$\max|\nabla\varphi(x,y)| \quad < \quad |\nabla c(x,y)|. \quad (5.47)$$

This means that the spatial carrier $c(x,y)$ may have many mathematical forms, but it must be high frequency to fulfill the requirement in Equation (5.47). To illustrate this thesis, now we explicitly develop the synchronous product $I(x,y)\exp[-ic(x,y)]$:

$$I(x,y)\exp[-ic(x,y)] = a(x,y)\exp[-ic(x,y)] + (1/2)b(x,y)\exp[i\varphi(x,y)]$$

$$+ (1/2)b(x,y)\exp[-i\varphi(x,y)-i2c(x,y)].$$

$$(5.48)$$

Assuming Equation (5.47) is fulfilled, the searched analytic signal is the only low-frequency term in Equation (5.48); the remaining terms are high-frequency ones spectrally displaced at the spatial carrier frequency $c(x,y)$ and twice the carrier frequency $2c(x,y)$, respectively. Thus, applying low-pass filtering to the synchronous product, we have

$$(1/2)b(x,y)\exp[i\hat{\varphi}(x,y)] = LPF\{I(x,y)\exp[-ic(x,y)]\}, \quad \forall(x,y)\in P, (5.49)$$

where $LPF\{\cdot\}$ stands for symmetric low-pass filtering. This means that we can apply the regularizing LPFs previously analyzed in this chapter to demodulate this general spatial carrier interferogram. For instance, using first-order regularization,

$$U[f(x,y)] = \sum_{(x,y)\in P} \left\{ |f(x,y)-I(x,y)e^{-ic(x,y)}|^2 \right.$$

$$\left. +\eta|f(x,y)-f(x-1,y)|^2 + \eta|f(x,y)-f(x,y-1)|^2 \right\}. \quad (5.50)$$

Once again, the filtered field $f(x,y)$ is a complex-valued (analytic) signal, the complex-valued demodulating carrier is $\exp[-ic(x,y)]$, and the searched modulated phase $\varphi(x,y)$ can be estimated by computing the angle of $f(x,y)$. Finally, recall that the bandwidth of the regularized low-pass filtering is controlled by the parameter η.

The main advantage of moving the desired spectral lobe toward the spectral origin is that, independently of the mathematical form of the spatial carrier, the interferometric method will be the same: just an LPF to keep the searched analytic signal $(1/2)\,b(x,y)\,\exp[i\varphi(x,y)]$ and filter out the rest of the spectral space. For instance, $c(x,y)$ may be a conic spatial carrier, widely used in corneal topography [16,17], or the 2-D pixelated carrier used for dynamic interferometry [18–20].

Example 5.4: Regularized Demodulation of Placido Images (Conic Spatial Carrier) for Corneal Topography

A conic spatial carrier $c(x,y) = \omega_0 \sqrt{x^2 + y^2}$ produces fringe data having uniformly distributed concentric rings. This conic carrier is widely used in corneal topography, in which concentric rings are projected over a human cornea to produce the so-called Placido gauging images [16,17]. Following our mathematical model, a conic carrier interferogram is given by

$$I(x,y) = a(x,y) + b(x,y)\cos[\varphi(x,y) + \omega_0(r - r_0)], \quad \forall (x,y) \in P,$$

$$r - r_0 = \sqrt{(x - x_0)^2 + (y - y_0)^2}, \tag{5.51}$$

where $\omega_0 = [\text{radians/pixel}]$ and (x_0, y_0) determine, respectively, the radial frequency and center location of the Placido rings, and $P = P(x,y)$ delimits the region with valid, well-defined interferometric data.

Proceeding according to Equation (5.50), the first step is to displace the interferometric data spectrally by means of the synchronous product $I(x,y)\exp[i\omega_0(r - r_0)]$, and then we isolate the searched analytic signal by applying a (regularized according to our choice) LPF. In this example, we used first-order regularization with $\eta = 5$. The main steps of the regularized conic carrier phase demodulation process are presented in Figure 5.11.

Note from Panel 5.11d that, unlike linear carrier interferometry, the conic carrier does not produce spectral separation between the searched analytic signal $(b/2)\exp(i\varphi)$ and its complex conjugate $(b/2)$ $\exp(-i\varphi)$. Instead, their spectral information fully overlaps within a circle of radius ω_0. Also, note the series of wider and less-bright spectral rings of radius $n\omega_0$; these correspond to higher-order harmonics caused by the binary profile of the Placido fringes [16,17].

Panel 5.11e corresponds to the Fourier spectrum of the synchronous product $I(x,y)\exp[i\omega_0(r - r_0)]$. Now, the searched analytic signal $(1/2)\, b(x,y)\, \exp[i\varphi(x,y)]$ is located at the spectral origin; the background signal $a(x,y)$ is modulated by the carrier $\exp[-i\omega_0(r - r_0)]$, so it is centered on a circle with radius

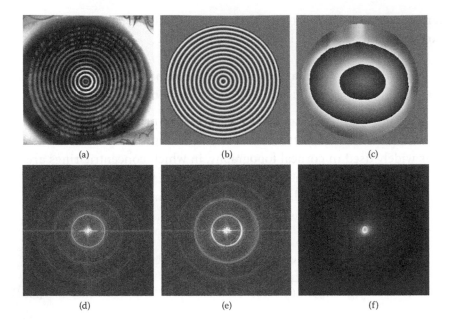

FIGURE 5.11   (a) Phase-modulated Placido target imaged over a human cornea; (b) conic carrier with the parameters $[\omega_0,(x_0,y_0)]$ estimated from the data; (c) wrapped estimated phase as obtained from its regularized synchronous demodulation. (d)–(f) Fourier spectra of the original data $I(x,y)$, the synchronous product $I(x,y)\exp[i\omega_0(r-r_0)]$, and the searched analytic signal $(1/2)b(x,y)\exp[i\varphi(x,y)]$, respectively.

ω_0; the conjugate $(1/2)\,b(x,y)\,\exp[-i\varphi(x,y)]$ is modulated by $\exp[-i2\omega_0(r-r_0)]$, so it is located around the circle of radius $2\omega_0$. Thus, the demodulated phase information is found by minimizing the following functional:

$$U[f(x,y)]=\sum_{(x,y)\in P}\{|f(x,y)-I(x,y)e^{-i\omega_0\rho}|^2$$

$$+\eta|f(x,y)-f(x-1,y)|^2+\eta|f(x,y)-f(x,y-1)|^2\}. \quad (5.52)$$

We can see that this functional uses $c(x,y)=\omega_0\sqrt{x^2+y^2}=\omega_0\rho$ as a spatial carrier, and the membrane-filtering potential remains the same. The Fourier spectrum of the searched analytic signal after the regularized low-pass filtering, $F\{(1/2)\,b(x,y)\,\exp[i\varphi(x,y)]\}$, is shown in panel 5.11f. Note that the complex conjugate signal and the high-frequency distorting harmonics have been removed.

5.7 TEMPORAL REGULARIZATION IN PHASE-SHIFTING INTERFEROMETRY

To this point, we have been working under the assumption that the interferometric modulating phase $\varphi(x,y)$ remains static (except for the temporal phase shift introduced). In this static measuring phase situation, we do not gain anything by replacing the convolution filter $h(t)$, a PSA, into a temporal regularized filter. If, however, the measuring interferometric phase is dynamic, $\varphi(x,y,t)$, and one needs to estimate the measuring phase at every temporal sample of the interferometric data, then a regularized temporal filter will be a much better choice.

To illustrate the temporal-regularized phase demodulation method more clearly, we need to consider the spectral behavior of standard (convolution-based) temporal PSAs. As shown in Section 5.3, temporal PSAs can be described as quadrature linear filters, which are spectrally displaced LPFs with a bandpass region centered at ω_0 and null spectral response at least at $\omega = \{0, -\omega_0\}$. Furthermore, in Section 5.5 we also showed that synchronous demodulation is applicable by spectrally displacing the interferometric data $I(x,y,t)\exp[i\omega_0 t]$ and afterward applying a conventional LPF (centered at the spectral origin). Thus, we conclude that both approaches illustrated in Figure 5.12 are equivalents.

The main motivation for regularizing temporal phase-shifting interferometry is because we may have dynamic temporal phase variations. In this case, the phase data are not static, as in standard phase-shifted interferometry, but vary with time: $\varphi(x,y,t)$. In other words, in standard (convolution-based) interferometry, the measuring phase is constant $\varphi(x,y)$ over time, except for the piston phase carrier introduced, $\omega_0 t$. But in dynamic

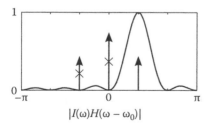
$|I(\omega)H(\omega - \omega_0)|$

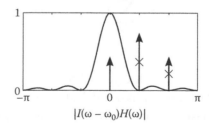
$|I(\omega - \omega_0)H(\omega)|$

FIGURE 5.12 Alternative approaches for the design of temporal phase-shifting algorithms. On the left, a quadrature linear filter was applied to the original data (centered at ω_0). On the right, a conventional low-pass linear filter (centered at the spectral origin) was applied to spectrally displaced interferometric data $I(x,y,t)\exp[i\,\omega_0 t]$.

temporal phase-shifted interferometry, the modulating phase may vary at each temporal sample. We may put this difference mathematically as

$I(x,y,t) = a(x,y) + b(x,y)\cos[\varphi(x,y) + \omega_0 t],$ Static Temporal Interferometry

$I(x,y,t) = a(x,y) + b(x,y)\cos[\varphi(x,y,t) + \omega_0 t],$ Dynamic Temporal Interferometry

$$(5.53)$$

We insist that in dynamic temporal interferometry, the measuring phase is not a fixed real number but instead varies over time. Of course, the maximum temporal variations of the modulating phase function must be bounded by the temporal carrier frequency ω_0:

$$\max\left|\frac{\partial\varphi(x,y,t)}{\partial t}\right| < \omega_0. \qquad (5.54)$$

For example, assume an object under study is dynamically moving during the time that three phase-shifted interferograms are taken with a phase step of ω_0. So, our interferometric data are given by

$$I(t) = \{a + b\cos[\varphi(0)]\}\delta(t) + \{a + b\cos[\varphi(1) + \omega_0]\}\delta(t-1)$$
$$+ \{a + b\cos[\varphi(2) + 2\omega_0]\}\delta(t-2). \qquad (5.55)$$

where $\varphi(0) \neq \varphi(1) \neq \varphi(2)$, and the spatial dependence was omitted for clarity.

Based in our previous analysis, assume that we have N temporal interferometric data. We may produce a spectrally centered analytic signal by means of the synchronous product $I(x,y,t)\exp(-i\omega_0 t)$. Thus, one may formulate the first-order temporal regularized filtering as

$$U[f(x,y,n)] = \sum_{n=0}^{N-1}\{|f(x,y,n) - I(x,y,n)e^{-i\omega_0 n}|^2 + \eta|f(x,y,n) - f(x,y,n-1)|^2\}.$$

$$(5.56)$$

The complex filtered field $f(x,y,n)$ is now formed by N temporal samples, and its phase will be the searched interferometric phase signal at each phase step, $\varphi(x,y,n)$. Note that in this energy functional the summation is over the temporal coordinate, not the spatial one. So, to emphasize its temporal character, we may omit the 2-D spatial coordinates:

$$U[f(n)] = \sum_{n=0}^{N-1}\{|f(n) - I(n)e^{-i\omega_0 n}|^2 + \eta|f(n) - f(n-1)|^2\}, \quad N \geq 3. \quad (5.57)$$

The minimizer of this functional leads to a set of N linear equations per pixel (x, y). The phase of $\{f(0), f(1), \ldots, f(N-1)\}$ at each phase step is known by inverting an $N \times N$ matrix only once and using this inverted matrix over all the temporal samples at each pixel $\{(x, y, 0),(x, y, 1), \ldots, (x, y, N-1)\}$ of the interferograms. As usual, the estimated phase $\hat{\varphi}(x, y, n)$ is calculated at each temporal plane n from the analytic filtered field $f(x, y, n)$ as

$$\hat{\varphi}(x,y,n) \bmod 2\pi = \tan^{-1}\left[\frac{Im[f(x,y,n)]}{Re[f(x,y,n)]}\right], \forall(x,y) \in P(x,y), \quad (5.58)$$

where the operators $Im[\cdot]$ and $Re[\cdot]$ take the imaginary and real part of its argument, respectively. From Equations (5.57) and (5.58), it should be noted that following this temporal regularization approach we need at least three phase steps to obtain a modulating-phase estimation. In the next section, we discuss another regularized method to work with single-image closed-fringes interferograms.

As summarized in Box 5.3, understanding spatial or temporal phase demodulation algorithms as quadrature linear filters allows us to take advantage of regularization techniques for enhanced reliability with respect to convolution-based filtering.

BOX 5.3   SYNCHRONOUS PHASE DEMODULATION WITH REGULARIZATION TECHNIQUES

- In general, a phase-carrier modulated interferogram $I(x,y,t)$ is given by

  $$I(x,y,t) = a(x, y) + b(x, y)\cos[\varphi(x, y) + c(x, y, t)],$$

  where $c(x,y,t)$ represents a *known* spatial, temporal, or spatiotemporal carrier.

- Following the synchronous demodulation method, the searched analytic signal is estimated as:

  $$A_0 \exp[i\varphi(x,y)] = LPF\{I(x,y,t)\exp[-ic(x,y,t)]\}.$$

- Here $LPF\{\cdot\}$ stands for a regularized low-pass filtering for enhanced reliability. For instance, considering a general spatial-carrier modulation and first-order regularization, we have:

  $$U[f(x,y)] = \sum_{(x,y)\in P}\left\{|f(x,y) - I(x,y)e^{-ic(x,y)}|^2 + \eta|f(x,y) - f(x-1,y)|^2 + \eta|f(x,y) - f(x,y-1)|^2\right\}.$$

> • Similarly, the first-order temporal-regularized synchronous demodulation is given by minimizing
>
> $$U[f(x,y,n)] = \sum_{n=0}^{N-1} \{|f(x,y,n) - I(x,y,n)e^{-i\omega_0 n}|^2 + \eta|f(x,y,n) - f(x,y,n-1)|^2\}.$$
>
> Note that in this functional, the summation is over the temporal coordinate not the spatial one.

5.8 REGULARIZED PHASE ESTIMATION OF SINGLE-IMAGE CLOSED-FRINGES INTERFEROGRAMS

To this point, we have assumed that our measured interferograms were codified with either a spatial or a temporal carrier. Now, we analyze the case of a single-image interferogram possibly having closed fringes. The regularized phase tracker (RPT) [7–9] is a nonlinear technique that allows phase demodulation of single-image closed-fringes interferograms; here, we show its main results.

Consider our mathematical model for a single-image closed-fringes interferogram:

$$I(x,y) = a(x,y) + b(x,y)\cos[\varphi(x,y)] \quad \forall (x,y) \in P(x,y). \quad (5.59)$$

The RPT method assumes that we have (or are able to obtain) a purely sinusoidal phase-modulated fringe pattern described in the following mathematical form:

$$I'(x,y) = \cos[\varphi(x,y)]. \quad (5.60)$$

To obtain this kind of signal, we must use normalization techniques. Basically, a normalized fringe pattern can be obtained by applying an isotropic 2-D *Hilbert* transform using a 2-D vortex operator, which suppresses the background signal $a(x,y)$ and compensates spatial variations of the contrast function $b(x,y)$. For more details, see Quiroga and Servin [21].

The RPT method is a sequential demodulation technique that assumes that the interferogram is spatially and locally monochromatic. This assumption implies that, in a small neighborhood around the pixel (x,y), one can approximate the modulating phase as a plane $p(\eta,\xi)$,

$$I'(\eta,\xi) = \cos[p(\eta,\xi)], \quad (5.61)$$

where $p(x, y)$ is a local phase-plane approximation given as

$$p(\eta, \xi) = \varphi_0(x, y) + u(x, y)(x - \eta) + v(x, y)(y - \xi), \quad \forall (\eta, \xi) \in \Gamma, \quad (5.62)$$

with Γ a given square neighborhood around (x, y). The functions $\varphi_0(x, y)$, $u(x, y)$, and $v(x, y)$ are respectively, the local (searched) phase and the local spatial frequencies of the fringe-pattern. In practice, the sizes of neighborhood Γ are usually from 5×5 to 15×15 pixels. Then, a cost function to estimate the parameters of the local phase-plane approximation is proposed. This cost function is given in the following way:

$$U[\varphi_0, u, v] = \sum_{(\eta, \xi) \in \Gamma} \{\cos[p(\eta, \xi)] - I'(\eta, \xi)\}^2 + \{I'_x(\eta, \xi) + u \sin[p(x, y)]\}^2$$

$$+ \{I'_y(\eta, \xi) - v \sin[p(x, y)]\}^2 + \eta[\hat{\varphi}(\eta, \xi) - p(\eta, \xi)]^2 m(\eta, \xi).$$

$$(5.63)$$

Here, $\varphi_0(x, y)$ is the estimated phase. The data term (the first term) makes consistent the approximation plane with the data in the least-squares sense. The other terms add restrictions to the local approximation; for example, the second and third terms constrain the quadrature of the solution with the derivative of the data. The spatial derivatives of the data may be computed as

$$I_x(x, y) = \frac{I(x - 1, y) - I(x + 1, y)}{2},$$

$$I_y(x, y) = \frac{I(x, y - 1) - I(x, y + 1)}{2}. \quad (5.64)$$

The fourth term reinforces for smooth solutions within the neighborhood Γ, and η is the regularization parameter. Finally, the regularization term uses an indicator function $m(\eta, \xi)$ that is 1 if site (η, ξ) has been previously demodulated and zero otherwise. To estimate the modulation phase map $\varphi(x, y)$, we minimize Equation (5.63) for each site (x, y) of the interferogram image. For instance, using the *steepest-descent* method, we have to solve

$$\hat{\varphi}_0^{k+1} = \hat{\varphi}_0^k - \tau \frac{\partial U}{\partial \varphi_0}[\hat{\varphi}_0^k, \hat{u}^k, \hat{v}^k],$$

$$\hat{u}^{k+1} = \hat{u}^k - \tau \frac{\partial U}{\partial u}[\hat{\varphi}_0^k, \hat{u}^k, \hat{v}^k], \quad (5.65)$$

$$\hat{v}^{k+1} = \hat{v}^k - \tau \frac{\partial U}{\partial v}[\hat{\varphi}_0^k, \hat{u}^k, \hat{v}^k].$$

Here, $(\hat{\phi}_0^k, \hat{u}^k, \hat{v}^k)$ represent the value of phase and local frequencies at iteration k. The parameter τ controls the speed of the steepest-descent method, and it must be small enough to ensure convergence. Because the steepest-descent method is iterative, it requires an initial starting point or seed. One possibility is to take $(\hat{\phi}_0, \hat{u}, \hat{v}) = (0,0,0)$ just for the first site and afterward take the previous estimated values $(\hat{\phi}_0^k, \hat{u}^k, \hat{v}^k)$ for the subsequent demodulating sites. This estimation process is effective only if the phase variation is locally quasi-monochromatic so it may be represented locally by a plane. That is applicable if the fringes are almost always "locally" open. To estimate the phase of closed fringes, the RPT method scans the image in such a way that it always (locally) sees open fringes, and the scanning strategy follows the fringes on almost isophase contours of the interferogram.

In Figure 5.13, we show an example of how the RPT method works. Figure 5.13a shows the single-image closed-fringes interferogram to demodulate. The isophase contours (the fringes' path) shown in Figure 5.13b are obtained by taking a binary threshold from the fringe pattern in Figure 5.13a. In Figures 5.13c–5.13e, we show the isophase path (fringe following) in the demodulation process. Finally, in Figure 5.13f the recovered phase is shown. One additional and attractive characteristic of the RPT method is

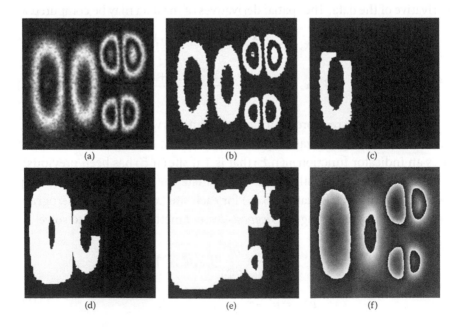

(a) (b) (c)

(d) (e) (f)

FIGURE 5.13 Several steps of the RPT demodulation method. Each panel of this figure is described in the text.

that it obtains the demodulated phase already unwrapped. However, in this figure the recovered phase was rewrapped for illustrative proposes (to compare the demodulated phase against the original interferogram).

Finally, note that the robustness of the RPT method is improved if the size of the neighborhood Γ is increased, but its maximum size is limited by the smoothness of the underlying phase surface, that is, the size of the larger window neighborhood where a local phase-plane approximation to the modulating phase remains valid. For more details, see Servin, Marroquin, and Cuevas [7,8] and Servin, Marroquin, and Quiroga [9].

5.9 REGULARIZED SPATIAL INTERPOLATION-EXTRAPOLATION IN INTERFEROMETRY

In most experimental situations, the modulating phase is continuous and smooth but because of experimental restraints (geometrical shadows, specular reflection, and so on), one cannot obtain valid interferometric data over the entire region under study. For this type of situation, the regularized techniques analyzed in this chapter allow us to interpolate or extrapolate the filtered interferogram by introducing the minor modifications discussed next.

Consider the first-order (rubber membrane) regularizer LPF for a general spatial carrier $c(x, y)$ given by [Equation (5.50), replicated here for the reader's convenience]

$$U[f(x,y)] = \sum_{(x,y)\in P_1} \{|f(x,y) - I(x,y)e^{-ic(x,y)}|^2 + \eta|f(x,y)$$

$$- f(x-1,y)|^2 + \eta|f(x,y) - f(x,y-1)|^2\}. \qquad (5.66)$$

As shown previously, the filtered field $f(x, y)$ that minimizes this energy functional will be a (complex-valued) analytic signal defined inside the bounding pupil $P_1(x, y)$, and the searched modulating phase $\varphi(x, y)$ can be estimated by computing its angle. The subscript in the bounding pupil $P_1(x, y)$ is to avoid confusion in the following step.

If we want to interpolate or extrapolate the filtered complex field, we need to modify the energy functional of the regularized filtering as

$$U[f(x,y)] = \sum_{(x,y)\in P_1} |f(x,y) - I(x,y)e^{ic(x,y)}|^2$$

$$+ \eta \sum_{(x,y)\in(P_1\cup P_2)} \{|f(x,y) - f(x-1,y)|^2 + |f(x,y) - f(x,y-1)|^2\},$$

$$(5.67)$$

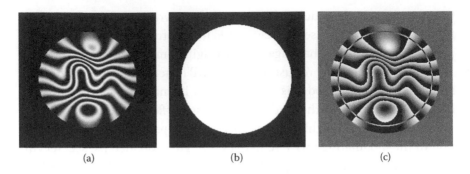

(a)   (b)   (c)

FIGURE 5.14   (a) Simulated fringe pattern bounded by a circular pupil $P_1(x,y)$. (b) Extended bounding pupil $[P_1(x,y) \cup P_2(x,y)]$, with a slightly larger radius because of $P_2(x,y)$, where we want to extrapolate the modulating phase. (c) Wrapped phase estimated with regularized first-order extrapolation; the inverse contrast ring was artificially added to indicate the boundary of the original pupil $P_1(x,y)$ where the original interferogram data reside.

where $P_1(x,y)$ is the pupil where we have valid fringe data, and $P_2(x,y)$ is the region where one desires to interpolate or extrapolate the fringe data. In this way, the filtered field will have a spatial support of $P_1(x,y) \cup P_2(x,y)$. For instance, if $P_1(x,y)$ is a circular pupil with radius r_1, the extropolation region $P_2(x,y)$ may be a donut outside and in contact with $P_1(x,y)$. An illustrative example is shown in Figure 5.14, for which we applied first-order spatially regularized low-pass filtering.

The advantage for doing interpolation or extrapolation using regularized potentials resides in restricting the behavior of the estimated smooth field. For instance, the analytic signal in the extrapolated region $P_2(x,y)$ can be estimated assuming that the modulating phase is continuous and smooth outside the data within $P_1(x,y)$. Also, recall that unlike typical averaging convolution filtering, regularized filtering prevents the mixing of valid data within $P_1(x,y)$ and invalid data outside $P_1(x,y)$.

5.10 REGULARIZATION IN LATERAL SHEARING INTERFEROMETRY

In this section, we analyze a regularization method for wavefront reconstruction in LSI. This may be considered a resumed version of the work presented in Servin, Malacara, and Marroquin [22] and Servin, Cywiak, and Davila [23].

Lateral shearing is a phase reconstruction method for wavefront analysis developed by Babcock [24]. Lateral shearing interferometers use a common-path configuration (see Figure 5.15), which makes them ideal

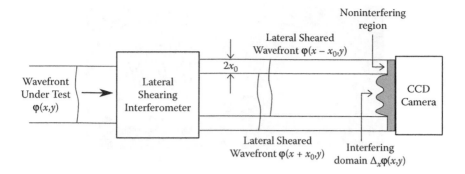

FIGURE 5.15 Schematics of a lateral shearing interferometer with displacement x_0 in the x direction. The elements of the figure are discussed in the text.

for testing setups on poorly stabilized environments and other adverse experimental conditions. Assuming a displacement x_0 in the x direction, an LSI produces two mutually displaced copies of the wavefront under test, $\varphi(x - x_0, y)$ and $\varphi(x + x_0, y)$. Considering $\varphi(x, y)$ to be spatially bounded by a pupil $P(x, y)$, the interference fringes are observed only in the intersection of $P(x - x_0, y)$ and $P(x + x_0, y)$.

The output signal in the lateral shearing interferometer can be modeled by the following formula:

$$I(x, y) = a(x, y) + b(x, y)\cos[\Delta_x \varphi(x, y)], \quad \forall (x, y) \in P(x, y) \quad (5.68)$$

where $a(x, y)$ and $b(x, y)$ are smooth functions that represent the background and contrast of the shearing interferogram, respectively; and using ensemble set notation, the modulating phase of a lateral shearing interferogram is given by:

$$\Delta_x \varphi(x, y) = [\varphi(x - x_0, y) - \varphi(x + x_0, y)][P(x - x_0, y) \cap P(x + x_0, y)],$$

$$\approx 2x_0 \frac{\partial \varphi(x, y)}{\partial x}[P(x - x_0, y) \cap P(x + x_0, y)]. \quad (5.69)$$

Note from Equation (5.69) that, unlike the other techniques analyzed in this chapter, the modulating phase is proportional (at first-order approximation) to the derivative of the searched wavefront $\varphi(x, y)$. Also, note that a laterally sheared interferogram has reduced spatial support because

$$[P(x + x_0, y) \cap P(x - x_0, y)] \subset P(x, y) \, \forall x_0 \neq 0. \quad (5.70)$$

An illustrative example is presented in Figure 5.16.

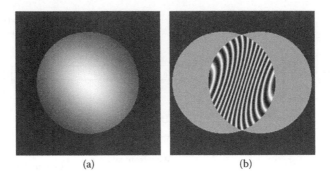

(a)   (b)

FIGURE 5.16   (a) Computer-simulated wavefront $\varphi(x,y)$ bounded by a circular pupil $P(x,y)$ and (b) laterally sheared interferogram with displacement in the x direction. Note that the interference fringes appear only in the overlapping region $[P(x-x_0,y) \cap P(x+x_0,y)]$.

5.10.1 Standard Method for Wavefront Estimation in Lateral Shearing Interferometry

Assuming that the amount of displacement introduced in the lateral shearing interferometer is known or can be estimated from the interferogram, one must be able (in principle) to demodulate the fringe pattern in $[P(x-dx,y) \cap P(x+dx,y)]$ to obtain $\Delta_x\varphi(x,y)$ by means of Fourier demodulation, linear PSAs, and any other appropriate phase demodulation process. For more details, see, for instance, Wyant [25] and Paez, Strojnik, and Mantravadi [26]. Thus, starting from some estimation of the modulating phase $\Delta_x\varphi(x,y)$,

$$\Delta_x\varphi(x,y) \approx 2x_0 \frac{\partial\varphi(x,y)}{\partial x}[P(x-dx,y) \cap P(x+dx,y)], \qquad (5.71)$$

the standard way of estimating the wavefront $\hat{\varphi}(x,y)$ is by simple integration along the x-axis:

$$\hat{\varphi}(x,y)[P(x-x_0,y) \cap P(x+x_0,y)] = \left\{ \frac{1}{2x_0} \int [\Delta_x\varphi(x,y)]dx \right\}. \qquad (5.72)$$

Note that Equation (5.71) is a good approximation for small values of the lateral displacement $2x_0$. However, in practice one often uses large values for $2x_0$ to increase the sensitivity of the lateral shearing interferometer

[22,26]. Furthermore, this simple integration approach also has the disadvantage of too much reduction of the support $[P(x-x_0,y) \cap P(x+x_0,y)]$ of the estimated phase $\hat{\varphi}(x,y)$ when the shearing distance $2x_0$ is significant.

Before proceeding further, we need to discuss how a lateral shearing interferometer can be described as a linear operator with input and output signals given by $\varphi(x,y)$ and $\Delta_x\varphi(x,y)$ respectively. As conventionally done, to simplify the exposition let us assume that we have an infinite bounding pupil or $P(x,y) = 1$ in the entire Euclidian plane. That is [from Equation (5.69)],

$$\Delta_x\varphi(x,y) = [\varphi(x-x_0,y) - \varphi(x+x_0,y)]. \qquad (5.73)$$

Now, the FTF for our lateral shearing operator $H_S(u,v)$ is found by taking the forward Fourier transform of Equation (5.73):

$$F[\Delta_x\varphi(x,y)] = [\exp(ix_0u) - \exp(-ix_0u)]F[\varphi(x,y)] = 2i\sin(x_0u)F[\varphi(x,y)]$$

$$(5.74)$$

and solving for the ratio between the output and the input signals:

$$H_S(u,v) = \frac{F[\Delta_x\varphi(x,y)]}{F[\varphi(x,y)]} = 2i\sin(x_0u). \qquad (5.75)$$

As we can see, the frequency response of a lateral shearing interferometer has null response at the spatial frequencies $u_n = n\pi/x_0$ for $n = \{0,1,2, \ldots\}$, so the information about the wavefront at these spatial frequencies is lost. This also means that one cannot implement a "direct" inverse filter to estimate the searched wavefront,

$$\hat{\Phi}(u,v) = H_S^{-1}(u,v)F[\Delta_x\varphi(x,y)] = \frac{1}{(2i)\sin(x_0u)}F[\Delta_x\varphi(x,y)], \quad (5.76)$$

because $H_S^{-1}(u,v)$ has poles (unstable behavior) at the spatial frequencies $u_n = n\pi/x_0$. This is illustrated in Figure 5.17.

Another nonregularized estimation method for $\hat{\varphi}(x,y)$ from the laterally sheared interferometric data $\Delta_x\varphi(x,y)$ is the least-squares fitting proposed separately by Fried [27] and Hudgin [28]. This is given by the

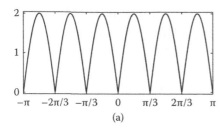 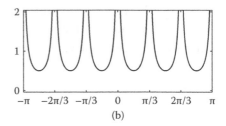

(a)   (b)

FIGURE 5.17   (a) Frequency transfer functions for the shearing operator $|H_S(u)|$, (b) "Direct" inverse filter $|H_S^{-1}(u)|$ with a lateral displacement $x_0 = 3$.

minimization of the following quadratic functional (adjusted to our current notation):

$$U[\hat{\varphi}(x,y)] = \sum_{(x,y)\in\mathbb{R}^2}\left[\hat{\varphi}(x,y)-\hat{\varphi}(x-1,y)-\frac{\Delta_x\varphi(x,y)}{2x_0}\right]^2 P(x,y)P(x-1,y).$$

(5.77)

Henceforward, we refer to this least-squares fit as the Fried-Hudgin operator. Once again, to simplify the analysis let us assume a large aperture in the bounding pupil, so $P(x,y) = 1$ for the entire image plane. Now, taking the derivative $\partial U/\partial\hat{\varphi} = 0$, we have

$$[2\hat{\varphi}(x,y)-\hat{\varphi}(x-1,y)-\hat{\varphi}(x+1,y)] = \frac{1}{2x_0}[\Delta_x\varphi(x,y)-\Delta_x\varphi(x+1,y)].$$

(5.78)

Taking the Fourier transform and solving for the FTF, results in the following:

$$|H_{FH}(u,v)| = \left|\frac{F[\hat{\varphi}(x,y)]}{F[\Delta_x\varphi(x,y)]}\right| = \frac{\{[1-\cos(u)]^2+\sin^2(u)\}^{1/2}}{|2x_0[2-2\cos(u)]|}.$$

(5.79)

Although minimizing the least-squares functional in Equation (5.77) is a well-posed inverse problem, it cannot recover the full bandwidth and support of the original (unsheared) wavefront $\varphi(x,y)$. Thus, one must use regularizing to solve these two problems and find the space of suitable solutions for $\hat{\varphi}(x,y)$. Nevertheless, this functional is our reference to compare with the spatially regularized approach presented next.

5.10.2 Regularized Methods for Wavefront Estimation in Lateral Shearing Interferometry

Now, let us analyze the regularized method to estimate the full wavefront under study $\hat{\varphi}(x, y)$ from its sheared interferometric data $\Delta_x \varphi(x, y)$. For ease of reference with our previous publications [22, 23], we refer to this inverse shearing method using the following convention: $\hat{\varphi}(x, y) = \Delta_x^{-1}[\Delta_x \varphi(x, y)]$, where $\Delta_x^{-1}[\cdot]$ denotes the inverse shearing operator.

The proposed first-order (rubber membrane) regularized quadratic functional is given by

$$U[\hat{\varphi}(x, y)] = \sum_{(x,y)\in\mathbb{R}} [\hat{\varphi}(x + x_0, y) - \hat{\varphi}(x - x_0, y) - \Delta_x \varphi(x, y)]^2 P(x + x_0, y) P(x - x_0, y)$$

$$+ \eta \sum_{(x,y)\in\mathbb{R}} [\hat{\varphi}(x, y) - \hat{\varphi}(x - 1, y)]^2 P(x, y) P(x - 1, y),$$

$$(5.80)$$

where, once again, the input and output signals are given by the demodulated shearing phase data $\Delta_x \varphi(x, y)$ and the recovered unsheared wavefront $\hat{\varphi}(x, y)$, respectively.

For illustrative purposes, let us consider the derivative with respect to $\hat{\varphi}(x, y)$ of only the first summation from Equation (5.80). This corresponds to the input-output quadratic fitting term:

$$\frac{1}{2} \frac{\partial U[\hat{\varphi}(x, y)]}{\partial \hat{\varphi}(x, y)}\bigg|_{\eta=0} = [\hat{\varphi}(x - 2x_0, y) - \hat{\varphi}(x, y) - \Delta_x(x - x_0, y)] P(x - 2x_0, y) P(x, y)$$

$$+ [\hat{\varphi}(x + 2x_0, y) - \hat{\varphi}(x, y) - \Delta_x(x + x_0, y)] P(x, y) P(x + 2x_0, y).$$

$$(5.81)$$

From Equation (5.81), we see that when estimating $\hat{\varphi}(x, y) = \Delta_x^{-1}[\Delta_x \varphi(x, y)]$ we have valid data whenever we are positioned within $[P(x - 2x_0, y) \cap P(x, y)]$ or within $[P(x, y) \cap P(x + 2x_0, y)]$. Thus, the spatial support $\text{supp}[\hat{\varphi}(x, y)]$ where we can calculate $\hat{\varphi}(x, y)$ with this technique is given by

$$\text{supp}[\hat{\varphi}(x, y)] = P(x, y) \cap [P(x - 2x_0, y) \cup P(x + 2x_0, y)]. \quad (5.82)$$

This contradicts the common belief that we are only allowed to estimate $\hat{\varphi}(x, y)$ where the LSI fringes are formed, as occurs in the standard integrating method [Equation (5.72)]. However, as illustrated in Figure 5.18,

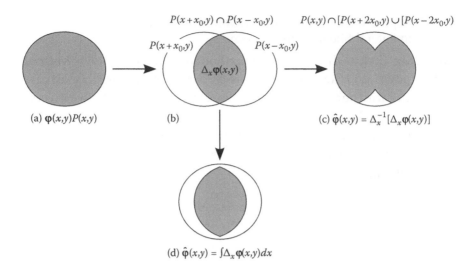

FIGURE 5.18 The gray area indicates the spatial support (a) for the original wavefront under study; (b) where shearing interference fringes appear; (c) where the estimated wavefront using the regularized approach is; and (d) for the recovered phase by simple integration of $\Delta_x \varphi(x, y)$.

the spatial support for this method also has some regions inside $P(x, y)$ where the wavefront $\hat{\varphi}(x, y)$ cannot be (directly) estimated.

Recall from the previous section that regularization allows interpolation or extrapolation of our estimated signals whenever the bounding pupil of the output contains the bounding pupil of the input. Thus, considering Equation (5.80), this means that once we obtain our wavefront estimation $\hat{\varphi}(x, y)$ in $[P(x + 2x_0, y) \cap P(x - 2x_0, y)]$, we can extrapolate this result to cover the entire bounding pupil $P(x, y)$ in a well-defined way given by the behavior of the linear regularizer chosen. Proceeding as previously, we can prove that the spectral response of our regularized wavefront estimation method is given by

$$H(u, v) = \frac{F[\text{Estimated } \hat{\varphi}(x, y)]}{F[\text{Sheared data } \Delta_x \varphi(x, y)]} = \frac{F[\hat{\varphi}(x, y)]}{F[\Delta_x \varphi(x, y)]}$$

$$= \frac{1}{(2i)\sin(x_0 u) + 2\eta[1 - \cos(u)]}. \tag{5.83}$$

In this case, one ordinarily uses small values for the regularizing parameter ($\eta < 1$) to preserve the spatial frequency content of the data. The next step is to explicitly introduce the spectral response of the lateral shearing operator $F[\Delta_x\varphi(x,y)]$ in Equations (5.79) and (5.83):

$$|H_{FH}(u,v)||H_{SI}(u,v)| = \frac{\{[1-\cos(u)]^2 + \sin^2(u)\}^{1/2}}{|2-2\cos(u)|}\left|\frac{\sin(x_0u)}{x_0}\right|,$$

$$|H(u,v)||H_{SI}(u,v)| = \left|\frac{\sin(x_0u)}{\sin(x_0u)+\eta[1-\cos(u)]}\right|. \tag{5.84}$$

As we can see, these spectral responses as a whole compensate for the spectral poles of the bare lateral shearing operator $\Delta_x\varphi(x,y)$, and consequently the linear system that estimates $\hat{\varphi}(x,y)$ has a well-behaved (stable) behavior. Their FTFs are illustrated in Figure 5.19. Note that the Fried-Hudgin technique falls faster in the high-frequency region than the proposed regularized method.

Finally, we present the straightforward generalization of the regularized wavefront estimation method for 2-D, orthogonally sheared data. This is given by

$$U[\hat{\varphi}(x,y)] = \sum_{(x,y)\in\mathbb{R}^2}\left\{[\hat{\varphi}(x+x_0,y)-\hat{\varphi}(x-x_0,y)-\Delta_x\hat{\varphi}(x,y)]^2\,P(x-x_0,y)P(x+x_0,y)\right.$$

$$+\sum_{(x,y)\in\mathbb{R}^2}[\hat{\varphi}(x,y+y_0)-\hat{\varphi}(x,y-y_0)-\Delta_y\hat{\varphi}(x,y)]^2\,P(x,y-y_0)P(x,y+y_0)$$

$$+\eta\sum_{(x,y)\in\mathbb{R}^2}[\hat{\varphi}(x,y)-\hat{\varphi}(x-1,y)]^2\,P(x,y)P(x-1,y)$$

$$+\eta\sum_{(x,y)\in\mathbb{R}^2}[\hat{\varphi}(x,y)-\hat{\varphi}(x,y-1)]^2\,P(x,y)P(x,y-1). \tag{5.85}$$

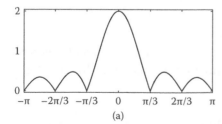 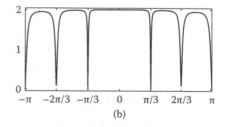

FIGURE 5.19 Frequency transfer function of the interferometer plus wavefront recovery system for (a) the Fried-Hudgin least-squares fitting technique and (b) the regularized method with $\eta = 0.01$.

**BOX 5.4   REGULARIZED PHASE ESTIMATION
METHODS IN INTERFEROMETRY**

In this chapter, we reviewed the following implementations of classical regularization techniques for phase estimation in interferometry:

- First- and second-order regularized low-pass filtering to improve the signal-to-noise ratio while preventing the mixture of valid fringe data with background signals (outside the bounding pupil where no interferometric data is available);
- Regularized quadrature linear filtering for phase demodulation of spatial or temporal carrier interferograms;
- Regularized phase-tracking demodulation for closed-fringes single-image interferograms;
- Interpolation or extrapolation of interferometric data following well-defined behavior using first-order (rubber membrane) and second-order (metallic thin-plate) regularizing potentials; and
- Wavefront reconstruction by regularizing a previously ill-posed inverse problem in lateral shearing interferometry with large lateral displacements.

Using this regularized estimator, our recovered wavefront now resides within the ensemble given by

$$\text{supp}[\hat{\varphi}(x,y)] = P(x,y) \cap \{[P(x-x_0,y) \cup P(x+x_0,y)]$$

$$\cup [P(x,y-y_0)P(x,y+y_0)]\}. \tag{5.86}$$

Nevertheless, just as in the previously analyzed case, it is easy to extrapolate or interpolate to cover the entire pupil $P(x,y)$. With this ends our brief review of regularized wavefront estimation methods for LSI. For more details, see Servin et al. [22,23] and Paez et al. [26].

Finally, in Box 5.4 we summarize the different applications of classical regularization techniques for phase estimation discussed in this chapter.

5.11 CONCLUSIONS

In this chapter, we reviewed several well-known phase demodulation techniques in interferometry. We showed that traditional approaches based on convolution filtering have some limitations that may be lifted using regularization techniques. To that end, we demonstrated how to use spatial, temporal, and combined spatial-temporal regularization techniques for phase demodulation of interferometric data.

When reviewing spatial interferometry we have shown that convolution filters mix up valid and invalid data (located respectively inside and outside of the interferometer's pupil); this translates into degradation of the demodulated phase near the boundaries of the pupil. On the other hand, regularization techniques allow us to demodulate the phase map under study preserving well-separated the valid and invalid data.

In temporal interferometry, we emphasized that only a single phase at the middle of the temporal sequence may be reliably known using convolution-based PSAs. In contrast, regularized temporal filtering allows us to demodulate the interferometric data for every instant within the temporal sequence and even to work without phase carriers (assuming prior normalization of the fringe pattern). We also proved that spatial regularization allows us to extrapolate or interpolate fringe data or estimated phase, which is particularly useful for wavefront reconstruction in LSI.

To summarize, regularized methods can be used in conjunction with (or as a substitute for) traditional convolution-based phase demodulation techniques.

REFERENCES

1. J. L. Marroquin. Deterministic interactive particle models for image processing and computer graphics. *Computer and Vision, Graphics and Image Processing* **55**, 408–417 (1993).
2. J. Bruning, D. Herriott, J. Gallagher, D. Rosenfeld, A. White, and D. Brangaccio. Digital wavefront measuring interferometer for testing optical surfaces and lenses. *Applied Optics* **13**, 2693–2703 (1974).
3. K. Freischlad and C. L. Koliopoulos. Fourier description of digital phase measuring interferometry. *Journal of the Optical Society of America A* **7**, 542–551 (1990).
4. Y. Surrel. Design of algorithms for phase measurement by use of phase stepping. *Optical Letters* **35**, 51–60 (1996).
5. M. Servin, J. Estrada, and J. Quiroga. The general theory of phase shifting algorithms. *Optics Express* **17**, 21867–21881 (2009).
6. A. Gonzalez, M. Servin, J. Estrada, and J. Quiroga. Design of phase-shifting algorithms by fine-tuning spectral shaping. *Optics Express* **19**, 10692–10697 (2011).
7. M. Servin, J. Marroquin, and F. Cuevas. Demodulation of a single interferogram by use of a two-dimensional regularized phase-tracking technique. *Applied Optics* **36**, 4540–4548 (1997).

8. M. Servin, J. Marroquin, and F. Cuevas. Fringe-follower regularized phase tracker for demodulation of closed-fringe interferograms. *Journal of the Optical Society of America A* **18**, 689–695 (2001).

9. M. Servin, J. Marroquin, and J. Quiroga. Regularized quadrature and phase tracking from a single closed-fringe interferogram. *Journal of the Optical Society of America A* **21**, 411–419 (2004).

10. D. Malacara, M. Servin, and Z. Malacara. *Interferogram Analysis for Optical Testing.* Boca Raton, FL: Taylor & Francis, CRC Press, 2005.

11. D. Malacara. *Optical Shop Testing.* 3rd ed. New York: Wiley-Interscience, 2007.

12. J. Schwider, R. Burow, K. E. Elssner, J. Grzanna, R. Spolaczyk, and K. Merkel. Digital wave-front measuring interferometry: Some systematic error sources. *Applied Optics* **22**(21), 3421–3432 (1983).

13. P. Hariharan, B. F. Oreb, and T. Eiju. Digital phase-shifting interferometry: A simple error-compensating phase calculation algorithm. *Applied Optics* **26**(13), 2504–2506 (1987).

14. Y. Ichioka and M. Inuiya. Direct phase detecting system. *Applied Optics* **11**, 1507–1514 (1972).

15. M. Takeda, H. Ina, and S. Kobayashi. Fourier-transform method of fringe-pattern analysis for computer-based topography and interferometry. *Journal of the Optical Society of America* **72**, 156–160 (1982).

16. M. Servin. Synchronous phase-demodulation of concentric-rings Placido mires in corneal topography and wave-front aberrometry (theoretical considerations). 2012. Available at http://arxiv.org/abs/1203.1886.

17. M. Servin. Digital interferometric demodulation of Placido mires applied to corneal topography. 2012. Available at http://arxiv.org/abs/1204.2210.

18. J. Millerd, N. Brock, J. Hayes, M. North-Morris, M. Novak, and J. C. Wyant. Pixelated phase-mask dynamic interferometer. *Proceedings SPIE* **5531**, 304–314 (2004).

19. J. M. Padilla, M. Servin, and J. C. Estrada. Harmonics rejection in pixelated interferograms using spatio-temporal demodulation. *Optics Express* **19**, 19508–19513 (2011).

20. J. M. Padilla, M. Servin, and J. C. Estrada. Synchronous phase-demodulation and harmonic rejection of 9-step pixelated dynamic interferograms. *Optics Express* **20**, 11734–11739 (2012).

21. J. A. Quiroga and M. Servin. Isotropic n-dimensional fringe pattern normalization. *Optics Communications* **224**, 221–227 (2003).

22. M. Servin, D. Malacara, and J. Marroquin. Wave-front recovery from two orthogonal sheared interferograms. *Applied Optics* **35**, 4343–4348 (1996).

23. M. Servin, M. Cywiak, and A. Davila. Extreme shearing interferometry: Theoretical limits with practical consequences. *Optics Express* **15**, 17805–17818 (2007).

24. H. W. Babcock. The possibility of compensating astronomical seeing. *Publications of the Astronomical Society of the Pacific* **65**, 229 (1953).

25. J. Wyant. Use of an AC heterodyne lateral shear interferometer with real-time wave-front correction systems. *Applied Optics* **14**, 2622–2626 (1975).
26. M. Strojnik, G. Paez, and M. Mantravadi. Lateral shear interferometers. In *Optical Shop Testing*, edited by D. Malacara. New York: Wiley, 2007.
27. D. L. Fried. Least-squares fitting a wave-front distortion estimate to an array of phase difference measurements. *Journal of the Optical Society of America* **67**, 370–375 (1977).
28. R. H. Hudgin. Wave-front reconstruction for compensated imaging. *Journal of the Optical Society of America* **67**, 375–378 (1977).

CHAPTER 6

Local Polynomial Phase Modeling and Estimation

Gannavarpu Rajshekhar

Beckman Institute
Urbana, Illinois, USA

Sai Siva Gorthi

Indian Institute of Science
Bangalore, India

Pramod Rastogi

Swiss Federal Institute of Technology
Lausanne, Switzerland

6.1 INTRODUCTION

As discussed in the previous chapters, reliable estimation of phase plays a crucial role in the application of optical interferometric techniques. This chapter introduces a phase estimation approach based on local polynomial phase modeling. Essentially, the approach relies on approximating the phase as a local polynomial, which transforms phase measurement into a parameter estimation problem. The approach is primarily discussed in the context of digital holographic interferometry (DHI), which is a prominent optical measurement technique with wide applicability in the areas of experimental mechanics, material characterization, and nondestructive testing.

Initially, we provide a brief description about DHI in Section 6.2. The idea is to introduce the working methodology and salient features of this interferometric technique in a concise manner. The basic concepts of the local polynomial phase-modeling approach are discussed in Section 6.3. Subsequently, the prominent phase estimation techniques based on this approach, such as maximum likelihood estimation (MLE; Section 6.4), cubic phase function (CPF; Section 6.5), high-order ambiguity function (HAF; Section 6.6), and phase-differencing operator (Section 6.7), are presented.

6.2 DIGITAL HOLOGRAPHIC INTERFEROMETRY

Holography is a technique for recording and reconstructing the amplitude and phase of an optical wavefield and was invented by Dennis Gabor [1,2]. In holography, the interference of a wavefield scattered from a diffuse object, also known as the *object wave*, and a coherent background, called the *reference wave*, is recorded on a photographic plate. The recorded intensity pattern is known as the *hologram*. Subsequently, an optical reconstruction procedure is applied: The hologram is illuminated with the reference wave to obtain the object wave information from the hologram.

During reconstruction, the illuminating reference wave is diffracted by the hologram. The resulting undiffracted wave provides a direct current (DC) term, whereas the positive and negative first-order diffraction terms provide the virtual and real images of the object [3]. The virtual image pertains to a diverging wavefront and appears at the same position where the object was located during the recording of the hologram. In contrast, the real image corresponds to a converging wavefront and appears in the opposite direction from the photographic plate. Both images are often referred to as the *twin images* in holography.

The optical configuration in holography can be broadly classified in two categories: (1) the in-line configuration, used in the original setup of Gabor, in which the reference wave and object wave are located along the axis normal to the photographic plate, and (2) the off-axis configuration [4,5], in which the reference wave is impinged on the photographic plate at an angle with respect to the object wave. For in-line holography, the DC term and the twin images are superimposed and hence difficult to separate. Comparatively, the off-axis configuration provides spatial separability of the three components.

An important development in the field came with the application of the charged-coupled device (CCD) camera as the recording medium, which led to the birth of digital holography [6]. This enabled digital recording and processing of holograms, thus removing the need for complex and

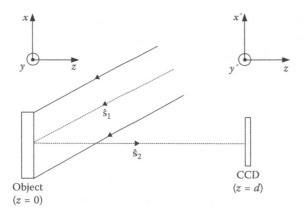

FIGURE 6.1   Coherent illumination of object.

time-consuming chemical processing steps involved with the use of the conventional photographic plates. In addition, the reconstruction is performed numerically, which enables direct retrieval of the complex object wave. Further, digital processing offers exciting opportunities, such as aberration compensation, reconstructing multiple object planes from a single hologram, numerical shearing, and so on.

In recent decades, DHI, which combines the principles of digital holography and optical interferometry, has emerged as a prominent technique for metrological applications. The technique offers noninvasive operability, whole-field measurement capability, and high resolution. It has been applied for deformation analysis [7–9], measurement of displacement derivatives [10–13], surface contouring [14–16], and observing refractive index variations in transparent media [17–19].

To gain a better understanding of DHI, we explore its application deformation analysis. Consider Figure 6.1, in which a diffuse object i minated by a coherent beam. The illuminated surface is represe the plane xy. The CCD camera, which records the hologram, is a distance d from the object. The CCD surface is denoted by th The direction of the light incident on the object is represente vector $\hat{s}_1$, whereas the observation direction toward the CC the unit vector $\hat{s}_2$.

To analyze the displacement caused by deforming th grams, corresponding to the object states before an are recorded in DHI. For the nondeformed state, and object waves at the CCD as $R(x',y')$ and $O(x',y')$, respectively. The hologram (i.e., the intensity recorded at the CCD camera because of the superposition of the two waves) is given as

$$I = |R+O|^2$$

$$= I_0 + RO^* + R^*O \tag{6.1}$$

where $I_0 = |R|^2 + |O|^2$ and $*$ denotes the complex conjugate. Here, the spatial coordinate (x',y') is omitted for the sake of brevity.

The next step is the reconstruction of the object wave from the hologram using numerical methods. Some of the prominent reconstruction techniques in digital holography are based on the Fresnel transform [20], convolution approach [21], angular spectrum technique [22], and wavelets [23]. In this section, we focus only on the Fresnel transform because it is a widely used reconstruction method in digital holography. Accordingly, the object wave is reconstructed at the xy-plane using the Fresnel transform as [20]

$$\Gamma_0(x,y) = \frac{j}{\lambda d} \exp\left[\frac{-j2\pi d}{\lambda}\right] \int_{-\infty}^{\infty} \int_{-\infty}^{\infty} R(x',y') I(x',y')$$

$$\times \exp\left[\frac{-j\pi}{\lambda d}((x-x')^2 + (y-y')^2)\right] dx' dy' \tag{6.2}$$

litude of the object wave before defor-
ide the integral because they do not
Equation (6.2) is modified as

$$(x',y')\exp\left[\frac{-j\pi}{\lambda d}(x'^2 + y'^2)\right]$$

$$y')\Big] dx' dy' \tag{6.3}$$

zed on a rectangular grid consti-
words, if the CCD has $N_x \times N_y$

pixels, with pixel sizes $\Delta x'$ and $\Delta y'$ along horizontal and vertical directions, we have

$$\Gamma_0(m\Delta x, n\Delta y) = \sum_{q=0}^{N_y-1} \sum_{p=0}^{N_x-1} R(p\Delta x', q\Delta y')I(p\Delta x', q\Delta y')$$

$$\times \exp\left[\frac{-j\pi}{\lambda d}((p\Delta x')^2 + (q\Delta y')^2)\right] \quad (6.4)$$

$$\times \exp\left[j2\pi\left(\frac{pm}{N_x} + \frac{qn}{N_y}\right)\right]$$

where $m \in [0, N_x - 1]$ and $n \in [0, N_y - 1]$. Here, Δx and Δy are the spatial resolutions in the xy-plane and are given as

$$\Delta x = \frac{\lambda d}{N_x \Delta x'} \quad (6.5)$$

$$\Delta y = \frac{\lambda d}{N_y \Delta y'} \quad (6.6)$$

This discretized form can be efficiently implemented using an inverse fast Fourier transform (FFT) algorithm [24].

For the moment, we consider the nondiscretized form of the complex amplitude in Equation (6.3). Using Equation (6.1), the complex amplitude can be expressed as

$$\Gamma_0(x, y) = \int_{-\infty}^{\infty} \int_{-\infty}^{\infty} [RI_0 + R^2 O^* + |R|^2 O]\exp\left[\frac{-j\pi}{\lambda d}(x'^2 + y'^2)\right]$$

$$\times \exp\left[\frac{j2\pi}{\lambda d}(xx' + yy')\right]dx'dy' \quad (6.7)$$

In Equation (6.7), the contributions arising from the terms RI_0, $R^2 O^*$, and $|R|^2 O$ constitute the DC part, the real image, and the virtual image, respectively. As discussed previously, for an off-axis configuration, the three components are spatially separable because of the inclination of the reference beam with respect to the optical axis of the object. Hence, focusing on the contribution from the term $|R|^2 O$, the complex amplitude before deformation can be written as

$$\Gamma_1(x,y) = a_1(x,y)\exp[j\psi(x,y)] \tag{6.8}$$

Here, $\psi(x,y)$ is the random phase corresponding to the scattered object wave.

Next, we consider the hologram obtained after deforming the object. Proceeding with the numerical reconstruction and other steps as discussed, the complex amplitude after deformation can be expressed as

$$\Gamma_2(x,y) = a_1(x,y)\exp[j(\psi(x,y) + \phi(x,y))] \tag{6.9}$$

where $\phi(x,y)$ is the phase change in the object wave caused by deformation and is usually referred to as the *interference phase* in DHI.

Subsequently, the interference field signal is obtained by multiplying the postdeformation complex amplitude of the object wave with the complex conjugate of predeformation complex amplitude. In other words, we have

$$\Gamma(x,y) = \Gamma_2(x,y)\Gamma_1^*(x,y)$$

$$= a(x,y)\exp[j\phi(x,y)] \tag{6.10}$$

where the amplitude $a(x,y) = a_1^2(x,y)$. The real part of $\Gamma(x,y)$ provides the fringe pattern in DHI.

The phase $\phi(x,y)$ is related to the displacement $\vec{d}$ experienced by the deformed object as

$$\phi(x,y) = \frac{2\pi}{\lambda}\vec{d}\cdot(\hat{s}_2 - \hat{s}_1) \tag{6.11}$$

Equation (6.11) signifies the importance of reliable phase estimation in DHI for deformation analysis because the quantity of interest (i.e., the object displacement) is directly dependent on the interference phase. To this end, observing that the interference field $\Gamma(x,y)$ is a complex signal, the phase $\phi(x,y)$ can be derived from the argument of $\Gamma(x,y)$ by applying an arctan function:

$$\phi_w(x,y) = \arg\{\Gamma(x,y)\}$$

$$= \tan^{-1}\left[\frac{Im\{\Gamma(x,y)\}}{Re\{\Gamma(x,y)\}}\right] \tag{6.12}$$

with *Im* and *Re* denoting the imaginary and real parts of a complex number, respectively. Because the principal values of the inverse function lie in the interval $(-\pi, \pi]$, the output of the arctan function, that is, $\phi_w(x, y)$ in Equation (6.12), lies between $\pm\pi$. As $\phi_w(x, y)$ is confined within a finite interval, it is usually known as *wrapped* phase. Mathematically, the true phase $\phi(x, y)$ and the wrapped phase $\phi_w(x, y)$ are related as [25] follows:

$$\phi_w(x, y) = \phi(x, y) + 2\pi n(x, y) \tag{6.13}$$

where $n(x, y)$ is an integer function that forces $-\pi < \phi_w(x, y) \leq \pi$. In light of Equation (6.13), we can observe that the wrapped phase has discontinuities whenever the true phase value lies outside the given interval.

For all practical applications, the true phase value is required, which necessitates the estimation of $\phi(x, y)$ from the wrapped function $\phi_w(x, y)$. This procedure is referred to as *phase unwrapping* and usually involves adding integral multiples of 2π to the wrapped phase to obtain the continuous phase distribution. An excellent introduction on the topic was provided by Ghiglia [25], and some of the prominent phase-unwrapping methods have been discussed [26–28]. Note that the phase-unwrapping operation is error prone in the presence of noise. Further, it adds an extra step in the phase estimation procedure, thereby increasing the overall computational burden. These limitations are addressed by the phase estimation techniques based on the local polynomial phase-modeling approach, which are discussed in the ensuing sections.

To summarize, the basic aspects of DHI were covered in this section. Nevertheless, we would like to emphasize that the topics covered in this section are by no means exhaustive; they merely provide the requisite background and context for presenting the local polynomial phase-modeling approach in the next section. For an elaborate and detailed treatment of DHI, refer to the work by Kreis [3].

6.3 PRINCIPLE

As discussed in the previous section, the complex interference field in DHI can be expressed as

$$\Gamma(x, y) = a(x, y) \exp[j\phi(x, y)] + \eta(x, y) \tag{6.14}$$

where $\phi(x, y)$ is the interference phase, and $\eta(x, y)$ is the noise, assumed to be white and Gaussian with zero mean and variance σ_η^2. The size of the complex signal is denoted as $N_x \times N_y$ samples, that is $x \in [1, N_x]$ and

$y \in [1, N_y]$. From a mathematical perspective, the interference field $\Gamma(x, y)$, as a two-dimensional (2-D) signal, can be visualized as a matrix whose columns and rows are represented by x and y.

As the phase $\phi(x, y)$ in DHI is, in general, a continuous function of the spatial coordinates, it can be modeled arbitrarily closely as a polynomial of sufficient degree, in accordance with the Weierstrass approximation theorem [29]. The main reason for choosing this model is that polynomials have a simple form, and their properties are well known. In the literature, several polynomial-fitting models [30–34] have been proposed for interferometric applications. However, there are certain limitations to these approaches. In particular, when approximating the phase globally as a polynomial, the degree would be high if the phase has rapid variations. Accordingly, the associated number of parameters to be computed is also large, resulting in more computational effort. In addition, for high-degree polynomials, overfitting could cause significant estimation errors [34].

These limitations can be addressed by considering *local* polynomial phase modeling [35]. In other words, the phase is approximated as a polynomial within a finite region or segment, as opposed to the whole length of the signal. As a result, even rapidly varying phase can be approximated by a low-degree polynomial with reasonably high accuracy. This is illustrated in Box 6.1.

BOX 6.1 LOCAL POLYNOMIAL PHASE MODELING

Consider a fifth-degree polynomial phase as follows:

$$\phi(y) = 10^{-6} y^2 (y - 45)(y - 75)(y - 105) \quad \forall y \in [1, 100]$$

The phase $\phi(y)$ in radians is shown in Figure 6.2a. For local modeling, the phase is divided into four segments, each containing 25 data samples. Within each segment, the phase is fitted with a third-degree polynomial. The original phase and the corresponding polynomial fit for the four segments are shown in Figures 6.2b–6.2e.

From these figures, it is evident that the fifth-degree polynomial phase is approximated reasonably well with a lower- (third-) degree polynomial within each segment. Thus, local polynomial phase modeling avoids the need for using high-degree polynomials, thereby overcoming associated drawbacks such as overfitting.

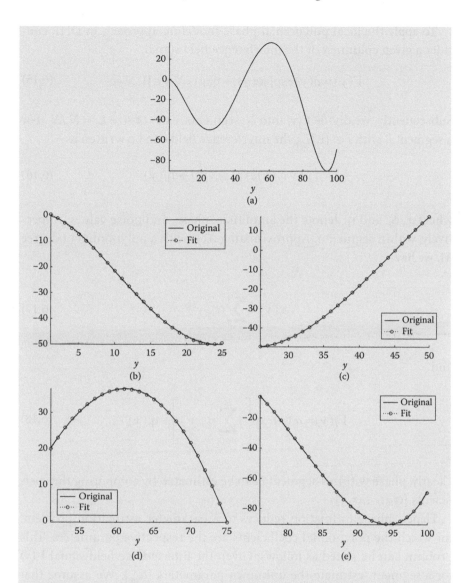

FIGURE 6.2 (a) Fifth-degree polynomial phase in radians. Third-degree polynomial fit for (b) first, (c) second, (d) third, and (e) fourth segments.

To apply the local polynomial phase-modeling approach in DHI, consider a given column x of the interference field signal,

$$\Gamma(y) = a(y)\exp[j\phi(y)] + \eta(y) \quad \forall\, y \in [1, N_y] \tag{6.15}$$

Subsequently, we divide $\Gamma(y)$ into N_s segments, each of size $L_s = N_y/N_s$. For a segment i, with $i \in [1, N_s]$, the interference field can be written as

$$\Gamma_i(y) = a_i(y)\exp[j\phi_i(y)] + \eta_i(y) \tag{6.16}$$

where a_i, ϕ_i, and η_i denote the amplitude, phase, and noise values, respectively, within segment i. Approximating $\phi_i(y)$ with a polynomial of degree M, we have

$$\phi_i(y) = \sum_{m=0}^{M} \alpha_{im} y^m \tag{6.17}$$

and

$$\Gamma_i(y) = a_i(y)\exp\left[j\sum_{m=0}^{M} \alpha_{im} y^m\right] + \eta_i(y) \tag{6.18}$$

Clearly, phase within a segment i can be estimated by computing the coefficients $[\alpha_{i0}, \dots, \alpha_{iM}]$.

Hence, phase estimation reduces to a parameter estimation problem, for which the polynomial coefficients are the respective parameters. This problem can be stated as follows: Given the interference field signal $\Gamma_i(y)$ for a segment, estimate the unknown parameters $\{\alpha_{im}\}$. We assume that the polynomial degree M is known.

By solving the previously mentioned parameter estimation problem, one can obtain the coefficients $[\hat{\alpha}_{i0}, \dots, \hat{\alpha}_{iM}]$, where the superscript $\wedge$ indicates the estimated value. Using these coefficients, the phase within a segment i is computed. Further, by varying i in $[1, N_s]$, the phase estimates for all N_s segments within a given column x are obtained.

The next step is to ensure that the phase varies smoothly between adjacent segments. This is performed using an *offsetting* operation. To elucidate the working of this operation, consider two adjacent segments i and $i + 1$, wherein the polynomial coefficients have been estimated as $[\hat{\alpha}_{i0},\ldots,\hat{\alpha}_{iM}]$ and $[\hat{\alpha}_{(i+1)0},\ldots,\hat{\alpha}_{(i+1)M}]$, respectively, using one of the techniques mentioned in the next sections. The corresponding phases in the two segments are given, respectively, as

$$\hat{\phi}_i(y) = \sum_{m=0}^{M} \hat{\alpha}_{im} y^m \quad \forall y \in [1, L_s] \tag{6.19}$$

$$\hat{\phi}_{i+1}(y) = \sum_{m=0}^{M} \hat{\alpha}_{(i+1)m} y^m \quad \forall y \in [1, L_s] \tag{6.20}$$

Note that the phases in both segments are computed on the same interval $1 \leq y \leq L_s$. For a continuous phase, the values of $\hat{\phi}_i(y)$ at $y = L_s + 1$ and $\hat{\phi}_{i+1}(y)$ at $y = 1$ should be identical. Hence, we compute the offset as

$$\gamma_{i+1} = \hat{\phi}_i(L_s + 1) - \hat{\phi}_{i+1}(1) \tag{6.21}$$

Using the offset γ_{i+1}, the phase in the segment $i + 1$ is modified through an assignment:

$$\hat{\phi}_{i+1}(y) \leftarrow \hat{\phi}_{i+1}(y) + \gamma_{i+1} \tag{6.22}$$

Because the offset is essentially a constant value for a given segment, this assignment can also be equivalently expressed in terms of the zero-order coefficient. Thus, the zero-order coefficient is modified as

$$\hat{\alpha}_{(i+1)0} \leftarrow \hat{\alpha}_{(i+1)0} + \gamma_{i+1} \tag{6.23}$$

By varying $i \in [1, N_s - 1]$ in Equation (6.21), we obtain the requisite offsets for the respective segments. It is interesting to note that the offsets are required only for the second segment and beyond. This means that the zero-order coefficient for the first segment remains unchanged, and only those for segments $i \geq 2$ are modified. The offsetting operation provides a smoothly varying phase for a given column.

We would like to point out that the scheme presented is not the only way to implement the offsetting operation. A 50% overlapping segment strategy [36] can also be used for the offsetting operation, with the offsets computed from estimated phase values near the middle of each segment; it provides better estimation accuracy, albeit with a more complicated scheme and at the cost of a higher computational effort. For the sake of simplicity, we would confine the implementation of the offsetting operation to the one described and urge those interested to look into the suggested reference.

It needs to be emphasized that the local polynomial phase model is applied in a one-dimensional (1-D) sense; that is, one column is analyzed segment-wise at a time. As a result, the offsetting operation provides an estimate that is continuous only along one dimension (i.e., with respect to y in this case). By repeating the phase estimation procedure for all columns, we obtain a 2-D estimate that varies continuously with y. Thus, the next step is to ensure that the phase also varies smoothly along the other dimension, that is, along x. This is achieved via a *stitching* operation, by which we perform the 1-D unwrapping of the estimate with respect to x. This operation is as simple as the 1-D unwrapping of an ideal wrapped phase. Thus, the offsetting and stitching operations provide the overall continuous phase distribution.

The local polynomial phase-modeling approach for phase estimation in DHI can be summarized in the following steps:

1. Consider the interference field signal $\Gamma(y)$ for a given column x, as in Equation (6.15).

2. Divide the column into i segments to obtain $\Gamma_i(y)$, as in Equation (6.16).

3. Within each segment, the phase $\phi_i(y)$ is modeled as a polynomial of degree M using Equation (6.17).

4. The respective polynomial coefficients $\alpha_{i0}, \ldots, \alpha_{iM}$ are estimated from the complex signal $\Gamma_i(y)$ using a parameter estimation technique.

5. The phase within a segment is estimated using Equation (6.19).

6. These steps are repeated for all segments, and subsequently for all columns, in conjunction with the offsetting and stitching operations, to obtain the overall phase distribution.

Note that the approach has been presented by taking into account the interference field signal for a given column x. However, without any loss of generality, the approach can also be applied by analyzing the signal along a given row y and proceeding with the subsequent steps.

The crucial part in the local polynomial phase-modeling approach lies in solving the parameter or coefficient estimation problem. Hence, the definitive features of this approach, such as the estimation accuracy and computational complexity, are decided by the technique selected to solve the respective problem. In the following sections, we outline several techniques for solving the corresponding parameter estimation problem and discuss their application in the local polynomial phase-modeling approach for reliable phase estimation.

6.4 MAXIMUM LIKELIHOOD ESTIMATION

To this point, we explored the basic principle of the local polynomial phase-modeling approach. For convenience, the mathematical form of the interference field signal and the phase within a segment i for a given column x are reiterated as follows:

$$\Gamma_i(y) = a_i(y)\exp\left[j\sum_{m=0}^{M}\alpha_{im}y^m \right] + \eta_i(y) \qquad (6.24)$$

and

$$\phi_i(y) = \sum_{m=0}^{M}\alpha_{im}y^m \qquad (6.25)$$

Our objective is to estimate the respective polynomial coefficients. The MLE algorithm [37] offers a solution to the coefficient estimation problem. The algorithm involves the maximization of a multivariate objective function to compute the polynomial coefficients within a given segment i. The corresponding objective function in M variables can be expressed as [38]

$$U(x_1,\dots,x_M) = \left| \sum_{y=1}^{L_s}\Gamma_i(y)\exp\left[-j\sum_{m=1}^{M}x_m y^m \right] \right| \qquad (6.26)$$

From the given objective function, the first and other higher-order coefficients are estimated by tracing the peak of the function. Effectively, we have

$$[\hat{\alpha}_{i1},\ldots,\hat{\alpha}_{iM}] = \arg\max_{x_1,\ldots,x_M} U(x_1,\ldots,x_M) \tag{6.27}$$

With the estimates $[\hat{\alpha}_{i1},\ldots,\hat{\alpha}_{iM}]$ known, we perform a peeling operation, by which the contribution of the higher-order coefficients is removed:

$$\Gamma_i^0(y) = \Gamma_i(y)\exp\left[-j\sum_{m=1}^{M}\hat{\alpha}_{im}y^m\right]$$

$$\approx a_i(y)\exp[j\alpha_0] + \eta_i^0(y) \tag{6.28}$$

with

$$\eta_i^0(y) = \eta_i(y)\exp\left[-j\sum_{m=1}^{M}\hat{\alpha}_{im}y^m\right].$$

Subsequently, the zero-order coefficient is estimated by using the argument operator:

$$\hat{\alpha}_{i0} = \arg\left\{\frac{1}{L_s}\sum_{y=1}^{L_s}\Gamma_i^0(y)\right\}$$

$$= \tan^{-1}\left[\frac{Im\left\{\sum_{y=1}^{L_s}\Gamma_i^0(y)\right\}}{Re\left\{\sum_{y=1}^{L_s}\Gamma_i^0(y)\right\}}\right] \tag{6.29}$$

Using these coefficients, the phase can be estimated as

$$\hat{\phi}_i(y) = \sum_{m=0}^{M}\hat{\alpha}_{im}y^m \tag{6.30}$$

It is clear from Equation (6.27) that the MLE technique essentially involves an optimization problem. This optimization could be easily performed using the gradient-based techniques [39] if the multivariate objective function being maximized were convex. However, in general, because of the presence of noise, the objective function can have many local maxima and is thus not convex.

Consequently, a maximum likelihood solution can be implemented in two steps. Initially, an exhaustive multidimensional grid search is performed to find the vicinity of the global maximum. The spatial coordinates of the peak thus located provide the *coarse* estimates of the polynomial coefficients. In the next step, we refine our search using the coarse estimates as initial values in a suitable optimization algorithm. For our analysis, we use the Nelder-Mead simplex method [40] to implement the optimization routine. As a result, we obtain the *refined* estimates of the polynomial coefficients.

To gain better insights about the MLE technique, we illustrate its application in Box 6.2.

Next, we focus on applying the MLE technique in the local polynomial phase-modeling approach for phase estimation in DHI [35]. To understand this, consider a simulated interference field signal as follows:

$$\Gamma(x, y) = \exp[j\phi(x, y)] + \eta(x, y) \tag{6.31}$$

where the phase is given as

$$\phi(x, y) = 72\pi \left[\left(\frac{x-128}{256} \right)^2 + \left(\frac{y-128}{256} \right)^2 \right] \tag{6.32}$$

The size of the signal is $N \times N$ samples, with $N = 256$. The noise standard deviation $\sigma_\eta = 0.1$. The phase $\phi(x, y)$ is shown in Figure 6.4a. The real part of $\Gamma(x, y)$, which constitutes a fringe pattern, is shown in Figure 6.4b.

For phase estimation, we proceed by dividing each column of $\Gamma(x, y)$ into four segments, that is, $N_s = 4$. Thus, the length of a segment is given as $L_s = 64$. Within each segment, we approximate the phase as a quadratic polynomial with degree $M = 2$. Subsequently, we apply the MLE technique for estimating the corresponding quadratic coefficients. From these coefficients, the phase $\hat{\phi}_i(y)$ is computed using Equation (6.30) for the segment. This process is applied for all segments within a given column to obtain the phase estimate corresponding to that column.

BOX 6.2   MAXIMUM LIKELIHOOD ESTIMATION

Consider a quadratic phase signal as follows:

$$\Gamma(y) = \exp[j\phi(y)] + \eta(y)$$

with

$$\phi(y) = \alpha_0 + \alpha_1 y + \alpha_2 y^2 \quad \forall y \in [1,128]$$

The coefficients are given as $[\alpha_0, \alpha_1, \alpha_2] = [0.9, 0.2, 0.004]$, and the noise standard deviation $\sigma_\eta = 0.1$. The phase is shown in Figure 6.3a.

To estimate the quadratic coefficients, a maximum likelihood estimation (MLE) technique is applied. Using Equation (6.26), the bivariate objective function to be maximized is given as

$$U(x_1, x_2) = \left| \sum_{y=1}^{128} \Gamma(y) \exp[-j(x_1 y + x_2 y^2)] \right|$$

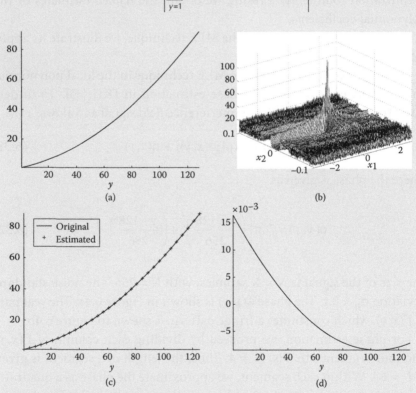

FIGURE 6.3   (a) Simulated phase $\phi(y)$ in radians. (b) Objective function $U(x_1, x_2)$. (c) Original phase $\phi(y)$ and estimated phase $\hat{\phi}(y)$ in radians. (d) Estimation error in radians.

This function is shown in Figure 6.3b. A dominant peak is clearly visible in the figure.

Subsequently, the first- and second-order coefficients are estimated by locating the maximum of the objective function. In other words, using Equation (6.27), we have

$$[\hat{\alpha}_1, \hat{\alpha}_2] = \arg\max_{x_1, x_2} U(x_1, x_2)$$

As previously discussed, a two-dimensional (2-D) grid search is initially performed to find the peak. The spatial coordinates corresponding to the peak provide the unrefined or coarse estimates of the respective coefficients. These are obtained as $[\hat{x}_1, \hat{x}_2] = [0.1885, 0.0040]$.

Using these estimates as initial values, a Nelder-Mead simplex method is applied to obtain the refined estimates $[\hat{\alpha}_1, \hat{\alpha}_2] = [0.2005, 0.0040]$. Note that the simplex method is applicable for minimization problems; hence, the given maximization problem is transformed into minimization of the negative objective function.

Further, we estimate the zero-order coefficient using the peeling operation of Equation (6.28) and taking the argument, as in Equation (6.29). The quadratic coefficients thus estimated using the MLE algorithm are given as $[\hat{\alpha}_0, \hat{\alpha}_1, \hat{\alpha}_2] = [0.8831, 0.2010, 0.0040]$. Finally, from these coefficients, the phase is estimated using Equation (6.30). The original and the estimated phases are shown in Figure 6.3c. The corresponding estimation error is shown in Figure 6.3d.

This analysis and the associated results explain the working of the MLE technique for polynomial phase estimation.

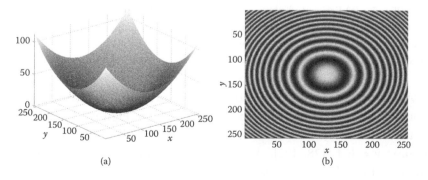

(a)   (b)

FIGURE 6.4   (a) Original phase $\phi(x, y)$ in radians. (b) Simulated fringe pattern.

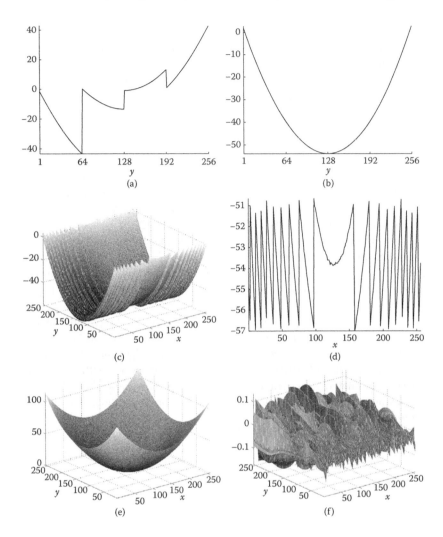

FIGURE 6.5   (a) Estimated phase in radians for the four segments along column $x = 100$. (b) Phase along the same column after offsetting operation. (c) Estimated phase for all columns. (d) Phase along a row $y = 100$. (e) Estimated phase $\hat{\phi}(x, y)$ in radians after stitching operation. (f) Estimation error in radians.

As an example, consider a particular column $x = 100$, for which the estimated phase using the MLE technique is shown in Figure 6.5a. Evidently, phase discontinuities are observed at the edges of adjacent segments, that is, near $y = [64, 128, 192]$. As discussed previously, this arises because phase within all segments is computed on the same interval $1 \le y \le 64$. Hence, we need to add appropriate offsets at the edges to maintain phase continuity. Applying the offsetting operation using Equations (6.21)–(6.23), the continuous phase estimate obtained for the given column is shown in Figure 6.5b.

Proceeding in a similar fashion, we obtain the phase estimates for all columns. This is shown in Figure 6.5c. From the figure, it is visible that the phase is continuous for each column, but uneven variations are seen between adjacent columns. To elaborate this, we plot a given row $y = 100$ in Figure 6.5d. The noticeable phase discontinuities in the figure arise because of the use of the arctan function for computing the zero-order coefficient, as shown in Equation (6.29). In the offsetting operation, the zero-order coefficients for all segments except the first are suitably modified, as in Equation (6.23), to provide a continuous phase within a given column. However, along the other dimension (i.e., for a given row), the wrapping effect of the arctan function in the zero-order coefficient evaluation needs to be addressed. This is accomplished through the phase-stitching operation, by which we perform a 1-D unwrapping along the rows, as discussed previously. The output of this operation provides the continuous 2-D phase estimate $\hat{\phi}(x, y)$, which is shown in Figure 6.5e. Normally, the stitching operation adds a constant shift to the overall estimated phase because of the effect of unwrapping. Hence, throughout the entire chapter, we remove the respective minimum values from the original and estimated phase distributions so that they can be compared with respect to the same base level (i.e., the origin). Finally, the phase estimation error is shown in Figure 6.5f.

This analysis illustrates the basic steps associated with the local polynomial phase-modeling approach for phase estimation. We can clearly visualize that the accuracy of phase estimation using this approach is directly dependent on the accuracy with which the polynomial coefficients are estimated. Also, any technique for coefficient estimation from a segment is performed N_s times within a column and then for N_x columns, that is, a total of $N_s \times N_x$ runs. Hence, the computational efficiency of the overall procedure is strongly governed by the computational requirements of the coefficient estimation technique.

An extremely important point to note about the MLE technique is that it requires a large amount of computations for maximizing the multivariate objective function. For instance, the technique involves peak detection on a 2-D grid for a quadratic phase signal, a three-dimensional (3-D) grid search for a cubic phase signal, and so on. As a result, for increasing polynomial orders, the computational cost involved with the MLE becomes prohibitively high; thus, the technique is primarily suitable for quadratic phase estimation only. This constitutes a major drawback for the MLE technique and limits its application for many practical situations.

It is clear that a computationally less-intensive approach for estimating the coefficients is usually desired for polynomial phase estimation. As such, the next sections describe some prominent techniques to address this problem.

6.5 CUBIC PHASE FUNCTION

In the previous section, we saw that the practical applicability of the MLE technique is confined to quadratic phase signals. Now, we extend the possibilities beyond second-order polynomial phase estimation using the CPF technique [41, 42]. As the name suggests, the technique is applicable for third-order polynomial phase signals.

For applying the CPF technique, we model the interference phase within a segment i as A cubic phase polynomial. Effectively, consider the signal

$$\Gamma_i(y) = a_i(y)\exp[j\phi_i(y)] + \eta_i(y) \quad \forall \, y \in [-(L_s-1)/2, (L_s-1)/2] \quad (6.33)$$

with the phase given by

$$\phi_i(y) = \alpha_{i0} + \alpha_{i1}y + \alpha_{i2}y^2 + \alpha_{i3}y^3 \quad (6.34)$$

where the signal length L_s is odd. The CPF for this signal is defined as [42]

$$CP(y,\Omega) = \sum_{m=0}^{(L_s-1)/2} \Gamma_i(y+m)\Gamma_i(y-m)\exp[-j\Omega m^2] \quad (6.35)$$

For understanding how the CPF method works, the concept of instantaneous frequency rate (IFR) has to be explained. The IFR is defined as the second-order derivative of the phase, that is,

$$IFR(y) = \frac{\partial^2 \phi_i(y)}{\partial y^2}$$

$$= 2(\alpha_{i2} + 3\alpha_{i3}y) \quad (6.36)$$

To explore the relationship between the CPF method and the IFR, consider the term $\Gamma_i(y + m) \, \Gamma_i(y - m)$ in Equation (6.35). For the sake of simplicity, if we assume a slowly varying amplitude and ignore the effect of noise, we have

$$\Gamma_i(y+m)\Gamma_i(y-m) \approx a^2(y)\exp[j\{2(\alpha_{i0}+\alpha_{i1}y+\alpha_{i2}y^2+\alpha_{i3}y^3)$$

$$+2(\alpha_{i2}+3\alpha_{i3}y)m^2\}] \qquad (6.37)$$

In Equation (6.37), the right-hand side comprises a complex signal whose phase has two components: (1) the term $2(\alpha_{i0}+\alpha_{i1}y+\alpha_{i2}y^2+\alpha_{i3}y^3)$, which is invariant to m and is a constant for a given y, and (2) the term $2(\alpha_{i2}+3\alpha_{i3}y)m^2$, which is quadratic in m, and the corresponding quadratic coefficient is essentially the IFR.

Hence, the CPF in Equation (6.35) can now be expressed as

$$CP(y,\Omega) \approx \sum_m a^2(y)\exp[j\{2(\alpha_{i0}+\alpha_{i1}y+\alpha_{i2}y^2+\alpha_{i3}y^3)$$

$$+2(\alpha_{i2}+3\alpha_{i3}y)m^2\}]\times\exp[-j\Omega m^2]$$

$$\approx a^2(y)\exp[j2(\alpha_{i0}+\alpha_{i1}y+\alpha_{i2}y^2+\alpha_{i3}y^3)] \qquad (6.38)$$

$$\sum_m \exp[-j\{\Omega-IFR(y)\}m^2]$$

Equation (6.38) implies that the magnitude of the CPF will be maximum when $\Omega = IFR(y)$. In other words, the IFR at any given y can be estimated as

$$IFR(y) = \arg\max_\Omega |CP(y,\Omega)| \qquad (6.39)$$

The main motivation for IFR estimation using CPF is to compute the values of the polynomial coefficients. This is because the IFR is dependent on the second- and third-order coefficients, as evident from Equation (6.36). Obviously, if the IFR is known at two positions of y, say y_1 and y_2, the coefficients α_{i2} and α_{i3} can be determined. To this effect, we compute the CPFs at y_1 and y_2 and estimate the corresponding IFRs as

$$IFR(y_1) = \arg\max_\Omega |CP(y_1,\Omega)|$$

$$= 2(\hat{\alpha}_{i2}+3\hat{\alpha}_{i3}y_1) \qquad (6.40)$$

and

$$IFR(y_2) = \arg\max_{\Omega} |CP(y_2, \Omega)|$$

$$= 2(\hat{\alpha}_{i2} + 3\hat{\alpha}_{i3} y_2) \tag{6.41}$$

Clearly, by solving Equation (6.40) and Equation (6.41), the estimates of the second- and third-order polynomial coefficients (i.e., $\hat{\alpha}_{i2}$ and $\hat{\alpha}_{i3}$) can be obtained. Note that the accuracy of estimating the two coefficients depends on the choice of y_1 and y_2, that is, the positions at which the IFR is estimated. For minimum estimation error, the optimal values are given as $y_1 = 0$ and $y_2 = 0.11 L_s$ [42].

The next step is to estimate the first-order polynomial coefficient α_{i1}. To do so, we perform a peeling operation, by which the contributions arising from the second- and third-order coefficients are removed from $\Gamma_i(y)$:

$$\Gamma_i^1(y) = \Gamma_i(y) \exp[-j(\hat{\alpha}_{i2} y^2 + \hat{\alpha}_{i3} y^3)]$$

$$\approx a_i(y) \exp[j(\alpha_{i0} + \alpha_{i1} y)] + \eta_i^1(y) \tag{6.42}$$

with $\eta_i^1(y) = \eta_i(y) \exp[-j(\hat{\alpha}_{i2} y^2 + \hat{\alpha}_{i3} y^3)]$. Here, $\Gamma_i^1(y)$ is a single-tone signal whose frequency corresponds to α_{i1}. Hence, the first-order coefficient can be estimated by computing the discrete Fourier transform (DFT) of $\Gamma_i^1(y)$ and tracing the peak of the magnitude spectrum. Doing so, we have

$$X(\omega) = \sum_{y=-(L_s-1)/2}^{(L_s-1)/2} \Gamma_i^1(y) \exp[-j\omega y] \tag{6.43}$$

$$\hat{\alpha}_{i1} = \arg\max_{\omega} |X(\omega)| \tag{6.44}$$

For efficient implementation of the DFT, the FFT algorithm [24] is used. In addition, to improve the accuracy of estimating the single-tone frequency, we use the iterative frequency estimation by interpolation on Fourier coefficients (IFEIF) method [43].

Finally, we estimate the zero-order coefficient by another peeling operation, essentially removing the contribution arising from the first-order coefficient. Thus, we have

$$\Gamma_i^0(y) = \Gamma_i^1(y) \exp[-j\hat{\alpha}_{i1} y]$$

$$\approx a_i(y) \exp[j\alpha_{i0}] + \eta_i^0(y) \tag{6.45}$$

with $\eta_i^0(y) = \eta_i^1(y)\exp[-j\hat{\alpha}_{i1}y]$. Then, we obtain the zero-order coefficient as

$$\hat{\alpha}_{i0} = \arg\left\{ \frac{1}{L_s} \sum_{y=-(L_s-1)/2}^{(L_s-1)/2} \Gamma_i^0(y) \right\} \tag{6.46}$$

Once the cubic coefficients are obtained, the phase within a segment i can be estimated as

$$\hat{\phi}_i(y) = \hat{\alpha}_{i0} + \hat{\alpha}_{i1}y + \hat{\alpha}_{i2}y^2 + \hat{\alpha}_{i3}y^3 \tag{6.47}$$

The working of the CPF technique for phase estimation is illustrated in Box 6.3.

BOX 6.3 CUBIC PHASE FUNCTION

Consider a cubic phase signal with $L_s = 129$ samples as follows:

$$\Gamma(y) = \exp[j\phi(y)] + \eta(y)$$

where

$$\phi(y) = \alpha_0 + \alpha_1 y + \alpha_2 y^2 + \alpha_3 y^3 \quad \forall y \in [-64, 64]$$

and the coefficients $[\alpha_0, \alpha_1, \alpha_2, \alpha_3] = [0.7, 0.4, 0.006, -0.0002]$. The noise standard deviation $\sigma_\eta = 0.2$. The phase is shown in Figure 6.6a.

Using Equation (6.35), the cubic phase function (CPF) is computed at $y_1 = 0$ and $y_2 = 0.11L_s$. The magnitudes of the CPF at these positions are shown in Figures 6.6b and 6.6c. In each of these figures, a distinct peak is visible that corresponds to the instantaneous frequency rate for the specific position. By tracing these peaks, the two instantaneous frequency rate (IFR) estimates are determined. Then, we estimate the second- and third-order coefficients from the obtained IFRs by solving Equations (6.40) and (6.41). As a result, we obtain $[\hat{\alpha}_2, \hat{\alpha}_3] = [0.0060, -0.0002]$.

Then, for estimating α_1, a peeling operation is performed using Equation (6.42) to obtain $\Gamma^1(y)$. The Fourier transform $X(\omega)$ of the resulting single-tone signal is computed, and the magnitude spectrum $|X(\omega)|$ is shown in Figure 6.6d. The visible peak corresponds to $\hat{\alpha}_1$. Using the iterative frequency estimation by interpolation on Fourier coefficients (IFEIF method), we obtain $\hat{\alpha}_1 = 0.3999$. After this, the zero-order coefficient is estimated using Equations (6.45) and (6.46).

Overall, the estimated cubic coefficients are given as $[\hat{\alpha}_0, \hat{\alpha}_1, \hat{\alpha}_2, \hat{\alpha}_3] = [0.6637, 0.3999, 0.0060, -0.0002]$. From these coefficients, the phase is obtained using Equation (6.47). The original phase $\phi(y)$ and the estimated phase $\hat{\phi}(y)$ in radians are shown in Figure 6.6e. The corresponding estimation error is shown in Figure 6.6f.

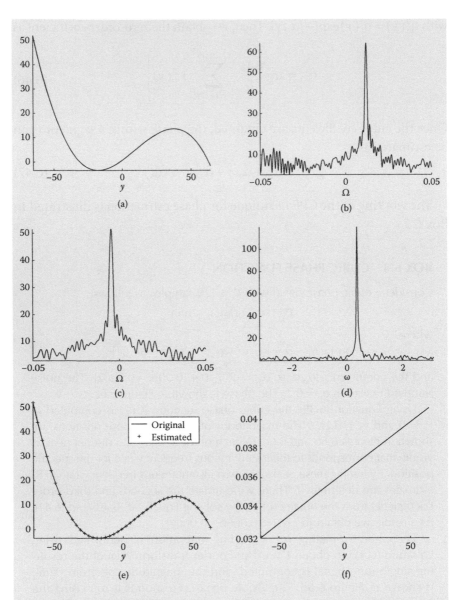

FIGURE 6.6 (a) Simulated phase $\phi(y)$ in radians. (b) Cubic phase function magnitude $|CP(y_1,\Omega)|$ at $y = y_1$. (c) Cubic phase function magnitude $|CP(y_2,\Omega)|$ at $y = y_2$. (d) Magnitude spectrum $|X(\omega)|$. (e) Original phase $\phi(y)$ and estimated phase $\phi(y)$ in radians. (f) Estimation error in radians.

This analysis explains how the CPF technique works for cubic phase estimation.

It needs to be emphasized that the CPF technique relies on solving multiple 1-D maximization problems to estimate the cubic coefficients. For instance, two 1-D maximizations are required to estimate the IFRs in Equations (6.40) and (6.41) for obtaining the second- and third-order coefficients, and subsequently another 1-D maximization is required in Equation (6.44) to estimate the first-order coefficient. Hence, when compared to the MLE technique, which requires a 3-D maximum search to estimate the cubic coefficients, the CPF technique breaks down the problem of coefficient estimation into multiple 1-D maximum searches. As a result, the computational requirements for the CPF technique are significantly lower than those for the MLE technique for cubic phase estimation.

Next, we discuss the application of the CPF technique in the local polynomial phase-modeling approach for phase estimation in DHI [44,45]. Consider an interference field signal

$$\Gamma(x,y) = \exp[j\phi(x,y)] + \eta(x,y) \tag{6.48}$$

where the size of the signal is 512×512 samples, and the noise standard deviation $\sigma_\eta = 0.1$. The phase is simulated as

$$\phi(x,y) = 15\left[\left(1+\sin\left(\frac{\pi x}{512}\right)\right)\left(1+\sin\left(\frac{\pi y}{512}\right)\right)-1\right] \tag{6.49}$$

followed by removal of its minimum value. The resulting $\phi(x,y)$ is shown in Figure 6.7a, and the real part of $\Gamma(x,y)$ (i.e., the fringe pattern) is shown in Figure 6.7b.

We divide each column of $\Gamma(x,y)$ into $N_s = 4$ segments, with each segment consisting of 128 samples, and use a cubic phase approximation. Subsequently, the cubic coefficients for each segment are estimated using the CPF technique, and the phase is constructed using these coefficients. Note that the CPF technique operates on odd signal lengths; hence, we only consider 127 samples of $\Gamma_i(y)$ in segment i by removing the first (or last) sample in the computation of the CPF. This process is repeated for all segments in a given column and is followed by the offsetting operation. The original and the estimated phases for a given column $x = 100$ are shown in Figure 6.7c. Continuing in a similar manner for all columns and applying the phase-stitching operation, the overall 2-D estimated phase $\hat{\phi}(x,y)$ is shown in Figure 6.7d. The corresponding estimation error is shown in Figure 6.7e.

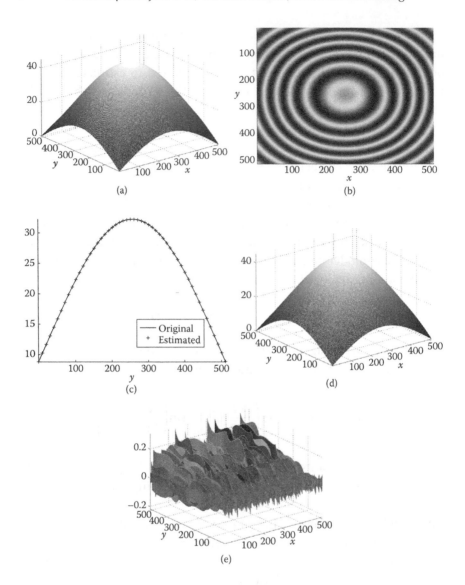

FIGURE 6.7 (a) Simulated phase $\phi(x, y)$ in radians. (b) Simulated fringe pattern. (c) Original versus estimated phase for a given column $x = 100$. (d) Estimated phase $\hat{\phi}(x, y)$ in radians. (e) Estimation error in radians.

Further, we analyze the practical utility of the CPF technique for a DHI experiment. A circularly clamped object was subjected to central loading, and two holograms were digitally recorded before and after deformation. The complex amplitudes were obtained using the numerical reconstruction based on the discrete Fresnel transform, as discussed in Section 6.2. The interference field signal was then obtained

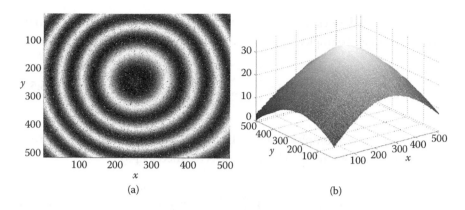

FIGURE 6.8 (a) Experimental fringe pattern. (b) Estimated phase in radians.

by multiplying the postdeformation complex amplitude with the conjugate of the predeformation complex amplitude. The real part of the interference field, which constitutes the experimental fringe pattern, is shown in Figure 6.8a. Applying the CPF technique as described and using $N_s = 4$, the interference phase was estimated and is shown in Figure 6.8b.

In this section, we explored the fundamentals of the CPF technique for cubic phase estimation. Although the technique is computationally efficient when compared to the MLE technique, its scope, by definition, is solely limited to cubic phase modeling only. In the next section, we show how to circumvent this limitation.

6.6 HIGH-ORDER AMBIGUITY FUNCTION

In the previous sections, we presented two techniques for reliable coefficient estimation in the local polynomial phase-modeling approach: (1) the MLE technique, which is practically confined to quadratic phase signals, and (2) the CPF technique, which is applicable only for cubic phase signals. Hence, there is a need for a generalized technique that is suitable for arbitrary-order polynomial phase signals. This is addressed by the HAF technique [38, 46].

As stated in Section 6.3, we model the interference field as an Mth-degree polynomial phase signal within a segment i for a given column x,

$$\Gamma_i(y) = a_i(y) \exp\left[j \sum_{m=0}^{M} \alpha_{im} y^m \right] + \eta_i(y) \tag{6.50}$$

The HAF is suited for estimating the polynomial coefficients from the signals of the form $\Gamma_i(y)$. To implement this technique, for any integer q we define

$$\Gamma_i^{(\dagger q)}(y) = \begin{cases} \Gamma_i(y), & q \text{ even} \\ \Gamma_i^*(y), & q \text{ odd} \end{cases} \tag{6.51}$$

where $*$ denotes complex conjugation. Subsequently, the high-order instantaneous moment (HIM) operator is defined as [46]

$$P_M[\Gamma_i(y);\tau] = \prod_{q=0}^{M-1} \left[\Gamma_i^{(\dagger q)}(y-q\tau)\right]^{\binom{M-1}{q}} \tag{6.52}$$

where τ is a positive number and is usually known as the *delay* parameter. In Equation (6.52), we used the notation

$$\binom{n}{r} = \frac{n!}{r!(n-r)!} \tag{6.53}$$

The HIM operator can also be expressed recursively as

$$P_M[\Gamma_i(y);\tau] = \begin{cases} \Gamma_i(y) & M = 1 \\ P_{M-1}[\Gamma_i(y);\tau]\{P_{M-1}[\Gamma_i(y-\tau);\tau]\}^* & M > 1 \end{cases} \tag{6.54}$$

The fundamental property of the HIM operator is that it transforms the polynomial phase signal of degree M into a complex sinusoid, that is, a single-tone signal. By assuming the amplitude is constant and ignoring the effect of noise for the sake of simplicity, the HIM operator reduces to

$$P_M[\Gamma_i(y);\tau] \approx a_i^{M-1} \exp[j\psi_i]\exp[j\omega_{iM}y] \tag{6.55}$$

where

$$\psi_i = (M-1)!\tau^{M-1}\alpha_{i(M-1)} - 0.5M!(M-1)\tau^M\alpha_{iM} \tag{6.56}$$

and the single-tone frequency is given as

$$\omega_{iM} = M!\tau^{M-1}\alpha_{iM} \tag{6.57}$$

Equation (6.57) signifies that by estimating the frequency associated with the HIM operator, the Mth-order coefficient α_{iM} can be obtained. Accordingly, we compute the DFT of $P_M[\Gamma_i(y);\tau]$, which is referred to as the HAF. The HAF is given as

$$P_M[\Gamma_i(y);\omega,\tau] = \sum_{y=1}^{L_s} P_M[\Gamma_i(y);\tau]\exp[-j\omega y] \qquad (6.58)$$

It is evident that the frequency can be estimated by tracing the peak of the amplitude spectrum. So, we have

$$\hat{\omega}_{iM} = \arg\max_{\omega} |P_M[\Gamma_i(y);\omega,\tau]| \qquad (6.59)$$

As discussed in the previous section, we use the FFT algorithm for HAF computation and apply the IFEIF method [43] for frequency estimation. Then, using Equation (6.57), the coefficient α_{iM} is estimated as

$$\hat{\alpha}_{iM} = \frac{\hat{\omega}_{iM}}{M!\tau^{M-1}} \qquad (6.60)$$

The next step is to remove the contribution of the highest-order coefficient from $\Gamma_i(y)$ using a peeling operation as follows:

$$\Gamma_i^{M-1}(y) = \Gamma_i(y)\exp\left[-j\hat{\alpha}_{iM}y^M\right]$$

$$\approx a_i(y)\exp\left[j\sum_{m=0}^{M-1}\alpha_{im}y^m\right] + \eta_i^{M-1}(y) \qquad (6.61)$$

with $\eta_i^{M-1}(y) = \eta_i(y)\exp[-j\hat{\alpha}_{iM}y^M]$. As a result of the peeling operation, we obtain $\Gamma_i^{M-1}(y)$ as a polynomial phase signal with degree $M-1$. Hence, for this signal, the computation of HIM, that is, $P_{M-1}[\Gamma_i^{M-1}(y);\tau]$, and after that HAF, that is, $P_{M-1}[\Gamma_i^{M-1}(y);\omega,\tau]$, followed by the frequency estimation strategy, would provide us with the coefficient estimate $\hat{\alpha}_{i(M-1)}$.

This procedure of HAF computation for coefficient estimation and a subsequent peeling operation can be sequentially applied to estimate up to the first-order coefficient. With the estimates $[\hat{\alpha}_{i1},...,\hat{\alpha}_{iM}]$ known, the zero-order coefficient can be estimated using the following peeling operation and application of the argument operator:

$$\Gamma_i^0(y) = \Gamma_i(y) \exp\left[-j \sum_{m=1}^{M} \hat{\alpha}_{im} y^m \right] \tag{6.62}$$

$$\hat{\alpha}_{i0} = \arg\left\{ \frac{1}{L_s} \sum_{y=1}^{L_s} \Gamma_i^0(y) \right\} \tag{6.63}$$

Note that the HIM operator was defined using a constant-delay parameter τ. However, it has been shown [47] that by using different values of τ in the computation of HIM, the estimation accuracy can be significantly improved. Accordingly, by using variable τ, the HIM recursion relation in Equation (6.54) is modified as

$$P_1[\Gamma_i(y)] = \Gamma_i(y)$$

$$P_2[\Gamma_i(y); \tau_1] = P_1[\Gamma_i(y)]\{P_1[\Gamma_i(y-\tau_1)]\}^* \tag{6.64}$$

$$\vdots$$

$$P_M[\Gamma_i(y); \tau_{M-1}] = P_{M-1}[\Gamma_i(y); \tau_{M-2}]\{P_{M-1}[\Gamma_i(y-\tau_{M-1}); \tau_{M-2}]\}^*$$

where $\tau_k = (\tau_1, \tau_2, \ldots, \tau_k)$. The optimal values of these delay parameters to minimize the mean square error for coefficient estimation have been determined [47] and are highlighted in Table 6.1.

Further, the HAF in Equation (6.58) is modified as

$$P_M[\Gamma_i(y); \omega, \tau_{M-1}] = \sum_{y=1}^{L_s} P_M[\Gamma_i(y); \tau_{M-1}] \exp[-j\omega y] \tag{6.65}$$

TABLE 6.1 Optimal Delay Parameters $(\tau_1, \ldots, \tau_{M-1})$ as a Fraction of Signal Length L_s for Different Values of Polynomial Degree M

M	Delay Parameters as Fractions of L_s
2	0.5
3	0.12, 0.44
4	0.08, 0.18, 0.37
5	0.07, 0.19, 0.24, 0.31

The coefficient estimate $\hat{\alpha}_{iM}$ is now given as

$$\hat{\alpha}_{iM} = \frac{\hat{\omega}_{iM}}{M! \prod_{k=1}^{M-1} \tau_k} \tag{6.66}$$

Henceforth, we use only the modified forms of HIM and HAF for analysis because they provide more reliable estimates for the coefficients.

We can now summarize how the HAF technique works:

1. Assume a variable r and initialize it as $r = M$, that is, the highest order or degree. We have $\Gamma_i^r(y) = \Gamma_i(y)$ for $r = M$.

2. Compute the HIM $P_r[\Gamma_i^r(y); \tau_{r-1}]$ using Equation (6.64).

3. Compute the HAF $P_r[\Gamma_i^r(y); \omega, \tau_{r-1}]$ using Equation (6.65) and determine the frequency $\hat{\omega}_{ir}$ corresponding to the spectral peak.

4. Estimate the coefficient α_{ir} using Equation (6.66).

5. Perform a peeling operation $\Gamma_i^{r-1}(y) = \Gamma_i^r(y) \exp[-j\hat{\alpha}_{ir} y^r]$.

6. Substitute $r = r - 1$. If $r \geq 1$, go to step 2.

7. Estimate the zero-order coefficient using Equations (6.62) and (6.63).

8. Compute the phase from the estimated coefficients using Equation (6.19).

This work flow indicates that the HAF provides a generalized approach for estimating the coefficients from any arbitrary-order polynomial phase signal. This is a significant advancement in contrast to the previous techniques, which were limited to only the second or third orders.

Another important feature of HAF, as already mentioned, is that it operates by transforming a polynomial phase signal into a complex sinusoid, and the coefficient estimation boils down to single-tone frequency estimation. This facilitates the use of the highly reliable and computationally efficient spectral analysis methods [48], such as IFEIF, which is applied here, or others such as the estimation of signal parameters by rotational invariance technique (ESPRIT) [49], multiple signal classification (MUSIC) technique [50], and others. Consequently, the HAF technique can be implemented with less computational effort in comparison with the MLE or CPF techniques.

To gain better understanding of the HAF technique, we illustrate its application for polynomial phase estimation in Box 6.4.

BOX 6.4   HIGH-ORDER AMBIGUITY FUNCTION

Consider the following signal:

$$\Gamma(y) = \exp[j\phi(y)] + \eta(y)$$

with quartic phase

$$\phi(y) = \alpha_0 + \alpha_1 y + \alpha_2 y^2 + \alpha_3 y^3 + \alpha_4 y^4$$

where the coefficients $[\alpha_0,\alpha_1,\alpha_2,\alpha_3,\alpha_4]$ = [0.7, 0.3, 0.006, 0.0005, -4×10^{-6}], and the noise standard deviation σ_η = 0.1. The simulated phase is shown in Figure 6.9a.

To estimate the quartic coefficients, we apply the high-order ambiguity function (HAF) technique. Initially, for $M = 4$ (i.e., the highest order), we compute the HAF $P_4[\Gamma(y);\omega,\tau_3]$ using Equation (6.65). The corresponding spectrum $|P_4[\Gamma(y);\omega,\tau_3]|$ is shown in Figure 6.9b. We observe a distinct peak, which corresponds to the coefficient α_4. For the sake of clarity, all HAF plots in this example are shown on a frequency-scaled axis, that is, $\omega/(M!\tau_1\tau_2\cdots\tau_{M-1})$ instead of ω, so that the peak location directly provides the coefficient estimate.

Once $\hat{\alpha}_4$ is obtained, we proceed with the peeling operation to reduce the polynomial degree by 1 and subsequently obtain the HAF for $M = 3$. This process of HAF computation, coefficient estimation using peak detection, and subsequent peeling is continued until we obtain the first-order coefficient. The HAF spectra for the decreasing polynomial degree M are shown in Figures 6.9c–6.9e; the peaks correspond to the third-, second-, and first-order coefficients, respectively. Finally, the zero-order coefficient is estimated using Equations (6.62) and (6.63). In this manner, we obtain the estimates $[\hat{\alpha}_0,\hat{\alpha}_1,\hat{\alpha}_2,\hat{\alpha}_3,\hat{\alpha}_4]$=[0.6596,0.3016,0.0060,0.0005,-3.9995×10^{-6}]. The phase estimated using these coefficients is shown against the original phase in Figure 6.10a, and the estimation error is shown in Figure 6.10b.

This analysis highlights the functioning of the HAF technique for polynomial phase estimation.

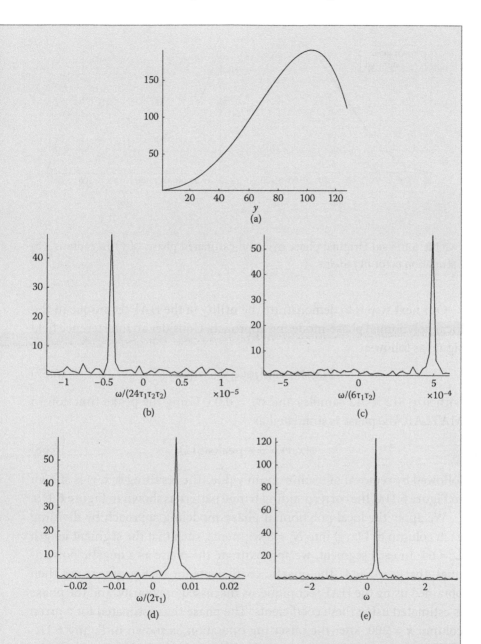

FIGURE 6.9 (a) Simulated phase $\phi(y)$ in radians. HAF spectrum for (b) $M = 4$, (c) $M = 3$, (d) $M = 2$, and (e) $M = 1$.

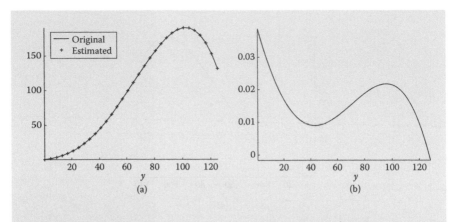

FIGURE 6.10 (a) Original phase $\phi(y)$ and estimated phase $\hat{\phi}(y)$ in radians. (b) Estimation error in radians.

Our next step is to demonstrate the utility of the HAF technique in the local polynomial phase-modeling approach. Consider an interference field signal as follows:

$$\Gamma(x,y) = \exp[j\phi(x,y)] + \eta(x,y) \tag{6.67}$$

with size 512×512 samples and $\sigma_\eta = 0.05$. Using the peaks function in MATLAB, the phase is simulated as

$$\phi(x,y) = 3 \times \text{peaks}(512) \tag{6.68}$$

followed by removal of its minimum value. The resulting $\phi(x, y)$ is shown in Figure 6.11a. The corresponding fringe pattern is shown in Figure 6.11b.

We apply the local polynomial phase-modeling approach by dividing each column of $\Gamma(x, y)$ into $N_s = 8$ segments, such that the segment length $L_s = 64$. In each segment, we approximate the phase as a quartic polynomial, that is, $M = 4$. The quartic coefficients for each segment are then obtained using the HAF technique, as discussed previously, and the phase is estimated using these coefficients. The phase thus estimated for a given column $x = 200$, after the offsetting operation, is shown in Figure 6.11c. Then, we continue this process with all columns and after that apply the stitching operation to obtain the phase estimate $\hat{\phi}(x, y)$, which is shown in Figure 6.11d. The estimation error is shown in Figure 6.11e.

It is interesting to observe that the simulated phase $\phi(x, y)$, as shown in Figure 6.11a, contains rapid variations. For signals with such rapidly varying phase, it often becomes necessary to use a polynomial approximation

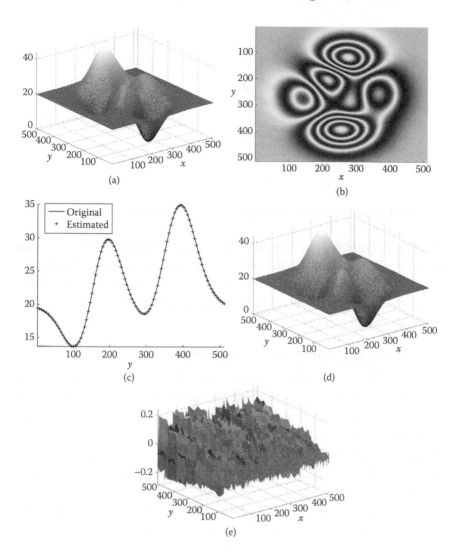

FIGURE 6.11 (a) Simulated phase $\phi(x, y)$ in radians. (b) Simulated fringe pattern. (c) Original versus estimated phase for a given column $x = 200$. (d) Estimated phase $\hat{\phi}(x, y)$ in radians. (e) Estimation error in radians.

beyond the second or third order. Hence, for these signals, the use of the MLE technique with quadratic phase approximation or the CPF technique with cubic phase approximation could lead to erroneous estimates. In contrast, the HAF is highly suitable in this scenario because it can be easily extended to higher polynomial orders with sufficient accuracy.

We also explore the practical utility of the HAF technique in DHI. In Figure 6.12a, we show an experimental fringe pattern, obtained in DHI.

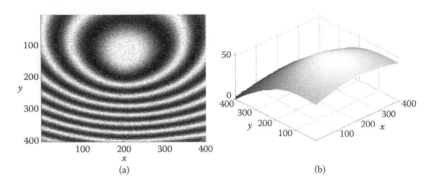

FIGURE 6.12   (a) Experimental fringe pattern. (b) Estimated phase in radians.

The local polynomial phase-modeling approach using the HAF technique was applied. For the analysis, we used $N_s = 4$ (i.e., four segments) and $M = 3$, that is, cubic phase approximation. The estimated phase using the method is shown in Figure 6.12b. These results affirm that the HAF technique has good potential for phase estimation in DHI.

6.7 PHASE-DIFFERENCING OPERATOR

In the techniques discussed so far, we can observe that all relied on 1-D processing only. In other words, we initiate the local polynomial phase-modeling approach by considering a given column (or row) and then model the interference field as a 1-D polynomial within a localized region, called the segment. It can be intuitively deduced that a 2-D approach would provide better generalization because the interference field is essentially a 2-D signal.

Another advantage of the 2-D processing is the enhanced robustness against noise because the samples along both dimensions are considered for phase estimation. To address this important issue, we present a 2-D local polynomial phase-modeling approach [51].

Let us reconsider the interference field signal in DHI as

$$\Gamma(x, y) = a(x, y) \exp[j\phi(x, y)] + \eta(x, y) \tag{6.69}$$

In our approach, we proceed by modeling the phase as a local 2-D polynomial. So, we divide $\Gamma(x, y)$ into multiple rectangular blocks B_{uv}, with $u \in [1, U]$ along x and $v \in [1, V]$ along y. The size of each block is $L_x \times L_y$ samples. This structure is shown in Figure 6.13.

Within a given block B_{uv}, the interference field signal can be expressed as

$$\Gamma_{uv}(x, y) = a_{uv}(x, y) \exp[j\phi_{uv}(x, y)] + \eta_{uv}(x, y) \tag{6.70}$$

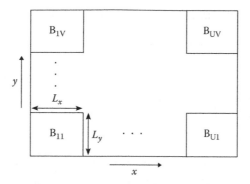

FIGURE 6.13   A 2-D block representation.

Approximating the phase as a 2-D polynomial of degree M, we have

$$\phi_{uv}(x,y) = \sum_{(k,l)\in I} c(k,l)x^k y^l \tag{6.71}$$

where $I = \{0 \le k,l \text{ and } 0 \le k + l \le M\}$. Effectively, the phase can be seen as composed of $M + 1$ layers such that

$$\phi_{uv}(x,y) = \underbrace{\{c(0,0)\}}_{\text{layer 1}} + \underbrace{\{c(0,1)y + c(1,0)x\}}_{\text{layer 2}}$$

$$+\cdots+\underbrace{\{c(0,M)y^M + c(1,M-1)xy^{M-1} +\cdots+ c(M-1,1)x^{M-1}y + c(M,0)x^M\}}_{\text{layer } M+1}$$

$$\tag{6.72}$$

where the layer i indicates a bivariate polynomial of degree $i - 1$ with i coefficients. Thus, once again, to retrieve phase, we have to solve the parameter estimation problem, with the parameters being the respective coefficients for each layer.

The bivariate coefficients are estimated using the 2-D phase-differencing operator [52, 53]. For the block B_{uv}, initially we define

$$\Gamma_{uv}^{(\dagger(p+q))} = \begin{cases} \Gamma_{uv}, & p+q \text{ even} \\ \Gamma_{uv}^*, & p+q \text{ odd} \end{cases} \tag{6.73}$$

The 2-D phase-differencing operator is now given as

$$PD_{x^{(P)},y^{(M-P-1)}}[\Gamma_{uv}(x,y)]$$

$$= \prod_{q=0}^{M-P-1}\left\{\prod_{p=0}^{P}\{\Gamma_{uv}^{(\dagger(p+q))}(x+p\tau_x,y+q\tau_y)\}^{\binom{P}{p}}\right\}^{\binom{M-P-1}{q}} \quad (6.74)$$

where $*$ denotes the complex conjugate, P is a parameter that varies in $[0, M - 1]$, and $\tau_x = L_x/(P + 1)$ and $\tau_y = L_y/(M - P)$ denote the delay parameters. We can observe that these equations closely resemble Equations (6.51) and (6.52). In fact, the 2-D phase-differencing operator can be viewed as the 2-D generalization of the HIMs.

The primary feature of the phase-differencing operator is that it transforms a polynomial phase signal of degree M into a 2-D complex sinusoid. Accordingly, ignoring the scaling factor and noise term for the sake of simplicity, we have

$$PD_{x^{(P)},y^{(M-P-1)}}[\Gamma_{uv}(x,y)] \approx \exp[j(\omega_x x + \omega_y y)] \quad (6.75)$$

where the frequencies are related to the coefficients as

$$\omega_y = (-1)^{M-1}P!(M-P)!\tau_x^P\tau_y^{M-P-1}c(P,M-P) \quad (6.76)$$

$$\omega_x = (-1)^{M-1}(P+1)!(M-P-1)!\tau_x^P\tau_y^{M-P-1}c(P+1,M-P-1) \quad (6.77)$$

From these equations, it is clear that the coefficients $c(P, M - P)$ and $c(P + 1, M - P - 1)$ can be estimated by finding the spatial frequencies corresponding to the phase-differencing operator. This is achieved by computing the 2-D DFT of $PD_{x^{(P)},y^{(M-P-1)}}[\Gamma_{uv}(x,y)]$ and tracing the spectral peak. Hence, we compute the DFT as follows:

$$X_{PM}(\omega_1,\omega_2) = \sum_y\sum_x PD_{x^{(P)},y^{(M-P-1)}}[\Gamma_{uv}(x,y)]\exp[-j(\omega_1 x + \omega_2 y)]$$

$$(6.78)$$

For efficient implementation of the DFT, we use the 2-D FFT algorithm. Interestingly, $X_{PM}(\omega_1,\omega_2)$ provides the 2-D extension of the HAF. The next step is to find the location of the maximum of the amplitude spectrum, which is provided as

$$\hat{\omega}_x, \hat{\omega}_y = \arg\max_{\omega_1, \omega_2} |X_{PM}(\omega_1, \omega_2)| \tag{6.79}$$

Subsequently, the coefficients are estimated as

$$\hat{c}(P, M-P) = \frac{\hat{\omega}_y}{(-1)^{M-1} P!(M-P)! \tau_x^P \tau_y^{M-P-1}} \tag{6.80}$$

$$\hat{c}(P+1, M-P-1) = \frac{\hat{\omega}_x}{(-1)^{M-1}(P+1)!(M-P-1)! \tau_x^P \tau_y^{M-P-1}} \tag{6.81}$$

By varying $P \in [0, M-1]$ in Equations (6.80) and (6.81), we obtain the coefficient estimates $[\hat{c}(0,M), \hat{c}(1, M-1), \ldots, \hat{c}(M-1,1), \hat{c}(M,0)]$ corresponding to the layer $M+1$ in Equation (6.72). Subsequently, we remove the contribution of these coefficients from the interference field signal $\Gamma_{uv}(x, y)$ using a peeling operation to reduce the polynomial degree by 1 as

$$\Gamma_{uv}^{M-1}(x, y) = \Gamma_{uv}(x, y) \exp\left[-j \sum_{P=0}^{M} \hat{c}(P, M-P) x^P y^{M-P}\right] \tag{6.82}$$

In Equation (6.82), $\Gamma_{uv}^{M-1}(x, y)$ is effectively a 2-D polynomial phase signal of degree $M-1$ (i.e., with M layers of coefficients). Hence, by successive application of the phase-differencing operator followed by a peeling operation, we estimate the coefficients right up to the second layer. From the resulting signal $\Gamma_{uv}^1(x, y)$, we obtain the zero-order coefficient of the first layer by removing the contribution of the second-layer coefficients and using the argument operator. Effectively, we have

$$\Gamma_{uv}^0(x, y) = \Gamma_{uv}^1(x, y) \exp[-j(\hat{c}(0,1) y + \hat{c}(1,0) x)] \tag{6.83}$$

and

$$\hat{c}(0,0) = \arg\left\{\frac{1}{L_x L_y} \sum_{y=1}^{L_y} \sum_{x=1}^{L_x} \Gamma_{uv}^0(x, y)\right\} \tag{6.84}$$

Finally, the phase is estimated using these coefficients:

$$\hat{\phi}_{uv}(x, y) = \sum_{(k,l) \in I} \hat{c}(k,l) x^k y^l \tag{6.85}$$

Following this procedure, the phase is estimated for all blocks B_{uv} shown in Figure 6.13.

The next step is to ensure that the phase varies smoothly between adjacent blocks. As discussed previously for the 1-D case, we need an offsetting operation to maintain phase continuity. As such, we consider the phase $\phi_{uv}(x, y)$ within an arbitrary block B_{uv}. The size of the block is $L_x \times L_y$ samples; hence, $x \in [1, L_x]$ and $y \in [1, L_y]$. To ensure phase continuity along x near the end of B_{uv} and the beginning of B_{u+1v}, the difference between the phase values of $\hat{\phi}_{uv}$ at $x = L_x + 1$ and $\hat{\phi}_{u+1v}$ at $x = 1$ should be minimal. Hence, in a least-squares sense, we can compute the offset γ_{u+1v} as the mean of the difference values. In other words, we have

$$\gamma_{u+1v} = \frac{1}{L_y} \sum_{y=1}^{L_y} (\hat{\phi}_{uv}(L_x + 1, y) - \hat{\phi}_{u+1v}(1, y)) \tag{6.86}$$

and use the assignment

$$\hat{\phi}_{u+1v}(x, y) \leftarrow \hat{\phi}_{u+1v}(x, y) + \gamma_{u+1v} \tag{6.87}$$

By varying $u \in [1, U - 1]$, the requisite offsets along x can be computed for all blocks. Similarly, an equivalent offsetting operation is performed along y to obtain the overall continuous phase distribution. Note that because of the offsetting operations in both dimensions, the phase continuity is ensured along both x and y; consequently, the stitching operation is not required. Thus, the 2-D approach removes the need for any unwrapping procedure.

At this point, we would like to emphasize that the implementation of the offsetting operation is not limited to the method described here. As the 2-D local polynomial phase-modeling approach is still in a nascent stage, we have presented only the basic method; nevertheless, other methodologies could be explored in the future.

For better understanding, we demonstrate the working of the phase-differencing operator technique in Box 6.5.

Next, we discuss the application of the 2-D local polynomial phase-modeling approach using a phase-differencing operator for phase estimation in DHI. Consider an interference field signal as follows:

$$\Gamma(x, y) = \exp[j\phi(x, y)] + \eta(x, y) \tag{6.88}$$

BOX 6.5   2-D PHASE-DIFFERENCING OPERATOR

Consider the following two-dimensional (2-D) signal:

$$\Gamma(x,y) = \exp[j\phi(x,y)] + \eta(x,y) \quad \{(x,y) \in [0,255]\}$$

with a bivariate quadratic phase, given as

$$\phi(x,y) = c(0,0) + c(0,1)y + c(1,0)x + c(0,2)y^2 + c(1,1)xy + c(2,0)x^2$$

where the coefficients $c(0,0) = 0$, $[c(0,1),c(1,0)] = [0.3, 0.2]$ and $[c(0,2),c(1,1),c(2,0)] = [-0.003, -0.002, 0.001]$, and the noise standard deviation $\sigma_\eta = 0.2$. This equation indicates that the phase can be viewed as having three layers. The simulated phase is shown in Figure 6.14a.

For the phase-differencing operator method, we proceed by estimating the coefficients of the highest layer. Accordingly, we compute the phase-differencing operator $PD_{x^{(P)},y^{(M-P-1)}}[\Gamma(x,y)]$ using Equation (6.74) for $P = 0$ and $M = 2$. The discrete Fourier transform (DFT) $X_{02}(\omega_1,\omega_2)$ is then obtained using Equation (6.78), and the corresponding amplitude spectrum is shown in Figure 6.14b. We estimate the frequencies by locating the clearly visible peak in the figure and then obtain the coefficient estimates $\hat{c}(0,2)$ and $\hat{c}(1,1)$ using Equations (6.80) and (6.81). Next, we increment the value of P by 1 and compute $X_{12}(\omega_1,\omega_2)$ for $P = 1$ and $M = 2$. The corresponding spectrum is shown in Figure 6.14c. Using the same steps, we obtain the estimates $\hat{c}(1,1)$ and $\hat{c}(2,0)$. Note that the coefficient $\hat{c}(1,1)$ is estimated twice in this process, and we can select either value for the analysis. Thus, all coefficients corresponding to the third layer are estimated.

In the next step, we reduce the polynomial degree by 1 through a peeling operation, as described in Equation (6.82). Then, the phase-differencing operator is computed for $P = 0$ and $M = 1$. Further, the DFT is calculated, and the amplitude spectrum $|X_{01}(\omega_1,\omega_2)|$ is shown in Figure 6.14d. As shown, the peak detection provides the estimates $\hat{c}(0,1)$ and $\hat{c}(1,0)$. Finally, the zero-order coefficient is estimated using Equations (6.83) and (6.84). Overall, the estimated coefficients are $\hat{c}(0,0) = 0.0009$, $[\hat{c}(0,1),\hat{c}(1,0)] = [0.3000, 0.2000]$ and $[\hat{c}(0,2),\hat{c}(1,1),\hat{c}(2,0)] = [-0.0030,-0.0020,0.0010]$.

Using these coefficients, the phase is estimated as shown in Equation (6.85). The estimated phase $\hat{\phi}(x,y)$ is shown in Figure 6.14e, and the estimation error is shown in Figure 6.14f.

This analysis explains the basic aspects of applying the 2-D phase-differencing operator for polynomial phase estimation.

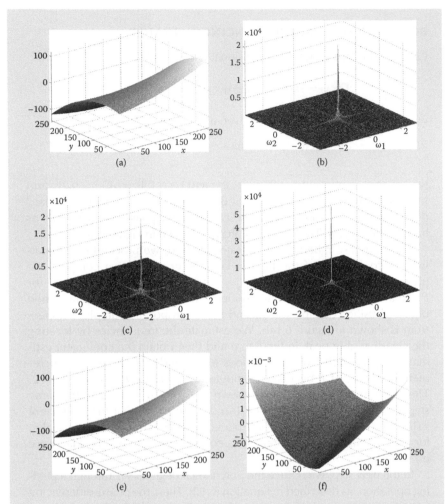

FIGURE 6.14   (a) Simulated phase $\phi(x,y)$ in radians. (b) $|X_{02}(\omega_1,\omega_2)|$ for $P = 0$, M = 2. (c) $|X_{12}(\omega_1,\omega_2)|$ for $P = 1$, $M = 2$. (d) $|X_{01}(\omega_1,\omega_2)|$ for $P = 0$, $M = 1$. (e) Estimated phase $\hat{\phi}(x,y)$ in radians. (f) Estimation error in radians.

where the signal size is $N \times N$ samples, with $N = 512$, and the noise standard deviation $\sigma_\eta = 0.3$. We simulate the phase as

$$\phi(x,y) = 30 \left\{ \exp\left[-15\left(\frac{(x-N/2)^2 + (y-N/2-50)^2}{N^2} \right) \right] \right.$$

$$\left. - \exp\left[-15\left(\frac{(x-N/2)^2 + (y-N/2+50)^2}{N^2} \right) \right] \right\} \quad (6.89)$$

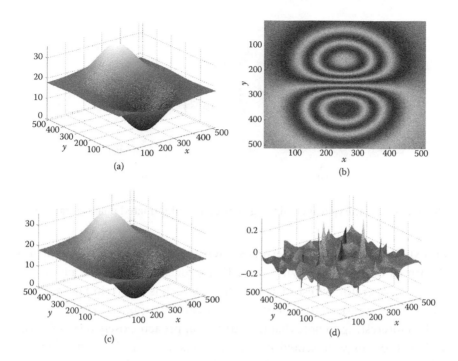

FIGURE 6.15 (a) Original phase $\phi(x, y)$ in radians. (b) Simulated fringe pattern. (c) Estimated phase $\hat{\phi}(x, y)$ in radians. (d) Estimation error in radians.

and remove the minimum value from it. The resulting phase distribution is shown in Figure 6.15a. The corresponding fringe pattern is shown in Figure 6.15b.

For applying the 2-D approach, we divide the interference field signal into four blocks along x (i.e., $U = 4$) and four blocks along y (i.e., $V = 4$). The size of each block is 128×128 samples. Within each block, we model the phase as a bivariate quartic polynomial, that is, with degree $M = 4$. The polynomial coefficients are estimated using the phase-differencing operator. Subsequently, the phase for the given block is estimated from these coefficients. This process is repeated for all blocks. Finally, the offsetting operation is used between adjacent blocks to obtain smooth phase distribution. The phase estimate $\hat{\phi}(x, y)$ is shown in Figure 6.15c. The estimation error is shown in Figure 6.15d. From these results, we can infer that the 2-D local polynomial phase-modeling approach is suitable for phase estimation in DHI.

Further, we highlight the utility of the 2-D approach for estimating the phase from experimental fringe patterns obtained in DHI. In Figure 6.16a,

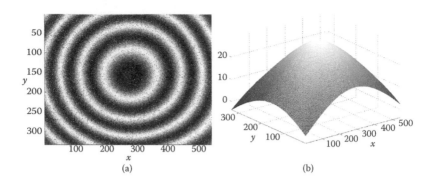

FIGURE 6.16 (a) Experimental fringe pattern. (b) Estimated phase in radians.

an experimental fringe pattern is shown that corresponds to a centrally deformed object. By applying the 2-D approach based on the phase-differencing operator, we estimate the phase, which is shown in Figure 6.16b. Clearly, the approach is suitable for practical DHI applications.

It is interesting to note that the current approach closely resembles the 1-D methodology presented in the previous sections. In particular, the 1-D segments are replaced by 2-D blocks, and the univariate polynomial phase approximation is substituted by a bivariate approximation. Further, the role of HAF for coefficient estimation in the previous section is now performed by the phase-differencing operator. Overall, this method helps us to visualize the basic concept of local polynomial phase modeling from a broader perspective because of the 2-D generalization.

6.8 CONCLUSIONS

In this chapter, we presented the local polynomial phase-modeling approach for phase estimation in the context of DHI. The basic principle of the approach was discussed in Section 6.3. The central idea is to approximate the phase as a polynomial within a small region as opposed to the entire signal span, thereby rendering the transformation of phase measurement into a parameter estimation problem, with the polynomial coefficients being the respective parameters. The direct implication of this transformation is that the fundamental traits of the approach are strongly dependent on the type of technique chosen to solve the parameter estimation problem.

Accordingly, some of the prominent coefficient estimation techniques for the local polynomial-modeling approach were explored. In Section 6.4, the MLE technique and its application in DHI were detailed. We saw that the high computational requirements associated with the multivariate optimization limit the practical application of this technique for quadratic phase modeling only. This limitation was relieved by the CPF technique discussed in Section 6.5, which allows cubic phase approximation at much lower computational effort. Subsequently, in Section 6.6, we generalized the approach for arbitrary-order polynomial phase modeling using HAF. The major advantage of this technique is that it allows local approximation of phase as a polynomial beyond the quadratic or cubic orders and provides a computationally efficient strategy for reliable estimation of the corresponding polynomial coefficients and thus phase. Finally, in Section 6.7 we extended the local polynomial phase-modeling approach to two dimensions using the phase-differencing operator. This technique can be considered the 2-D equivalent of the HAF.

The numerical simulation studies and the experimental cases presented in the chapter demonstrate the application potential of these techniques. Last but not least, we would like to emphasize that the idea of local polynomial phase modeling for estimating the phase distribution is still in its infancy and offers a tremendous scope of research possibilities in the future.

REFERENCES

1. D. Gabor. A new microscopic principle. *Nature* **161**, 777–778 (1948).
2. D. Gabor. Microscopy by reconstructed wave-fronts. *Proceedings of the Royal Society of London. Series A. Mathematical and Physical Sciences* **197**, 454–487 (1949).
3. T. Kreis. *Handbook of Holographic Interferometry: Optical and Digital Methods*. Weinheim: Wiley-VCH, 2005.
4. E. N. Leith and J. Upatnieks. Reconstructed wavefronts and communication theory. *Journal of the Optical Society of America* **52**, 1123–1128 (1962).
5. E. N. Leith and J. Upatnieks. Wavefront reconstruction with diffused illumination and three-dimensional objects. *Journal of the Optical Society of America* **54**, 1295–1301 (1964).
6. O. Schnars and W. Juptner. Direct recording of holograms by a CCD target and numerical reconstruction. *Applied Optics* **33**, 179–181 (1994).
7. U. Schnars. Direct phase determination in hologram interferometry with use of digitally recorded holograms. *Journal of the Optical Society of America* A **11**, 2011–2015 (1994).
8. U. Schnars and W. P. O. Juptner. Digital recording and reconstruction of holograms in hologram interferometry and shearography. *Applied Optics* **33**, 4373–4377 (1994).
9. H. J. Tiziani and G. Pedrini. From speckle pattern photography to digital holographic interferometry. *Applied Optics* **52**, 30–44 (2013).
10. C. Liu. Simultaneous measurement of displacement and its spatial derivatives with a digital holographic method. *Optical Engineering* **42**, 3443–3446 (2003).

11. C. Quan, C. J. Tay, and W. Chen. Determination of displacement derivative in digital holographic interferometry. *Optics Communication* **282**, 809–815 (2009).
12. G. Rajshekhar, S. S. Gorthi, and P. Rastogi. Strain, curvature, and twist measurements in digital holographic interferometry using pseudo-Wigner-Ville distribution based method. *Review of Scientific Instruments* **80**, 093107 (2009).
13. G. Rajshekhar, S. S. Gorthi, and P. Rastogi. Estimation of displacement derivatives in digital holographic interferometry using a two-dimensional space-frequency distribution. *Optics Express* **18**, 18041–18046 (2010).
14. C. Wagner, S. Seebacher, W. Osten, and W. Jüptner. Digital recording and numerical reconstruction of lensless Fourier holograms in optical metrology. *Applied Optics* **38**, 4812–4820 (1999).
15. I. Yamaguchi, T. Ida, M. Yokota, and K. Yamashita. Surface shape measurement by phase-shifting digital holography with a wavelength shift. *Applied Optics* **45**, 7610–7616 (2006).
16. A. Wada, M. Kato, and Y. Ishii. Multiple-wavelength digital holographic interferometry using tunable laser diodes. *Applied Optics* **47**, 2053–2060 (2008).
17. V. Kebbel, M. Adams, H. J. Hartmann, and W. Jüptner. Digital holography as a versatile optical diagnostic method for microgravity experiments. *Measurement Science & Technology* **10**, 893–899 (1999).
18. R. B. Owen and A. A. Zozulya. Comparative study with double-exposure digital holographic interferometry and a Shack-Hartmann sensor to characterize transparent materials. *Applied Optics* **41**, 5891–5895 (2002).
19. M. De Angelis, S. De Nicola, A. Finizio, G. Pierattini, P. Ferraro, S. Pelli, G. Righini, and S. Sebastiani. Digital-holography refractive-index-profile measurement of phase gratings. *Applied Physics Letters* **88** (2006).
20. U. Schnars and W. P. O. Juptner. Digital recording and numerical reconstruction of holograms. *Measurement Science & Technology* **13**, R85–R101 (2002).
21. T. H. Demetrakopoulos and R. Mittra. Digital and optical reconstruction of images from suboptical diffraction patterns. *Applied Optics* **13**, 665–670 (1974).
22. L. Yu and M. K. Kim. Wavelength-scanning digital interference holography for tomographic three-dimensional imaging by use of the angular spectrum method. *Optics Letters* **30**, 2092–2094 (2005).
23. M. Liebling, T. Blu, and M. Unser. Fresnelets: New multiresolution wavelet bases for digital holography. *IEEE Transactions on Image Processing* **12**, 29–43 (2003).
24. J. W. Cooley and J. W. Tukey. An algorithm for the machine computation of the complex Fourier series. *Mathematics of Computation* **19**, 297–301 (1965).
25. D. C. Ghiglia and M. D. Pritt. *Two-Dimensional Phase Unwrapping: Theory, Algorithms, and Software.* New York: Wiley-Interscience, 1998.
26. J. M. Huntley and C. R. Coggrave. Progress in phase unwrapping. *Proceedings SPIE* **3407**, 86–93 (1998).
27. X. Su and W. Chen. Reliability-guided phase unwrapping algorithm: A review. *Optics and Lasers in Engineering* **42**, 245–261 (2004).
28. G. Rajshekhar and P. Rastogi. Fringe analysis: Premise and perspectives. *Optics and Lasers in Engineering* **50**, iii–x (2012).

29. W. Rudin. *Principles of Mathematical Analysis.* 3rd ed. New York: McGraw-Hill, 1976.
30. M. P. Rimmer, C. M. King, and D. G. Fox. Computer program for the analysis of interferometric test data. *Applied Optics* **11**, 2790–2796 (1972).
31. J. Y. Wang and D. E. Silva. Wave-front interpretation with Zernike polynomials. *Applied Optics* **19**, 1510–1518 (1980).
32. C.-J. Kim. Polynomial fit of interferograms. *Applied Optics* **21**, 4521–4525 (1982).
33. A. Cordero-Dávila, A. Corínejo-Rodrfguez, and O. Cardona-Nuñez. Polynomial fitting of interferograms with Gaussian errors on fringe coordinates. 1: Computer simulations. *Applied Optics* **33**, 7339–7342 (1994).
34. J. Novak and A. Miks. Least-squares fitting of wavefront using rational function. *Optics and Lasers in Engineering* **43**, 40–51 (2005).
35. S. S. Gorthi and P. Rastogi. Piecewise polynomial phase approximation approach for the analysis of reconstructed interference fields in digital holographic interferometry. *Journal of Optics A: Pure and Applied Optics* **11**, 065405 (2009).
36. S. S. Gorthi and P. Rastogi. Windowed high-order ambiguity function method for fringe analysis. *Review of Scientific Instruments* **80**, 073109 (2009).
37. B. Boashash. Estimating and interpreting the instantaneous frequency of a signal. II. Algorithms and applications. *IEEE Proceedings* **80**, 540–568 (1992).
38. S. Peleg and B. Friedlander. The discrete polynomial-phase transform. *IEEE Transactions on Signal Processing* **43**, 1901–1914 (1995).
39. S. Boyd and L. Vandenberghe. *Convex Optimization.* Cambridge, UK: Cambridge University Press, 2004.
40. J. C. Lagarias, J. A. Reeds, M. H. Wright, and P. E. Wright. Convergence properties of the Nelder-Mead simplex method in low dimensions. *SIAM Journal on Optimization* **9**, 112–147 (1999).
41. P. O'Shea. A new technique for instantaneous frequency rate estimation. *IEEE Signal Processing Letters* **9**, 251–252 (2002).
42. P. O'Shea. A fast algorithm for estimating the parameters of a quadratic fm signal. *IEEE Transactions on Signal Processing* **52**, 385–393 (2004).
43. E. Aboutanios and B. Mulgrew. Iterative frequency estimation by interpolation on Fourier coefficients. *IEEE Transactions on Signal Processing* **53**, 1237–1242 (2005).
44. S. S. Gorthi, G. Rajshekhar, and P. Rastogi. Strain estimation in digital holographic interferometry using piecewise polynomial phase approximation based method. *Optics Express* **18**, 560–565 (2010).
45. S. S. Gorthi and P. Rastogi. Phase estimation in digital holographic interferometry using cubic-phase-function based method. *Journal of Modern Optics* **57**, 595–600 (2010).
46. B. Porat. *Digital Processing of Random Signals.* Englewood Cliffs, NJ: Prentice Hall, 1994.
47. S. Golden and B. Friedlander. A modification of the discrete polynomial transform. *IEEE Transactions on Signal Processing* **46**, 1452–1455 (1998).
48. P. Stoica and R. Moses. *Introduction to Spectral Analysis.* Upper Saddle River, NJ: Prentice Hall, 1997.

49. R. Roy and T. Kailath. Esprit—estimation of signal parameters via rotational invariance techniques. *IEEE Transactions on Acoustics, Speech, and Signal Processing* **37**, 984–995 (1989).
50. R. O. Schmidt. Multiple emitter location and signal parameter estimation. *IEEE Transactions on Antennas and Propagation* **34**, 276–280 (1986).
51. G. Rajshekhar and P. Rastogi. Fringe demodulation using the two-dimensional phase differencing operator. *Optics Letters* **37**, 4278–4280 (2012).
52. B. Friedlander and J. M. Francos. An estimation algorithm for 2-D polynomial phase signals. *IEEE Transactions on Image Processing* **5**, 1084–1087 (1996).
53. J. M. Francos and B. Friedlander. Two-dimensional polynomial phase signals: Parameter estimation and bounds. *Multidimensional Systems and Signal Processing* **9**, 173–205 (1998).

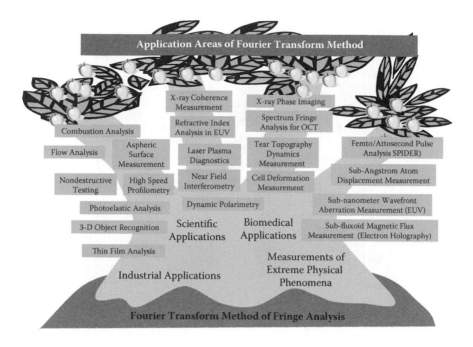

COLOR FIGURE 1.3 Application areas of Fourier transform method. (After M. Takeda, *Proceedings of SPIE 8011* (2011): 80116S; *AIP Conference Proceedings 1236* (2010): 445–448 with permission from SPIE.)

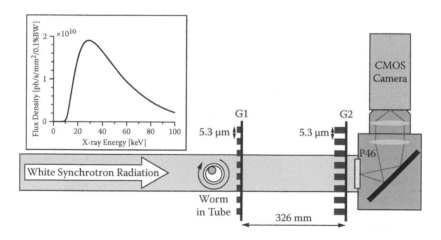

COLOR FIGURE 1.13 Experimental setup (side view) of four-dimensional X-ray phase tomography with a Talbot interferometer consisting of a phase grating (G1) and an amplitude grating (G2) and white synchrotron radiation. The inset shows the calculated flux density at the sample position. (After Momose, A., Yashiro, W., Harasse, S, and Kuwabara, H., *Optics Express* **19**(9) (2011): 8424–8432 with permission from OSA.)

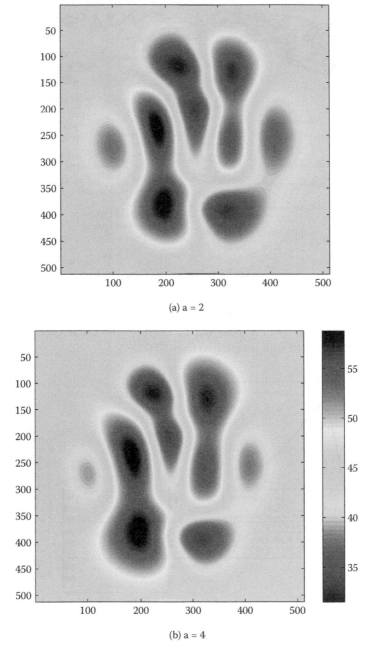

(a) a = 2

(b) a = 4

COLOR FIGURE 3.18   The 2-D CWT of the fringe pattern in Figure 3.12 with the 2-D Mexican hat with σ = 0.5. *(continued)*

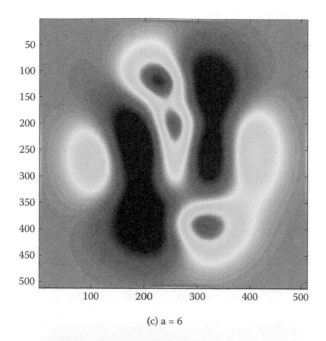

(c) a = 6

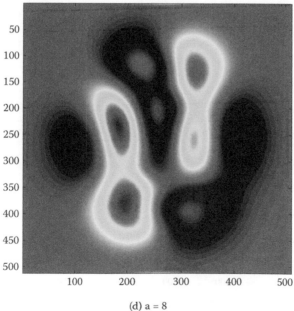

(d) a = 8

COLOR FIGURE 3.18   *(continued)* The 2-D CWT of the fringe pattern in Figure 3.12 with the 2-D Mexican hat with σ = 0.5. *(continued)*

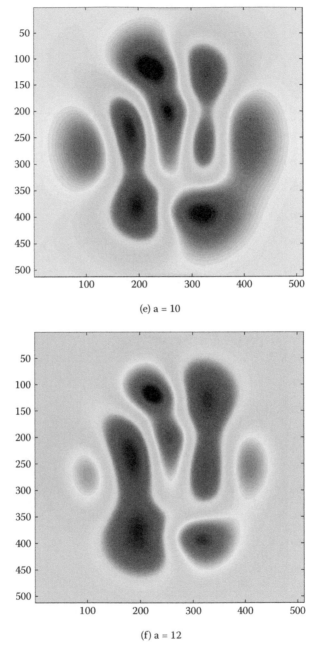

(e) a = 10

(f) a = 12

COLOR FIGURE 3.18   *(continued)* The 2-D CWT of the fringe pattern in Figure 3.12 with the 2-D Mexican hat with σ = 0.5. *(continued)*

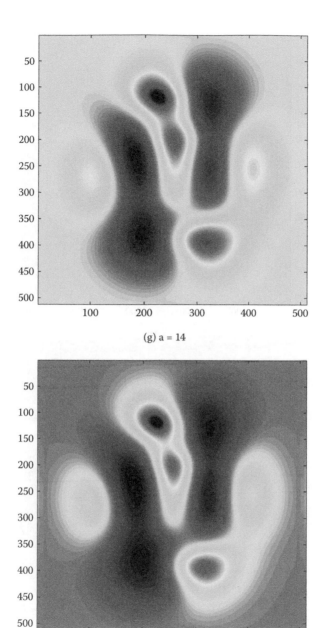

(g) a = 14

(h) a = 16

COLOR FIGURE 3.18   *(continued)* The 2-D CWT of the fringe pattern in Figure 3.12 with the 2-D Mexican hat with σ = 0.5. *(continued)*

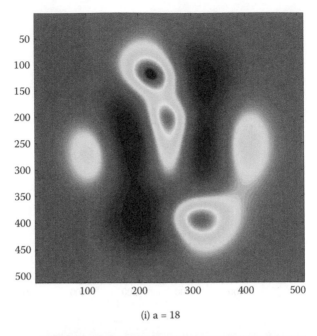

(i) a = 18

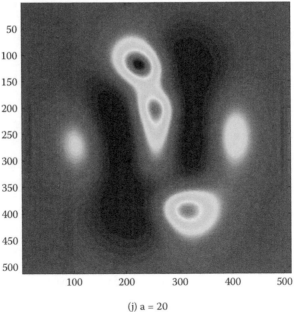

(j) a = 20

COLOR FIGURE 3.18   *(continued)* The 2-D CWT of the fringe pattern in Figure 3.12 with the 2-D Mexican hat with $\sigma = 0.5$.

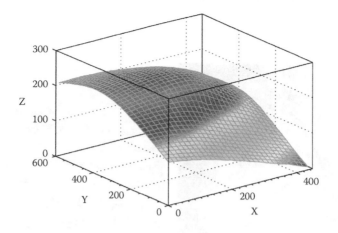

COLOR FIGURE 7.3 (d) unwrapped phase map using Goldstein [25] algorithm. (From A. Patil, R. Langoju, R. Sathish, and P. Rastogi. *Journal of the Optical Society of America A* **24**, 794–813, 2007.)

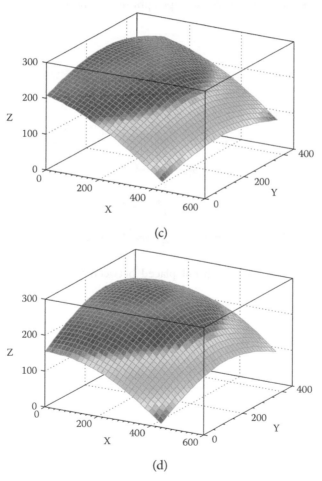

(c)

(d)

COLOR FIGURE 7.5 (c) φ_1 and (d) φ_2. (From A. Patil, R. Langoju, R. Sathish, and P. Rastogi. *Journal of the Optical Society of America A* **24**, 794–813, 2007.)

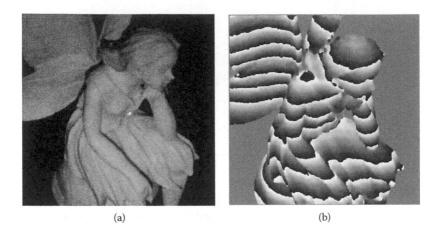

(a)   (b)

COLOR FIGURE 8.2   A complex wrapped phase distribution with context.

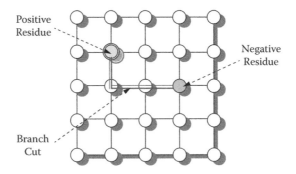

COLOR FIGURE 8.7   A branch cut placed between two residues that form a dipole pair.

CHAPTER 7

Signal-Processing Methods in Phase-Shifting Interferometry

Abhijit Patil

John F Welch Technology Centre
Bangalore, India

Rajesh Langoju

GE Global Research Centre
Bangalore, India

Pramod Rastogi

Swiss Federal Institute of Technology
Lausanne, Switzerland

7.1 INTRODUCTION

Previous chapters have introduced various spatial techniques for the extraction of phase information from a single frame. Among them, the Fourier transform technique has been most widely studied and applied in practical situations for fringe analysis, which involves transforming the intensity data into the Fourier domain. Fringe analysis using this technique involves the addition of a carrier fringe to the intensity data to separate the zero-order spectrum corresponding to the background intensity from the first-order spectrum. Note that it is the first-order spectrum that carries the phase information. The phase is then computed by processing only the first-order spectrum. The carrier fringes are added by tilting a wave front in an interferometer [1], resulting in a phase term, which is a linear function of the coordinates (x, y) to the actual phase φ. The intensity at (x, y) as a result of tilting is governed by the following equation:

$$I(x,y)=I_{dc}(x,y)\{1+\gamma(x,y)\cos[2\pi(xf_{cx}+yf_{cy})+\varphi(x,y)]\} \quad (7.1)$$

where (x, y) represent the spatial coordinates spanning the measurement domain, $I_{dc}(x, y)$ is the background intensity, $\gamma(x, y)$ is the fringe visibility (modulation intensity), and $\varphi(x, y)$ is the phase difference that carries the relevant information one is generally interested to measure. Equation (7.1) can be further expanded as

$$I(x,y)=I_{dc}(x,y)+\Psi(x,y)\exp[i2\pi(xf_{cx}+yf_{cy})]+\Psi^*(x,y)$$

$$\exp[-i2\pi(xf_{cx}+yf_{cy})] \quad (7.2)$$

where $\Psi(x,y)=\frac{1}{2}\gamma(x,y)\exp[i\varphi(x,y)]$, and $\{f_{cx}, f_{cy}\}$ represent carrier frequencies in x and y directions, respectively. The Fourier transform of Equation (7.2) gives

$$I_f(f_x,f_y)=P_0(f_x,f_y)+P_1(f_x-f_{cx},f_y-f_{cy})+P_1^*(f_x+f_{cx},f_y+f_{cy}) \quad (7.3)$$

where P_0 and P_1 represent the Fourier transforms of $I_{dc}(x, y)$ and $\Psi(x, y)$, respectively. From Equation (7.3), one can observe that $I_f(f_x, f_y)$ is a tri-modal function with peaks at $(-f_{cx}, -f_{cy})$, (f_{cx}, f_{cy}), and the origin. The function $P_1(f_x, f_y)$ can be isolated using a filter centered around (f_{cx}, f_{cy}). The carrier frequency is then removed by shifting $P_1(f_x-f_{cx}, f_y-f_{cy})$ to the origin to give $P_1(f_x, f_y)$. The inverse Fourier transform of $P_1(f_x, f_y)$ yields $\Psi(x, y)$. The phase difference $\varphi(x, y)$ can be obtained as

$$\varphi(x,y)=\tan^{-1}\left\{\frac{\Im[\Psi(x,y)]}{\Re[\Psi(x,y)]}\right\} \quad (7.4)$$

where $\Re[\Psi(x,y)]$ represents the real part and $\Im[\Psi(x,y)]$ represents the imaginary part of $\Psi(x, y)$.

In the literature, several methods have been proposed for Fourier analysis of the fringe pattern. However, these methods have their own limitations. For example, the Fourier transform is a global transform technique, so it is sensitive to the spatial variations or discontinuities in the fringe. Moreover, improper filtering of the zero- and negative first-order components will also affect the accuracy of the method. The spectral leakage between the first-order and zero-order Fourier components can be avoided by tilting one arm of the interferometer, thereby generating a higher carrier frequency. However, the increase in carrier frequency is limited by the detector response and bandwidth. Care should be taken that the frequencies of the signal phase $\varphi(x, y)$ should be within the passband of the filter centered on (f_{cx}, f_{cy}). In the presence of higher-order harmonics, care should be taken that harmonics must not fall in the filter passband. Moreover, the type of filter chosen will also affect the localization of the evaluated phase [2]. To address some of these concerns, a wavelet transform technique for fringe analysis has been proposed. This technique does not require a complex phase-unwrapping algorithm and thus allows direct phase demodulation [3]. However, this technique has met with limited success because of the need to add carrier fringes in both the horizontal and the vertical directions. If the fringes are added in the horizontal direction, only one-dimensional wavelet transform is possible. Moreover, the need to introduce an optimization algorithm to minimize the wavelet ridge is one of the potential bottlenecks in accessing its optimal performance.

Kemao [4–6] has demonstrated through several illustrations that a fringe map processed locally or block by block using the windowed Fourier transform (WFT) method removes artifacts in phase measurements as compared to the simple carrier-based Fourier transform method. The author has further shown an elegant comparison of the WFT method with various other spatial techniques that have been applied to phase measurements, such as windowed Fourier ridges, wavelet transform, and regularized phase tracking. Presently, WFT appears to be a promising tool among the spatial techniques for phase measurement.

7.2 TEMPORAL TECHNIQUES

An approach parallel to extracting the phase information spatially involves phase extraction by acquiring a number of intensity images with phase increments between successive frames. The phase increment corresponds to the relative change in the phase difference between the object and the reference beams. These phase increments can typically be obtained in various ways, such as by frequency modulation of laser diodes, by a ferroelectric liquid crystal and a combination of quarter- and half-wave plates, or by a piezoelectric device (lead zirconate titanate, PZT). Temporal techniques for extracting phase are known to offer advantages over the spatial techniques in terms of improved noise immunity, high-spatial-frequency resolution, and insensitivity to spatial variations in the detector response.

Various methods by which the phase shifts can be applied and the contemporary algorithms for extracting the phase information from temporal data sets are discussed in Patil and Rastogi [7]. In spite of the presence of a large number of phase-shifting elements, one device that has caught the attention of researchers is the piezoelectric device, commonly known as PZT. PZT is a ceramic perovskite material that shows a piezoelectric effect. Piezoelectricity results from the ability of certain crystals to generate a voltage in response to applied mechanical stress. Piezoelectric materials also show the opposite effect, called converse piezoelectricity, by which the application of an electrical field creates a mechanical deformation in the crystal. A deformation of about 0.1% of the original dimension can yield displacements on the order of nanometers.

Unfortunately, PZT is also one of the major sources of errors in phase measurement. In practice, this device exhibits a nonlinear response to the applied voltage and a mechanical vibration during its translation. In addition, the device has the characteristic of nonrepeatability and suffers from an aging effect. One of the most common errors during its usage is the so-called miscalibration error [8]. To make matters worse, the detector is known sometimes to be responsible for the presence of higher-order harmonics in the recorded fringe intensity [8,9]. Discussion of various contemporary algorithms aimed at minimizing the errors arising because of PZT movement to the applied voltages, and also those because of vibrations, stray reflections, temperature fluctuations, and quantization errors of the intensity, is beyond the scope of this chapter; an overview is presented in Patil and Rastogi [7].

The objective of this chapter is to introduce advanced temporal phase measurement methods that not only allow measurement of phase information in an optical setup involving a single PZT but also allow measurement of multiple phases at a pixel in an optical setup involving multiple PZTs. Examples of the two setups are phase measurement in holographic interferometry and multiple phase measurements in holographic moiré. Holographic interferometry basically involves comparison of two or more wave fields interferometrically, with at least one of them holographically recorded and reconstructed. The setup consists of two beams, the object beam and the reference beam. The object beam falls on the object, and the scattered wave front is collected on the recording medium. The reference beam falls directly on the recording medium. A phase shifter is typically placed in the path of the reference beam, and for each position of the phase shifter, data are recorded. Thus, series of phase-shifted images are solved to measure the quantity of interest. This technique offers the possibility to measure displacements, deformations, vibrations, and shapes of rough objects. However, this technique does not allow measurement of multiple information at a single pixel of the acquired data frame.

One possible way to accomplish this objective is to incorporate multiple PZTs in an optical setup. Rastogi [10, 11] has shown that when two PZTs are placed symmetrically across an object in a holographic moiré [12] configuration, two orthogonal displacement components can be determined simultaneously. Figures 7.1a and 7.1b show the optical setups for holographic interferometry and holographic moiré, respectively. In each of these setups, the first step involves estimating the phase steps at each pixel and subsequently solving the linear system of equations to extract the phase information at each pixel on the data frame.

This chapter focuses on signal-processing approaches for phase estimation. The chapter is organized as follows: Section 7.3 discusses the linear phase step estimation method based on a signal-processing approach for holographic interferometry. The assumption is that the movement of the PZT to the applied voltage is linear. Section 7.4 presents comparison of the signal-processing methods with other benchmarking algorithms. The section also discusses the extraction of phase for experimental data. Section 7.5 discusses dual PZTs in holographic moiré. Section 7.6 presents estimation of phase steps and phase information for holographic moiré setup.

In a practical scenario, the movement of the PZT to the applied voltage is nonlinear, and approaches mentioned in Sections 7.3 and 7.4 can produce erroneous results. To address this issue, we discuss nonlinear phase step estimation methods in Section 7.7 that could be applied for both the optical setups shown in Figure 7.1.

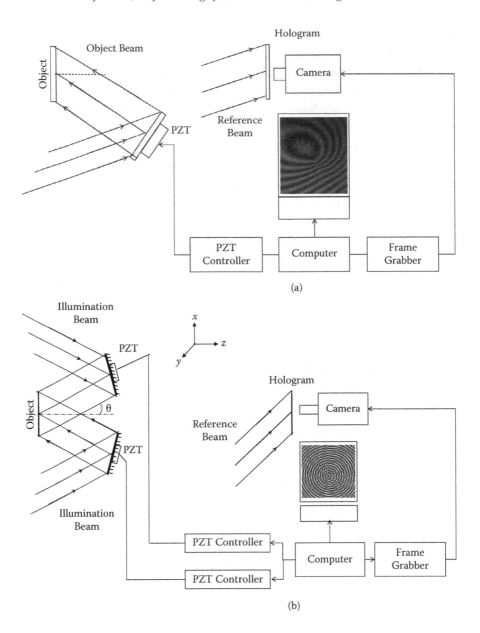

FIGURE 7.1 (a) Schematic of holographic interferometry. (b) Schematic of holographic moiré.

7.3 LINEAR PHASE STEP ESTIMATION METHODS

Let us first start the discussion of linear phase step estimation methods with a simple case of holographic interferometry involving a single PZT. The intensity acquired at any pixel (x, y) on the charge-coupled device (CCD) camera for the nth data frame is given by

$$I(x, y; n) = I_{dc} + \sum_{k=1}^{\kappa} a_k^* \exp[-jk(\varphi_1 + n\alpha)] + \sum_{k=1}^{\kappa} a_k \exp[jk(\varphi_1 + n\alpha)] \quad (7.5)$$

where $\{a_k = a_k^*\}$ are the complex Fourier coefficients of the kth-order harmonic, $j = \sqrt{-1}$ and I_{dc} are the local average value for intensity; and the pair (φ_1, α) represents the phase difference and phase shift applied, respectively. Let us assume that we have acquired N data frames. Now, Equation (7.5) can be rewritten as

$$I_n(x, y) = I_{dc} + \sum_{k=1}^{\kappa} \ell_k u_k^n + \sum_{k=1}^{\kappa} \ell_k^* (u_k^*)^n$$

$$\text{for } n = 0, 1, 2, \ldots, N-1. \quad (7.6)$$

where $\ell_k = a_k \exp(jk\varphi_1)$, $u_k = \exp(jk\alpha)$. Equation (7.6) represents a sum of complex valued sinusoidal signals with frequencies $\omega = 0, \alpha, \ldots, \kappa\alpha; -\alpha, \ldots, -\kappa\alpha$; by estimating the frequency content, the phase step α imparted to the PZT can be determined. So, the phase step estimation problem now boils down to a frequency estimation problem.

Frequency estimation is a common problem in the areas of signal processing and communication, and several efficient methods exist to estimate frequency. The next sections discuss in brief high-resolution frequency estimation methods such as the multiple signal classification (MUSIC) method [13] and the estimation of signal parameter via rotational invariance technique (ESPRIT) method [14] for determining linear phase step and phase information with high accuracy. These methods are called high-resolution methods because of their ability to retrieve frequency of complex exponentials in the presence of noise and limited data samples [15].

7.3.1 Multiple Signal Classification Method: root-MUSIC

Root-MUSIC is a subspace-based method [13] in which the first step involves the design of a covariance matrix from the acquired N data frames. In brief, let the covariance matrix be defined as $\mathbf{R_I} = E[\mathbf{I}^*(n)\mathbf{I}(n)]$, where

$I(n) = [I(n − 1)\ I(n − 2)\ \dots\ I(n − m)]$. Here, E is the expectation operator, and m is the covariance length. The next step is performing an eigenvalue decomposition of the autocovariance matrix that decomposes the matrix $\mathbf{R}_I$ into signal and noise subspaces as follows [16,17]:

$$\mathbf{R}_I = \underbrace{\mathbf{A}\mathbf{P}\mathbf{A}^c}_{\mathbf{R}_s} + \underbrace{\sigma^2\mathbf{I}}_{\mathbf{R}_\varepsilon} \tag{7.7}$$

where $\mathbf{R}_s$ and $\mathbf{R}_\varepsilon$ are the signal and noise contributions, respectively; $\mathbf{A}_{m\times 2\kappa+1} = [\mathbf{a}(\omega_1)\ \cdots\ \mathbf{a}(\omega_{2\kappa+1})]$, where, for instance, the element $\mathbf{a}(\omega_1)$ consists of an $m \times 1$ matrix with unity entries corresponding to I_{dc}; $(\cdot)^T$ and $(\cdot)^c$ represent the transpose and complex transpose of a matrix, respectively; $\mathbf{a}(\omega_2) = [1\ \exp(j\alpha)\ \cdots\ \exp(j\alpha_{(m-1)})]^T$; $\mathbf{I}$ is the $m \times m$ identity matrix; and the $\mathbf{P}_{(2\kappa+1)\times(2\kappa+1)}$ matrix is

$$\mathbf{P} = \begin{bmatrix} A_0^2 & 0 & \cdot & \cdot & 0 \\ 0 & A_1^2 & \cdot & \cdot & \cdot \\ \cdot & \cdot & \cdot & \cdot & \cdot \\ 0 & \cdot & \cdot & \cdot & A_{2\kappa}^2 \end{bmatrix} \tag{7.8}$$

Because $\mathbf{R}_I$ is positive semidefinite, its eigenvalues are nonnegative. The eigenvalues of $\mathbf{R}_I$ can be ordered as $\lambda_1 \geq \lambda_2 \geq \cdots \geq \lambda_{2\kappa+1} \geq \cdots \geq \lambda_m$. Let $\mathbf{S}_{m\times(2\kappa+1)} = [\mathbf{s}_1\ \mathbf{s}_2\ \cdots\ \mathbf{s}_{2\kappa+1}]$ be the orthonormal eigenvectors associated with $\lambda_1 \geq \lambda_2 \geq \cdots \geq \lambda_{2\kappa+1}$. The space spanned by $\{\mathbf{s}_1, \mathbf{s}_2, \dots, \mathbf{s}_{2\kappa+1}\}$ is known as *signal subspace*. The set of orthonormal eigenvectors $\mathbf{G}_{m\times m-(2\kappa+1)} = [\mathbf{g}_1\ \mathbf{g}_2\ \cdots\ \mathbf{g}_{m-(2\kappa+1)}]$ associated with eigenvalues $\lambda_{2\kappa+2} \geq \lambda_{2\kappa+3} \geq \cdots \geq \lambda_m$ spans a subspace known as the *noise subspace*. Because $\mathbf{A}\mathbf{P}\mathbf{A}^c \in \mathbb{C}^{m\times m}$ ($\mathbb{C}$ represents a complex matrix) has rank $2\kappa + 1$, noting that $(2\kappa + 1 < m)$, it has $(2\kappa + 1)$ nonzero eigenvalues, and the remaining $m − (2\kappa + 1)$ eigenvalues are zero. We can further write

$$\mathbf{R}_I\mathbf{G} = \mathbf{G} \begin{bmatrix} \lambda_{2\kappa+2} & 0 & \cdot & \cdot & 0 \\ 0 & \lambda_{2\kappa+3} & \cdot & \cdot & \cdot \\ \cdot & \cdot & \cdot & \cdot & \cdot \\ 0 & \cdot & \cdot & \cdot & \lambda_m \end{bmatrix} = \sigma^2\mathbf{G} = \mathbf{A}\mathbf{P}\mathbf{A}^c\mathbf{G} + \sigma^2\mathbf{G} \tag{7.9}$$

The last equality in Equation (7.9) means that $\mathbf{A}\mathbf{P}\mathbf{A}^c\mathbf{G} = 0$, and because $\mathbf{A}\mathbf{P}$ has a full column rank, we have $\mathbf{A}^c\mathbf{G} = 0$. This means that sinusoidals

$\{\mathbf{a}(\omega_k)\}_{k=1}^{2\kappa+1}$ are orthogonal to the noise subspace. This can be stated as $\sum_{k=m-(2\kappa+1)}^{m} \mathbf{a}^c(\omega)\mathbf{G}_k = 0$. Hence, true frequencies $\{\omega_k\}_{k=1}^{2\kappa+1}$ are the only solutions to the equation $\mathbf{a}^T(\omega)\hat{\mathbf{G}}\hat{\mathbf{G}}^c\mathbf{a}(\omega) = \|\hat{\mathbf{G}}^c\mathbf{a}(\omega)\|^2 = 0$ for $m > 2\kappa + 1$. Because, in practice, only the estimate $\hat{\mathbf{R}}_I$ of $\mathbf{R}_I$ is available, only the estimate $\hat{\mathbf{G}}$ of $\mathbf{G}$ can be determined. Thus, in root-MUSIC, we compute the frequencies (in our case, the phase steps) as angular positions of the $2\kappa + 1$ roots of the equation $\mathbf{a}^T(\omega)\hat{\mathbf{G}}\hat{\mathbf{G}}^c\mathbf{a}(\omega)$.

7.3.2 Multiple Signal Classification Method: spectral-MUSIC

A slight variant of root-MUSIC is spectral-MUSIC, in which a Fourier transform for the noise subspace is performed [15]. The term $\mathbf{a}_k^c\mathbf{G}_i$, $\forall i = (0, 1, 2, \ldots, m - 2\kappa - 2)$ and $k = (0, 1, 2, \ldots, 2\kappa)$, recalls the definition of the inner product of two complex sequences $\langle \mathbf{a},\mathbf{G}_i \rangle$, which is given by

$$\langle \mathbf{a},\mathbf{G}_i \rangle = \sum_{n=0}^{m-1} \mathbf{a}^c(n)\mathbf{G}_i(n). \tag{7.10}$$

From the previous discussion, it is evident that the signal vectors are orthogonal to the noise vectors, which yields an inner product of zero (because $\mathbf{a}_k$ and $\mathbf{G}_i$ are orthogonal). Hence, we can write

$$\langle \mathbf{a}_k,\mathbf{G}_i \rangle = \sum_{n=0}^{m-1} \mathbf{a}_k^c(n)\mathbf{G}_i(n) = 0 \; \forall i = (0,1,2,\ldots,m-2\kappa-2) \text{ and } k = (0,1,2,\ldots,2\kappa)$$

$$\tag{7.11}$$

Because $\mathbf{a}_k = \exp(j\omega_k)$ and from the definition of the inner product in Equation (7.11), we obtain

$$\begin{aligned}
\langle \mathbf{a}_k,\mathbf{G}_i \rangle &= \sum_{n=0}^{m-1} \exp(-jn\omega_k)\mathbf{G}_i(n) \\
&= \langle \mathbf{a},\mathbf{G}_i \rangle|_{\omega=\omega_k} \tag{7.12} \\
&= 0
\end{aligned}$$

Equation (7.12) is nothing but the N point discrete Fourier transform of the noise vector $\mathbf{G}_i$ evaluated at $\omega = \omega_k$. Hence, frequencies ω_k are the zeros of $\langle \mathbf{a}, \mathbf{G}_i \rangle$. This observation motivates us to form a power spectrum function $L_i[\exp(j\omega_x)]$, which is defined as

$$L_i[\exp(j\omega_x)] = \frac{1}{|\langle \mathbf{a}_x, \mathbf{G}_i \rangle|^2} \quad \omega_x = 0, 2\pi/NOF, \ldots, 2\pi(NOF-1)/NOF,$$

(7.13)

where $\mathbf{a}_x = [1, \exp(j\omega_x), \exp(j2\omega_x), \ldots, \exp(jm\omega_x)]^T$, and NOF is the number of frequencies generated. In Equation (7.13), we generate ω_x from $[0, 2\pi)$ in steps of $2\pi/NOF$, such that, at $\omega_x = \omega_k$, the inner product in Equation (7.13) is zero, which eventually manifests as a peak of infinite amplitude in the power spectrum. Because the noise eigenvectors have different eigenvalues, usually weighted averaging is performed for $m - (2\kappa + 1)$ noise eigenvectors. Hence, Equation (7.13) for the power spectrum can be written as [18]

$$L[\exp(j)] = \frac{1}{\sum_{i=0}^{m-2\kappa-2} \frac{1}{\lambda_{2\kappa+1+i}} |\langle \mathbf{a}_x, \mathbf{G}_l \rangle|^2} \quad \omega_x$$

$$= 0, 2\pi/NOF, \ldots, 2\pi(NOF-1)/NOF.$$

(7.14)

Examination of the prominent peaks in $L[\exp(j\omega)]$ gives the estimate of the frequencies present in the signal, which in turn gives the information relative to the phase step value applied to the PZT. It is important to emphasize that because the signal subspace is orthogonal to the noise subspace, we generate a set of signal vectors $\mathbf{a}_x$ for frequencies between $[0, 2\pi)$ and perform the inner product with all the noise eigenvectors $\mathbf{G}_i$ such that the resultant is zero. The vector $\mathbf{G}_i$ is obtained from the eigenvalue decomposition of matrix $\mathbf{R}_l$ in Equation (7.7). Usually, for a large data sample the noise subspace is larger than the signal subspace; hence, more accurate information can be extracted from the noise subspace.

Actually, if the number of data frames acquired is large, the phase steps applied can be easily obtained by taking the temporal discrete Fourier transform (DFT) of the acquired intensity data. However, in phase-shifting interferometry, acquiring a large number of data frames is not a regular practice because of the nonlinear response of the PZT to the applied voltage. So, a simple temporal Fourier transform of the intensity data cannot give accurate results with a limited number of data frames as the resolution in the Fourier transform is $2\pi/N$ rad. On the other hand, for the spectral-MUSIC method, given N data frames, the resolution is commonly beyond the limit of $2\pi/N$ and depends on the sample autovariance matrix, noise level, and number of frequencies *NOF* generated between the interval $[0, 2\pi)$. Hence, for the same number of data frames N, the frequency resolution is $2\pi/NOF$. Because the method performs the Fourier transform with enhanced resolution, the method is referred to as a superresolution Fourier transform (SRFT) method. The superresolution property of the method is demonstrated using the examples in Boxes 7.1 and 7.2. These illustrations

---

BOX 7.1   COMPARISON WITH DFT METHODS ($\kappa = 1$)

In Example 7.1, we considered a phase step of $\alpha = 0.5760$ rad (33°), $\kappa = 1$, and $N = 8$, and generated intensity data as shown in Equation (7.6). The discrete Fourier transform (DFT) of the intensity data is given by

$$\hat{I}(k) = \sum_{n=0}^{N-1} I_n \exp(-j2\pi nk/N); \quad k = 0, 1, 2, \ldots, N-1. \tag{7.15}$$

Figure 7.2a shows the spectrum obtained at an arbitrary (x, y) pixel using the DFT. Because the value of N is small and the separation between the successive phase step values is less than the DFT resolution limit, $2\pi/N$ and hence the phase steps could not be accurately retrieved from the plot. One possible solution by which the resolution in DFT can be increased is by padding zeros to the intensity signal in Equation (7.46). The DFT of the zero-padded signal $\hat{I}_{ZP}$ of length ZP is given by

$$\hat{I}_{ZP}(k) = \sum_{n=0}^{ZP-1} I_n \exp(-j2\pi nk/N); \quad k = 0, 1, 2, \ldots, N-1, \ldots, ZP-1. \tag{7.16}$$

Figure 7.2b shows the spectrum obtained using the zero-padded DFT method. For zero-padded DFT, the intensity signal is padded with 4,088 zeros such that the length of the zero-padded signal $ZP = 4,096$. From the plot, it can be observed that determining the phase step accurately with limited data frames is not possible.

Figure 7.2c shows the power spectrum obtained from the SRFT method using Equation (7.14). We observe that the peaks are defined clearly and with improved accuracy as compared to the plots obtained in Figures 7.2a and 7.2b. In this case, the peaks are observed as $\alpha = 0, \pm0.5768$ rad with an error of 0.1389%. This example proves that the spectral-MUSIC method can perform high-resolution phase step estimation even if the number of frames is limited.

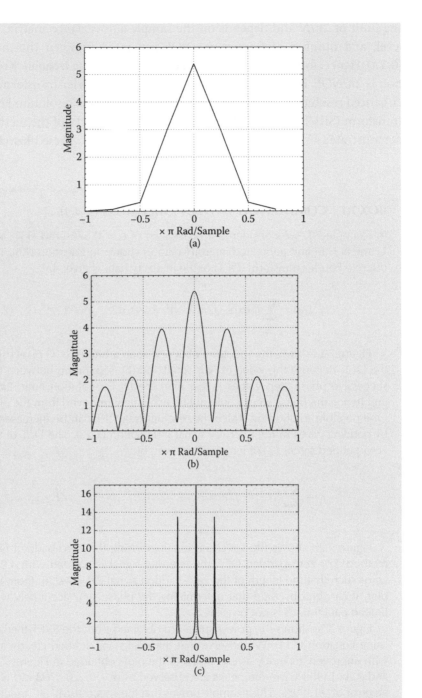

FIGURE 7.2 (a) Discrete Fourier transform. (b) Zero-padded discrete Fourier transform. (c) spectral-MUSIC method with κ = 1 harmonics. (From R. Langoju, A. Patil, and P. Rastogi. *Optics Express* **13**, 7160–7173, 2005.)

BOX 7.2   COMPARISON WITH DFT METHODS ($\kappa = 2$)

In a second example, we considered a phase step of $\alpha = 0.5760$ rad (33°), $\kappa = 2$, and $N = 12$, and generated intensity data as shown in Equation (7.6). Figures 7.3a and 7.3b show the power spectrum obtained from DFT and zero-padded discrete Fourier transform (DFT) methods for the second case ($\kappa = 2$). The plot in Figure 7.3c shows the power spectrum corresponding to frequencies 0, $\pm\alpha$, $\pm2\alpha$ obtained using Equation (7.14) for $NOF = 4096$, $N = 12$, and signal-to-noise ratio (SNR) = 60 dB. In this case the peaks are observed at 0, ±0.5768, and ±1.1508 rad, with percentage errors of 0.0%, 0.1389%, and 0.0955%, respectively.

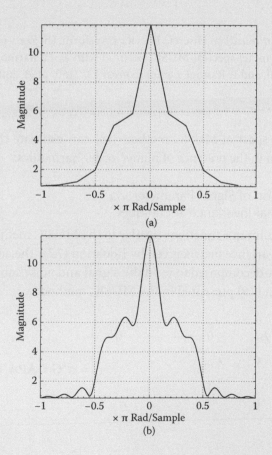

FIGURE 7.3   (a) Discrete Fourier transform. (b) Zero-padded discrete Fourier transform. (c) spectral-MUSIC method with $\kappa = 2$ harmonics. (From R. Langoju, A. Patil, and P. Rastogi. *Optics Express* **13**, 7160–7173, 2005.) *(continued)*

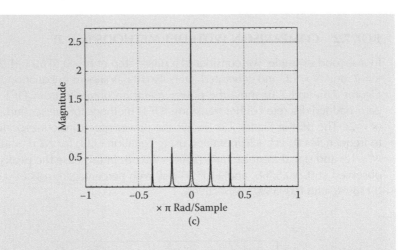

FIGURE 7.3 *(continued)* (a) Discrete Fourier transform. (b) Zero-padded discrete Fourier transform. (c) spectral-MUSIC method with $\kappa = 2$ harmonics. (From R. Langoju, A. Patil, and P. Rastogi. *Optics Express* **13**, 7160–7173, 2005.)

show that the spectral-MUSIC method [18] can estimate the phase step accurately even in the presence of higher-order harmonics.

7.3.3 Estimation of Signal Parameter via Rotational Invariance Technique

The ESPRIT technique functions similar to the MUSIC method by designing an autocovariance matrix given by Equation (7.7). The autocovariance matrix is eigen-decomposed to yield the signal and noise subspaces. Let us recall Equation (7.9) presented in the MUSIC method:

$$\mathbf{R}_I \mathbf{G} = \mathbf{G} \begin{bmatrix} \lambda_{f+1} & 0 & \cdot & \cdot & 0 \\ 0 & \lambda_{f+2} & \cdot & \cdot & \cdot \\ \cdot & \cdot & \cdot & \cdot & \cdot \\ 0 & \cdot & \cdot & \cdot & \lambda_m \end{bmatrix} = \sigma^2 \mathbf{G} = \mathbf{A}\mathbf{P}\mathbf{A}^c\mathbf{G} + \sigma^2 \mathbf{G}$$

(7.17)

The equality in Equation (7.17) means that $\mathbf{A}\mathbf{P}\mathbf{A}^c\mathbf{G} = 0$, and because $\mathbf{A}\mathbf{P}$ has full column rank, we have

$$\mathbf{A}^c\mathbf{G} = 0$$

(7.18)

From Equation (7.18), it can be seen that the sinusoidals $\{\mathbf{a}(\omega_k)\}_{k=0}^{f}$ are orthogonal to the noise subspace [15]. Let us set

$$
\Lambda =
\begin{bmatrix}
\lambda_1 - \sigma^2 & 0 & \cdot & \cdot & 0 \\
0 & \lambda_2 - \sigma^2 & \cdot & \cdot & \cdot \\
\cdot & \cdot & \cdot & \cdot & \cdot \\
0 & \cdot & \cdot & \cdot & \lambda_f - \sigma^2
\end{bmatrix}
\tag{7.19}
$$

Also, observing that $\lambda_1 \geq \lambda_2 \geq \cdots \geq \lambda_f \geq \sigma^2$ and $\lambda_{f+1} = \lambda_{f+2} = \cdots = \lambda_m = \sigma^2$, and from Equation (7.17), we obtain

$$
\mathbf{R}_I \mathbf{S} = \mathbf{S}
\begin{bmatrix}
\lambda_1 & 0 & \cdot & \cdot & 0 \\
0 & \lambda_2 & \cdot & \cdot & \cdot \\
\cdot & \cdot & \cdot & \cdot & \cdot \\
0 & \cdot & \cdot & \cdot & \lambda_f
\end{bmatrix}
= \mathbf{A}\mathbf{P}\mathbf{A}^c\mathbf{S} + \sigma^2\mathbf{S}.
\tag{7.20}
$$

We thus obtain

$$
\mathbf{S} = \mathbf{A}\underbrace{(\mathbf{P}\mathbf{A}^c\mathbf{S}\Lambda^{-1})}_{\Gamma}.
\tag{7.21}
$$

The concept of rotational invariance [15, 19] is applied to Equation (7.21). The basics of rotational invariance can be understood by the following illustrations: Let us construct matrices $\mathbf{A}_{1(m-1)\times f}$ and $\mathbf{A}_{2(m-1)\times f}$ from $\mathbf{A}$ in Equation (7.7) as

$$
\mathbf{A}_{1(m-1)\times f} = [\mathbf{I}_{(m-1)\times(m-1)} \quad \mathbf{0}_{(m-1)\times1}]\mathbf{A},
$$
$$
\mathbf{A}_{2(m-1)\times f} = [\mathbf{0}_{(m-1)\times1} \quad \mathbf{I}_{(m-1)\times(m-1)}]\mathbf{A}.
\tag{7.22}
$$

where $\mathbf{I}_{(m-1)\times(m-1)}$ is an identity matrix. The matrices $\mathbf{A}_1$ and $\mathbf{A}_2$ in Equation (7.22) are related by a unitary matrix $\mathbf{D}_{f\times f}$ in the following way:

$$
\mathbf{A}_{2(m-1)\times f} = \mathbf{A}_{1(m-1)\times f}
\underbrace{
\begin{bmatrix}
\exp(j\omega_1) & 0 & \cdot & \cdot & 0 \\
0 & \exp(j\omega_2) & \cdot & \cdot & \cdot \\
\cdot & \cdot & \cdot & \cdot & \cdot \\
0 & \cdot & \cdot & \cdot & \exp(j\omega_f)
\end{bmatrix}
}_{\mathbf{D}_{f\times f}}.
$$

$$
\tag{7.23}
$$

From Equation (7.23), we observe that the transformation in Equation (7.23) is rotation. This property plays a significant role in spectral estimation. Hence, using the same analogy as in Equation (7.23), we can derive the matrices $\mathbf{S}_1$ and $\mathbf{S}_2$ from the $\mathbf{S}$ matrix as

$$\mathbf{S}_{1(m-1)\times f} = [\mathbf{I}_{(m-1)\times(m-1)} \quad \mathbf{0}_{(m-1)\times 1}]\mathbf{S},$$

$$\mathbf{S}_{2(m-1)\times f} = [\mathbf{0}_{(m-1)\times 1} \quad \mathbf{I}_{(m-1)\times(m-1)}]\mathbf{S}. \tag{7.24}$$

We can thus represent the matrix $\mathbf{S}_2$ using Equations (7.21) and (7.24) as

$$\mathbf{S}_2 = \mathbf{A}_2\Gamma = \mathbf{A}_1\mathbf{D}\Gamma = \mathbf{S}_1\Gamma^{-1}\mathbf{D}\Gamma = \mathbf{S}_1\Upsilon. \tag{7.25}$$

where $\Upsilon = \Gamma^{-1}\mathbf{D}\Gamma$. It is important to note that because both matrices $\mathbf{S}$ and $\mathbf{A}$ in Equation (7.21) have full column rank, the matrix Γ is nonsingular. The matrices $\mathbf{A}_1$ and $\mathbf{A}_2$ in Equation (7.23) have full column rank (equal to f) because matrix $\mathbf{A}$ has a Vandermonde structure. Therefore, applying a similar analogy, we deduce from Equation (7.21) that matrices $\mathbf{S}_1$ and $\mathbf{S}_2$ have full rank. So, matrix Υ is uniquely given by

$$\Upsilon = (\mathbf{S}_1^c\mathbf{S}_1)^{-1}\mathbf{S}_1^c\mathbf{S}_2. \tag{7.26}$$

In Equation (7.26), Υ can be estimated from the available sample because Υ and $\mathbf{D}$ have the same eigenvalues. Finally, ESPRIT enables us to estimate the frequencies $(\omega_k)_{k=0}^f$ as the argument of $\hat{\omega}_k$, where $(\hat{\omega}_k)_{k=0}^f$ are the eigenvalues of the following consistent estimate of the matrix Υ:

$$\hat{\Upsilon} = (\hat{\mathbf{S}}_1^c\hat{\mathbf{S}}_1)^{-1}\hat{\mathbf{S}}_1^c\hat{\mathbf{S}}_2. \tag{7.27}$$

Because only the estimate of the covariance matrix in Equation (7.7) can be obtained, Equation (7.27) represents the estimated value of matrix Υ based on the estimated values $\hat{\mathbf{S}}_1$ and $\hat{\mathbf{S}}_2$ of $\mathbf{S}_1$ and $\mathbf{S}_2$, respectively. We infer from Equation (7.27) that ESPRIT has no problem in separating signal roots from noise roots, a problem usually encountered in other spectral estimation techniques, such as the MUSIC method [15].

The MUSIC method has been used to estimate frequencies from the noise subspace, based on the precept that sinusoidals are orthogonal to the *noise subspace*. In fact, MUSIC is more accurate than ESPRIT if a large number of data frames are present. But, such a case is unrealistic in the case of phase-shifting interferometry, for which we are constrained to work with a limited number of data frames to minimize the influence of noise and errors originating from the PZT itself. Moreover, the fact that also needs to be taken into account is that, in the case of a model mismatch, ESPRIT performs better than MUSIC [15]. In several communication applications (e.g., radar and sonar), the basic assumption is that the phase φ and the amplitudes $\{a_k\}$ are random. This in turn gives rise to a covariance matrix that, because it is identical in the least-squares sense, provides a more accurate estimation of phase steps. However, in the present application φ and $\{a_k\}$ are deterministic; hence, a true covariance matrix cannot be achieved.

7.4 EVALUATION OF PHASE DISTRIBUTION

Once the phase steps are obtained using approaches mentioned in Section 7.3, the next step involves estimating the phase distribution φ. The parameters ℓ_k can be solved using the linear Vandermonde system of equations obtained from Equation (7.6). The matrix thus obtained can be written as

$$
\begin{bmatrix}
\exp(j\kappa\alpha_0) & \exp(-j\kappa\alpha_0) & \exp[j(\kappa-1)\alpha_0] & \cdot & \dots & 1 \\
\exp(j\kappa\alpha_1) & \exp(-j\kappa\alpha_1) & \exp[j(\kappa-1)\alpha_1] & \cdot & \dots & 1 \\
\vdots & \vdots & \vdots & \vdots & \vdots & \vdots \\
\exp[j\kappa\alpha_{N-1}] & \exp[-j\kappa\alpha_{N-1}] & \exp[j(\kappa-1)\alpha_{N-1}] & \cdot & \dots & 1
\end{bmatrix}
$$

$$
\times
\begin{bmatrix}
\ell_\kappa \\
\ell_\kappa^* \\
\vdots \\
I_{dc}
\end{bmatrix}
=
\begin{bmatrix}
I_0 \\
I_1 \\
I_2 \\
\vdots \\
I_{N-1}
\end{bmatrix}
$$

(7.28)

where $\alpha_0, \alpha_1, \alpha_2, \dots, \alpha_{N-1}$ are phase steps in frames $I_0, I_1, I_2, \dots, I_{N-1}$, respectively. The phase φ is subsequently computed from the argument of ℓ_1. The next section discusses the performance of high-resolution methods with respect to other benchmarking algorithms.

TABLE 7.1   Categorization of the Conventional Benchmarking Phase-Shifting Algorithms and High-Resolution Methods According to Their Characteristics

	Calibration Error ε_1	Harmonics κ	Multiple PZTs H
	Group A	Group B	Group C
Carré [20]	✓	✗	✗
Hariharan [21]	✓	✗	✗
Schmit [22]	✓	✓	✗
Larkin [23]	✓	✓	✗
Surrel [8]	✓	✓	✗
de Groot [24]	✓	✓	✗
MUSIC [13]	✓	✓	✓
ESPRIT [14]	✓	✓	✓

7.4.1 Evaluation of Linear Phase Step Estimation Methods

Conventional single-PZT algorithms and high-resolution methods can be broadly classified into three groups, as shown in Table 7.1: A (effective in minimizing the linear calibration error in PZT), B (effective in handling the harmonics), and C (effective in accommodating multiple PZTs). Note that discussion of multiple PZT appears in Section 7.5. Reference numbers are given in the leftmost column of Table 7.1 to indicate sources for these methods.

This section presents a comparison of the performance of the conventional phase estimation algorithms in Table 7.1 with the ESPRIT method in the presence of noise and linear miscalibration errors. The algorithms proposed by Carré [20] and Hariharan [21] are not considered here as these algorithms are sensitive to higher-order harmonics and do not compensate for detector nonlinearity. We performed 1,000 Monte Carlo simulations at each signal-to-noise ratio (SNR) for SNR values from 10 to 80 dB and computed the mean squared error (MSE) in the estimation of phase φ at an arbitrary pixel point. Usually, the phase step applied to the PZT can be considered as $\alpha' = \alpha (1 + \varepsilon)$, where ε is the linear miscalibration error. Plots in Figure 7.4a show the MSE of φ for the case of zero calibration error. From the plots, it is evident that the performances of ESPRIT and the Surrel, Larkin, Schmit, and de Groot algorithms are similar. However, ESPRIT requires additional data frames to achieve the same performance as these algorithms. Unfortunately, during experiments there is always some magnitude of linear miscalibration error associated with the use of a PZT. Hence, we tested the robustness of these algorithms in the presence of linear miscalibration errors. Figures 7.4b–7.4d show the plots for $\varepsilon = 1.0\%$, $\varepsilon = 3.0\%$, and $\varepsilon = 5.0\%$, respectively. The plots

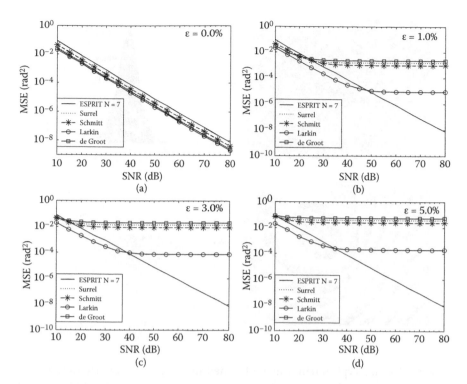

FIGURE 7.4 Comparison of ESPRIT with respect to the algorithms developed by Surrel, Schmit, Larkin, and de Groot for the computation of phase for different values of linear miscalibration errors: (a) $\varepsilon = 0.0\%$; (b) $\varepsilon = 1.0\%$; (c) $\varepsilon = 3.0\%$; and (d) $\varepsilon = 5.0\%$. (From A. Patil, R. Langoju, R. Sathish, and P. Rastogi. *Journal of the Optical Society of America A* **24**, 794–813, 2007.)

using methods proposed by Surrel, Larkin, Schmit, and de Groot show that as the miscalibration error increased, there was a certain bias in the estimation of phase even with an increase in the SNR. However, ESPRIT showed a continuous decrease in the MSE with an increase in the SNR. An important point to note here is that with ESPRIT, the MSE can be reduced further if additional data frames are used in the computation of phase. Moreover, algorithms designed by Surrel, Larkin, Schmit, and de Groot have been designed for particular phase step values, and even a slight deviation from these values has been seen to result in a considerable error in phase computation.

7.4.2 Phase Extraction Using ESPRIT: Experimental Results

This section presents an illustration in which ESPRIT was applied in holographic interferometry experiments. The holographic fringes were

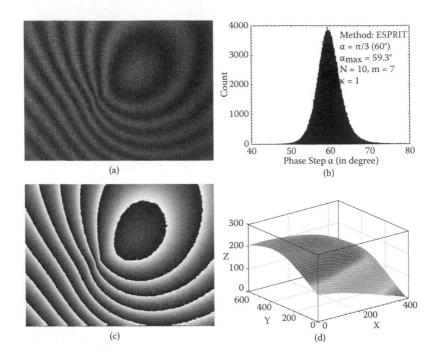

FIGURE 7.5 (a) Typical fringe map obtained in holographic interferometry; (b) histogram plot for phase step α; (c) wrapped phase obtained using ESPRIT; and (d; see also color insert) unwrapped phase map using Goldstein [25] algorithm. (From A. Patil, R. Langoju, R. Sathish, and P. Rastogi. *Journal of the Optical Society of America A* **24**, 794–813, 2007.)

captured on a 576 × 768 pixel CCD camera. Ten data frames were acquired with a phase step of π/3. Figure 7.5a shows a typical fringe map obtained after deforming the object. It is well known [26] that the retrieved phase steps at each pixel are not constant because of noise and other errors; hence, studying the distribution of the retrieved phase steps can be one way to compute the value of the phase step. We thus plotted a histogram to study the distribution of phase steps retrieved using the ESPRIT method. The histogram of the phase step obtained is shown in Figure 7.5b. The maximum value of the histogram (59.3°) was considered while computing the phase distribution. The wrapped phase was obtained by designing a Vandermonde system of equations mentioned in Equation (7.28). Figures 7.5c and 7.5d show the wrapped and unwrapped phase maps. Note that median phase filtering was applied to the wrapped phase map.

7.5 DUAL PZT IN HOLOGRAPHIC MOIRÉ

To recap, holographic moiré is essentially an incoherent superposition of two interferometric systems that result in the formation of a moiré pattern. Figure 7.1b illustrates a typical optical setup for holographic moiré [11]. In the optical setup, moiré carries information regarding the in-plane displacement component along the x-direction, and the carrier carries the information regarding the out-of-plane displacement component. However, in a conventional holographic moiré setup, the information carried by the carrier is lost if conventional phase-shifting algorithms are used. Thus, there are two drawbacks to conventional holographic moiré: First, simple phase shifting cannot be applied to holographic moiré, not even for measuring in real time the displacement information carried by moiré; second, the method is unable to obtain even optically simultaneous measurement of in-plane and out-of-plane displacement distributions. Incorporation of two phase-shifting devices in the holographic moiré configurations enables both drawbacks to be overcome. Two distinct phase steps are applied simultaneously to the two arms of the moiré setup. The method consists of acquiring data while voltages are simultaneously applied to the two PZTs.

The intensity for the N data frames acquired on the CCD camera at a pixel (x, y) for the nth phase step is given by

$$I(x,y;n) = I_{dc} + \sum_{k=1}^{\kappa} a_k^* \exp[-jk(\varphi_1 + n\alpha)] + \sum_{k=1}^{\kappa} a_k \exp[jk(\varphi_1 + n\alpha)]$$

$$+ \sum_{k=1}^{\kappa} b_k^* \exp[-jk(\varphi_2 + n\beta)] + \sum_{k=1}^{\kappa} b_k \exp[jk(\varphi_2 + n\beta)];$$

for $n = 0, 1, 2, \ldots, N-1$.

(7.29)

where a_k, a_k^*, b_k, and b_k^* are the complex Fourier coefficients of the kth-order harmonic, $j = \sqrt{-1}$; I_{dc} is the local average value for intensity; and pairs φ_1 and φ_2 and α and β represent phase differences and phase shifts, respectively, in the two arms of the holographic moiré. The coefficients a_k and b_k are in fact real and are related to the appropriate choice of phase origin at a point where the intensity reaches a maximum.

Let us rewrite Equation (7.29) in the following form:

$$I_n(x,y) = I_{dc} + \sum_{k=1}^{K} \ell_k u_k^n + \sum_{k=1}^{K} \ell_k^*(u_k^*)^n + \sum_{k=1}^{K} \wp_k v_k^n + \sum_{k=1}^{K} \wp_k^*(v_k^*)^n ;$$
$$\text{for } n = 0,1,2,\ldots,N-1.$$

(7.30)

where $\ell_k = a_k \exp(jk\varphi_1)$, $u_k = \exp(jk\alpha)$, $\wp_k = b_k\exp(jk\varphi_2)$, and $v_k = \exp(jk\alpha)$; $I_n(x,y)$ corresponds to the nth frame, where $n = 0$ refers to the data frame corresponding to the first phase-shifted holographic moiré intensity pattern. Equation (7.30) represents the complex valued sinusoidal signals where I_{dc}, α, ... , $\kappa\alpha$; $-\alpha$, ... , $-\kappa\alpha$; $-\beta$, ... , $-\kappa\beta$ represent the frequencies embedded in the signal, and by estimating them, the phase steps α and β imparted to the PZTs can be determined.

Fixed phase step algorithms in Table 7.1 (i.e., those proposed by Carré, Hariharan, Schmit, Larkin, Surrel, and de Groot) do not have the ability to accommodate dual PZTs. High-resolution methods, discussed in the previous section, seem to be well suited for applications involving dual PZTs in an optical configuration. Note that, using the mathematical framework presented in Section 7.3, one can estimate multiple phase steps. However, careful selection of phase steps is necessary for optimal performance. Patil et al. [27] provide information for optimizing the phase steps. It discusses the Cramér-Rao bound (CRB) for estimating phase steps in holographic moiré. The CRB provides valuable information on the potential performance of the estimators. The CRBs are independent of the estimation procedure, and the precision of the estimators cannot surpass the CRBs. In the cited reference, the CRB for the phase steps α and β as a function of SNR and N was derived. Although the phase steps α and β can take arbitrary values for pure-intensity signal, the article addresses a central question: What is the smallest difference between the phase steps α and β that can be retrieved reliably by an estimator as a function of SNR and N?

7.6 EVALUATION OF PHASE DISTRIBUTION IN HOLOGRAPHIC MOIRÉ

Once the phase steps are estimated using the approaches mentioned in Section 7.3, the next step involves measuring the phase distributions. The parameters ℓ_k and $\wp_k$ can be solved using the linear Vandermonde system of equations obtained from Equation (7.30). The matrix thus obtained can be written as

$$
\begin{bmatrix}
\exp(j\kappa\alpha_0) & \exp(-j\kappa\alpha_0) & \exp(j\kappa\beta_0) & \cdot & \exp[j(\kappa-1)\alpha_0] & \cdots & 1 \\
\exp(j\kappa\alpha_1) & \exp(-j\kappa\alpha_1) & \exp(j\kappa\beta_1) & \cdot & \exp[j(\kappa-1)\alpha_1] & \cdots & 1 \\
\vdots & \vdots & \vdots & \vdots & \vdots & \vdots & \vdots \\
\exp[j\kappa\alpha_{N-1}] & \exp[-j\kappa\alpha_{N-1}] & \exp[j\kappa\beta_{N-1}] & \cdot & \exp[j(\kappa-1)\alpha_{N-1}] & \cdots & 1
\end{bmatrix}
$$

$$
\times
\begin{bmatrix}
\ell_\kappa \\
\ell_\kappa^* \\
\wp_\kappa \\
\vdots \\
I_{dc}
\end{bmatrix}
=
\begin{bmatrix}
I_0 \\
I_1 \\
I_2 \\
\vdots \\
I_{N-1}
\end{bmatrix}
$$

(7.31)

where (α_0, β_0), (α_1, β_1), $(\alpha_2, \beta_2), \ldots, (\alpha_N, \beta_N)$ are phase steps in frames I_0, I_1, $I_2, \ldots, I_{N-1}$, respectively. The phases φ_1 and φ_2 are subsequently computed from the argument of ℓ_1 and $\wp_1$.

7.6.1 Holographic Moiré Experiments

A schematic of the holographic moiré experiments is given in Figure 7.1b and a brief description of the experiments is as follows: The light from the laser source (He-Ne operating at 532-nm wavelength) was split into two beams: the object beam OB and the reference beam RB. The reference beam fell directly on the holographic plate. The object beam was split into two beams, OB1 and OB2. The piezoelectric devices were placed in the path of the object beams. The OB1 and RB thus formed one holographic interferometry system, and OB2 and RB formed the other interferometry system. Holographic moiré is an incoherent superposition of two holographic interferometry systems.

The intensities of the object beams OB1 and OB2 were adjusted in such a way that equal scattered intensities from the object fell on the holographic plate. In the experiment, the object beam intensities falling on the holographic plates were 1.4 μw (OB1 = OB2 = 0.7 μw). The reference beam intensity was 9 μw. The exposure time for the holographic plate was selected as 3.5 s. The holographic plate, after exposure for 3.5 s, was developed using a wet chemical process and placed accurately in its original position. We now imparted two displacements—out of plane and in plane—to the object under study. A series of moiré fringe patterns was acquired while the phase steps were applied simultaneously to the two PZTs placed in the two arms of the optical system.

Figure 7.6a shows a typical moiré fringe pattern obtained experimentally. The fringe patterns corresponding to the two interferometer systems are shown in Figures 7.6b and 7.6c. In this experiment, the ESPRIT method was used to estimate the phase steps α and β. The phase steps

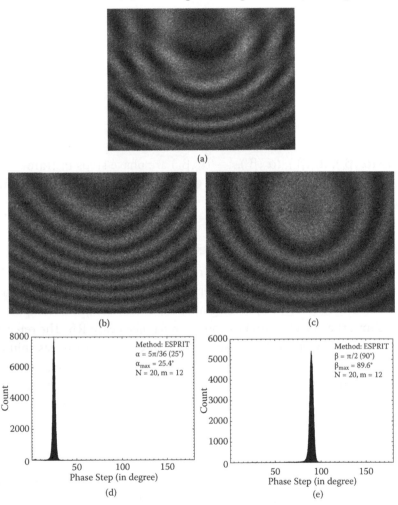

FIGURE 7.6   (a) A typical moiré fringe pattern in an experimental setup. (b) and (c) Corresponding fringe maps in the two arms of the interferometer obtained after the out-of-plane and in-plane displacements have been given to the object. (d) and (e) Corresponding phase steps estimated using ESPRIT for (b) and (c), respectively. (From A. Patil, R. Langoju, R. Sathish, and P. Rastogi. *Journal of the Optical Society of America A* **24**, 794–813, 2007.)

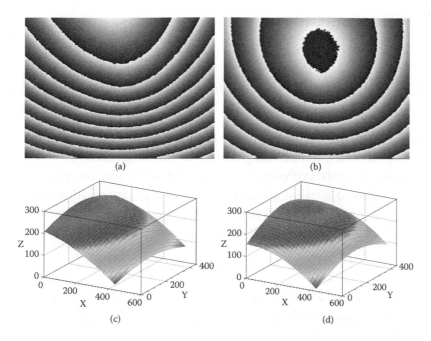

FIGURE 7.7 Wrapped phase distributions corresponding to phase (a) φ_1 and (b) φ_2. Unwrapped phase maps corresponding to wrapped phase (c; see also color insert) φ_1 and (d; see also color insert) φ_2. (From A. Patil, R. Langoju, R. Sathish, and P. Rastogi. *Journal of the Optical Society of America A* **24**, 794–813, 2007.)

obtained in Figures 7.6d and 7.6e (i.e., $\alpha = 25.4°$ and $\beta = 89.6°$) were used in the Vandermonde system of equations to compute the phase distributions φ_1 and φ_2. The wrapped phase maps corresponding to Figures 7.6b and 7.6c are shown in Figures 7.7a and 7.7b, respectively; the corresponding unwrapped phase maps are shown in Figures 7.7c and 7.7d, respectively.

7.7 NONLINEAR PHASE STEP ESTIMATION METHOD

The methods discussed in the previous sections assume that the movement of the PZT is linear. These methods are thus not effective in handling a nonlinear response of the PZT to the applied voltage. This section introduces a method that minimizes error in the measurement of phase because of the nonlinear response of the PZT to the applied voltage. Although a few traditional algorithms [24, 28, 29] have been proposed to handle nonlinearity in the PZT movement, these algorithms impose conditions such as the use of only specific phase step values to minimize the influence of particular order harmonics and nonlinearity of the phase shift.

The formulation for the intensity data recorded at a pixel (x, y) for $n = 0, 1, 2, \ldots, N - 1$ data frames in the presence of a nonlinear phase step is given by

$$\bar{I}_n = I_{dc}\left[1 + \sum_{k=1}^{\kappa} \gamma_k \cos(k\varphi + k\alpha_{n'})\right] + \eta_n, \tag{7.32}$$

where, for the nth data frame, I_{dc} is the local background intensity, κ is the number of harmonics, and γ_k represents the fringe visibility of the kth harmonic; φ and α'_n represent the phase difference and the phase shift, respectively; and η_n is the additive white Gaussian noise with mean zero and variance σ^2. For N data frames, the true phase shift α'_n caused by non-linear characteristics of the PZT can be represented as

$$\alpha'_n = n\alpha(1 + \epsilon_1) + \epsilon_2 (n\alpha)^2/\pi, \tag{7.33}$$

where α'_n is the polynomial representation of the phase step α, and ϵ_1, ϵ_2 are the linear and nonlinear error coefficients in the PZT response, respectively. Equation (7.33) can be further developed as

$$\alpha'_n = n\alpha_\epsilon + n^2\varpi, \tag{7.34}$$

where $\alpha_\epsilon = \alpha(1 + \epsilon_1)$ and $\varpi = \epsilon_2 \alpha^2/\pi$. The spectral-MUSIC method explained previously in Section 7.3.2 is primarily effective for linear phase steps between the acquired data frames. If this method is applied to the intensity data given in Equation (7.32), the phenomenon of spectral broadening and shifting of the peaks in the pseudospectrum can be observed. In controlled experimental conditions (i.e., for a high-SNR case), such broadening of peaks can be attributed to the nonlinearity of the phase steps. The example in Box 7.3 shows the broadening effect.

BOX 7.3   ILLUSTRATION OF BROADENING OF PEAKS

Figure 7.8 shows the pseudospectrum of the intensity data in the presence of linear and nonlinear phase steps between the data frames. The solid line in Figure 7.8 shows the pseudospectrum for a high signal-to-noise ratio (SNR) with nonlinear phase steps $\alpha = 0.2\pi$ rad (36°), $\epsilon_1 = 0.0$, and different values of ϵ_2. The dotted line shows the pseudospectrum with linear phase steps of $\alpha = 0.2\pi$ rad (36°) and $\epsilon_1 = \epsilon_2 = 0.0$. It is evident from the plot that nonlinearity in the phase shifts gave rise to a shift in the peaks of the spectrum.

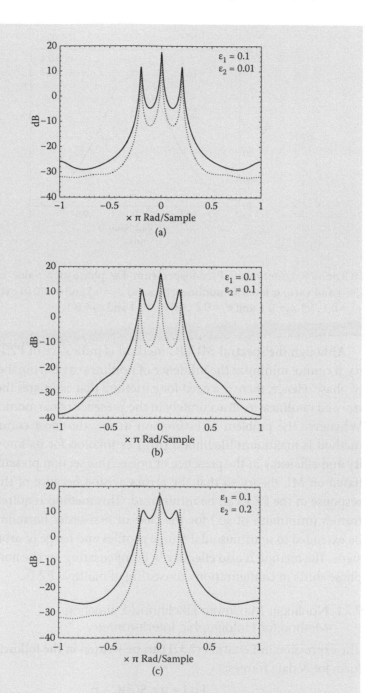

FIGURE 7.8 Pseudospectrum for phase step value $\alpha = 0.2\pi$ rad (36°) and various levels of nonlinearities: (a) $\epsilon_1 = 0.1$ and $\epsilon_2 = 0.01$; (b) $\epsilon_1 = 0.1$ and $\epsilon_2 = 0.1$; (c) $\epsilon_1 = 0.1$ and $\epsilon_2 = 0.2$; (d) $\epsilon_1 = 0.1$ and $\epsilon_2 = 0.4$. *(continued)*

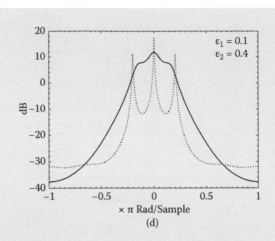

FIGURE 7.8 *(continued)* Pseudospectrum for phase step value $\alpha = 0.2\pi$ rad (36°) and various levels of nonlinearities: (a) $\epsilon_1 = 0.1$ and $\epsilon_2 = 0.01$; (b) $\epsilon_1 = 0.1$ and $\epsilon_2 = 0.1$; (c) $\epsilon_1 = 0.1$ and $\epsilon_2 = 0.2$; (d) $\epsilon_1 = 0.1$ and $\epsilon_2 = 0.4$.

Although the spectral-MUSIC method is indicative of PZT nonlinearity, it cannot minimize the influence of nonlinearity during the estimation of phase. Hence, there is a need for a method that estimates the phase step α_ϵ and nonlinearity ϖ accurately in the presence of harmonics and noise. Whenever the problem of estimation arises, the most commonly used method is maximum likelihood (ML) estimation for its known simplicity and efficiency in the presence of noise. This section presents a method based on ML theory so that the errors arising because of the nonlinear response of the PZT can be minimized. This method requires a few data frames (minimum of six) for the case of first-order harmonics and can be extended to nonsinusoidal fringe profiles and usage of arbitrary phase steps. The method is also effective in compensating for the nonlinearity of phase shifts in configurations consisting of multiple PZTs.

7.7.1 Nonlinear Maximum Likelihood Estimation Method for Holographic Interferometry

The expression in Equation (7.32) can be written in the following compact form for N data frames:

$$\bar{\mathbf{I}} = \mathbf{I} + \eta = \mathbf{S}(\xi)\mathbf{C} + \eta, \tag{7.35}$$

where $\mathbf{I} = [I_0 I_1 \dots I_{N-1}]^T$, $\eta = [\eta_0 \eta_1 \dots \eta_{N-1}]^T$, T is the transpose operator, and the matrix $\mathbf{S}(\xi)$ is expressed as $\mathbf{S}(\xi) = [1\, \mathbf{s}_1\, \mathbf{s}_1^* \dots \mathbf{s}_\kappa\, \mathbf{s}_\kappa^*]$, where

$\xi = \{\alpha_\epsilon, \omega\}$ is the parameter set, κ is the number of harmonics present in the signal, and each signal vector s_k is given by $s_k(n) = \exp(jk\alpha'_n)$, with $j = \sqrt{-1}$. The matrix C in Equation (7.35) is the coefficient matrix given by $C = [c_0 \, c_1 \, c_1^* \ldots c_\kappa \, c_\kappa^*]^T$ with $c_0 = I_{dc}$ and any element $c_k = I_{dc}\gamma_k \exp(jk\varphi)$. Because the noise is assumed to be additive white Gaussian with variance σ^2, the probability density function of the data vector $\bar{I}$, parameterized by ξ, which is given by $p(\bar{I};\xi)$, is

$$p(\bar{I};\xi) = \frac{1}{\pi^N \sigma^N} \exp\left\{-\frac{1}{2\sigma^2}[\bar{I} - S(\xi)C]^H[\bar{I} - S(\xi)C]\right\}, \qquad (7.36)$$

where H is the Hermitian transpose. The log-likelihood function of the data $L(\bar{I},\xi)$ is proportional to $p(\bar{I};\xi)$, which is given by

$$L(\bar{I},\xi) = \text{Constant} - \frac{1}{2\sigma^2}[\bar{I} - S(\xi)C]^H[\bar{I} - S(\xi)C] \qquad (7.37)$$

The ML estimate of ξ is obtained by maximizing the expression in Equation (7.37), which is equivalent to minimizing the expression $D(\xi) = [\bar{I} - S(\xi)C]^H[\bar{I} - S(\xi)C]$. Note that this expression is in the form of standard least squares if $S(\xi)$ is assumed to be a known matrix. Therefore, $D(\xi)$ is minimized over C by taking the complex gradient of $D(\xi)$ with respect to C and equating it to zero. We obtain

$$\frac{\partial D}{\partial C} = \frac{\partial D}{\partial C}\left[\bar{I}^H\bar{I} - \bar{I}^H SC - C^H S^H \bar{I} + C^H S^H SC\right]$$
$$= 2S^H\left[\bar{I} - SC\right] \qquad (7.38)$$

Equating Equation (7.38) with zero yields

$$C = \{[S(\xi)]^H S(\xi)\}^{-1} S(\xi)^H \bar{I} \qquad (7.39)$$

Substituting Equation (7.39) into the expression for $D(\xi)$, we obtain

$$\begin{aligned} D(\xi) &= \{\bar{I} - S[S^H S]^{-1} S^H \bar{I}\}^H \{\bar{I} - S[S^H S]^{-1} S^H \bar{I}\} \\ &= \{\bar{I}^H - I^H S[S^H S]^{-1} S^H\}\{\bar{I} - S[S^H S]^{-1} S^H \bar{I}\} \\ &= \bar{I}^H \bar{I} - \bar{I}^H S[S^H S]^{-1} S^H \bar{I}. \end{aligned} \qquad (7.40)$$

The estimation of nonlinear parameters in ξ is obtained by minimizing the expression for $\mathbf{D}(\xi)$ in Equation (7.40), which in turn leads to maximizing the following expression [30]:

$$\xi = \max \bar{\mathbf{I}}^{H} \mathbf{S}(\xi)\{[\mathbf{S}(\xi)]^{H}\mathbf{S}(\xi)\}^{-1}\mathbf{S}(\xi)^{H}\bar{\mathbf{I}}. \tag{7.41}$$

Because the vector ξ that appears in the matrix $\mathbf{S}(\xi)$ is nonlinearly related to $\bar{\mathbf{I}}$, the method is termed the nonlinear maximum likelihood (nonlinML) method. It can be observed from Equation (7.41) that a multidimensional search over the parameter space is required to obtain optimum ξ. Any global optimization algorithm, for instance the probabilistic global search Lausanne (PGSL), can be applied to obtain ξ. The advantage of PGSL is that this algorithm is a direct search algorithm, and it does not need any initial guess for the parameters. All we need are the bounds for each of the parameters. Because in several instances a priori information about the applied phase step is available, it is possible to provide good bounds for the parameters. Raphael and Smith [31] provide more details on PGSL.

Once the phase step response $\{\alpha_{\epsilon}, \varpi\}$ is estimated, the next step is to estimate the phase. This can be done by either a least-squares fit [32] or by solving a Vandermonde system of equations as follows:

$$\begin{bmatrix} 1 & 1 & \cdot & \cdot & 1 \\ \exp(j\kappa\alpha_1') & \exp(-j\kappa\alpha_1') & \cdot & \cdot & 1 \\ \exp(j\kappa\alpha_2') & \exp(-j\kappa\alpha_2') & \cdot & \cdot & 1 \\ \cdot & & \cdot & \cdot & \cdot \\ \exp(j\kappa\alpha_{N-1}') & & \cdot & \cdot & 1 \end{bmatrix} \begin{bmatrix} c_\kappa \\ c_\kappa^* \\ \cdot \\ \cdot \\ c_0 \end{bmatrix} = \begin{bmatrix} \bar{I}_0 \\ \bar{I}_1 \\ \cdot \\ \cdot \\ \bar{I}_{N-1} \end{bmatrix}$$

$$\tag{7.42}$$

Finally, the phase difference φ can be found from the argument of c_1 because $c_1 = \exp(j\varphi)$.

7.7.2 Evaluation of Nonlinear Phase Step Estimation Method

Finally, we simulated a fringe pattern for 512×512 pixels with a phase step of $\alpha_\epsilon = 60°$ and nonlinearity $\epsilon_2 = 0.2$ ($\varpi = 4°$). Additive noise with SNR = 25 dB was added to the signal. Because the process of finding the global minimum is computationally expensive, we randomly selected some 1,600 points on the image and estimated the phase step α_ϵ and ϖ at each of those

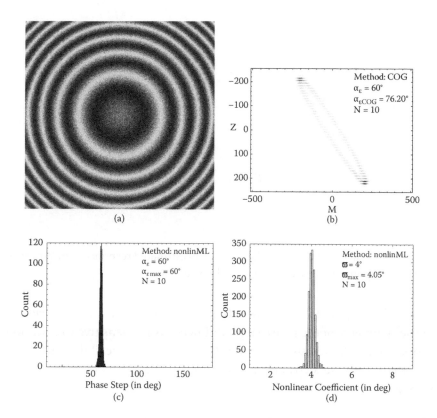

(a)

(b)

(c)

(d)

FIGURE 7.9 (a) Simulated fringe pattern. (b) Lattice site representation of the retrieved phase steps. (c) Histogram for retrieved phase steps α_ϵ. (d) Histogram for retrieved nonlinear coefficient ϖ.

pixels and plotted the histogram with the retrieved values. This method requires at least $2\kappa + 4$ data frames for minimizing the kth-order harmonic.

Figure 7.9a shows the simulated fringe pattern with additive random phase. According to Gutmann and Weber's lattice site representation of the retrieved phase steps, when nonlinear phase steps are applied, the lattice site representation of the phase steps will be in the form of an ellipse. The same is shown in Figure 7.9b, which proves the nonlinearity of the applied phase steps. Figures 7.9c and 7.9d show the histograms of the retrieved phase step α_ϵ and nonlinear coefficient ϖ, respectively. The value with maximum occurrences can be taken as the best phase step (α_ϵ) and nonlinearity (ϖ). In the present case, the maximum values are found at $\alpha_{\epsilon max} = 60°$ and $\varpi_{max} = 4.05°$. Figure 7.10a shows the actual and estimated responses, which show a good match, and Figure 7.10b shows the wrapped phase corresponding to the fringe shown in Figure 7.9a.

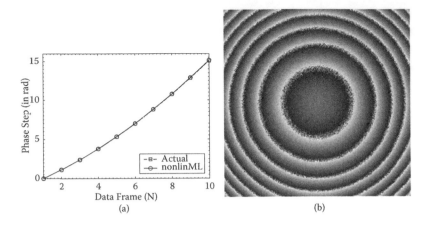

FIGURE 7.10   (a) Comparison of the actual and estimated responses. (b) Wrapped phase corresponding to fringe shown in Figure 7.9a.

The advantage of this method is that it does not impose any restriction on the symmetry of the phase steps and hence is not sensitive to common errors, such as hysteresis. Moreover, the method has flexibility in applying arbitrary phase steps and in minimizing the effects of higher-order harmonics.

7.7.3 Nonlinear Maximum Likelihood Estimation Method for Holographic Moiré

In this section, we extend the formulation in Equation (7.32) to a generalized log-likelihood function so that multiple phases in the presence of a nonlinear response of the PZTs can be estimated [33]. The recorded fringe intensity at any point (x, y) in the holographic moiré at the nth data frame is given by

$$\overline{I}_n = I_{dc1}\left[1 + \sum_{k=1}^{\kappa} \gamma_{1k} \cos(k\varphi_1 + k\alpha_n')\right] + I_{dc2}\left[1 + \sum_{k=1}^{\kappa} \gamma_{2k} \cos(k\varphi_2 + k\beta_n')\right] + \eta_n,$$

(7.43)

where I_{dc1} and I_{dc2} represent local background intensities, and γ_{1k} and γ_{2k} represent the fringe visibilities at the kth harmonic for each interferometric system. In Equation (7.43), the pair φ_1, φ_2 represents the phase differences and the pair α_n', β_n' represents phase shifts in the two systems. The term η_n represents additive random noise during the measurement with mean zero and variance σ^2. For N data frames, the true phase shifts $\{\alpha_n', \beta_n'\}$ applied to the PZTs because of nonlinear characteristics can be represented by the following:

$$\alpha_n' = n\alpha(1+\in_{11})+n^2 \in_{12} \alpha^2/\pi$$

$$\beta_n' = n\beta(1+\in_{12})+n^2 \in_{22} \beta^2/\pi,$$

(7.44)

where α_n', β_n' are the polynomial representations of the phase steps α and β with linear and nonlinear error coefficients $\{\in_{11}, \in_{21}\}$ and $\{\in_{12}, \in_{22}\}$, respectively. The expressions for α_n' and β_n' can be further modified as

$$\alpha_n' = n\alpha_\in + n^2\varpi^1$$

$$\beta_n' = n\beta_\in + n^2\varpi^2,$$

(7.45)

where $\alpha_\in = \alpha(1+\in_{11})$, $\beta_\in = \beta(1+\in_{21})$ and $\varpi^1 = \in_{12} \alpha^2/\pi$ and $\varpi^2 = \in_{22} \beta^2/\pi$. We apply the same analogy here as we did for the single-PZT case. For this, let us rewrite Equation (7.43) in the same format as Equation (7.35):

$$\bar{\mathbf{I}} = \mathbf{I} + \eta = \mathbf{S}(\xi)\mathbf{C} + \eta,$$

(7.46)

where $\mathbf{I} = [I_0 I_1 \ldots I_{N-1}]^T$, $\eta = [\eta_0 \eta_1 \ldots \eta_{N-1}]^T$, T is the transpose operator, and matrix $\mathbf{S}(\xi)$ is expressed as $\mathbf{S}(\xi) = [1 \, \mathbf{s}_\alpha \, \mathbf{s}_\alpha^* \, \mathbf{s}_\beta \, \mathbf{s}_\beta^* \ldots \mathbf{s}_{\kappa\alpha} \, \mathbf{s}_{\kappa\alpha}^* \, \mathbf{s}_{\kappa\beta} \, \mathbf{s}_{\kappa\beta}^*]$; where $\xi = \{\alpha_\in, \beta_\in, \varpi^1, \varpi^2\}$ is the parameter set; and signal vectors $\mathbf{s}_{k\alpha}$ and $\mathbf{s}_{k\beta}$ are given by $\mathbf{s}_{k\alpha}(n) = \exp(jk\alpha_n'), \mathbf{s}_{k\beta}(n) = \exp(jk\beta_n')$, with $j = \sqrt{-1}$. The matrix $\mathbf{C}$ in Equation (7.46) is the coefficient matrix given by $\mathbf{C} = [c_0 \, c_{11} \, c_{11}^* \, c_{21} \, c_{22}^* \ldots c_{1\kappa} \, c_{1\kappa}^* \, c_{2\kappa} \, c_{2\kappa}^*]^T$ with $c_0 = I_{dc1} + I_{dc2}$, and elements c_{1k}, c_{2k} are given by $c_{1k} = I_{dc1}\gamma_{1k} \exp(jk\varphi_1)$, $c_{2k} = I_{dc2}\gamma_{2k} \exp(jk\varphi_2)$, respectively. As discussed in the single-PZT case, the vector ξ that appears in the matrix $\mathbf{S}(\xi)$ is nonlinearly related to $\bar{\mathbf{I}}$. The estimation of nonlinear parameters in ξ is obtained by maximizing the following expression:

$$\xi = \max \bar{\mathbf{I}}^H \mathbf{S}(\xi)\{[\mathbf{S}(\xi)]^H \mathbf{S}(\xi)\}^{-1} \mathbf{S}(\xi)^H \bar{\mathbf{I}}.$$

(7.47)

One can use any global search method for the multigrid search. In the present analysis, the PGSL method was used for estimating phase steps and nonlinear coefficients. Once the phase step response is estimated, phase differences φ_1 and φ_2 are estimated by solving the Vandermonde system of equations for $\mathbf{C}$ as shown in Equation (7.42). Phases φ_1 and φ_2 can be obtained from the argument of c_{11} and c_{21} respectively. The next section discusses in detail the simulation results.

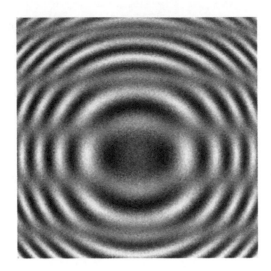

FIGURE 7.11   Typical moiré image.

7.7.4 Evaluation of Nonlinear Phase Step Estimation Method for Holographic Moiré

We simulated a moiré fringe pattern for 512×512 pixels with a phase step of $\alpha_\varepsilon = 30°$, $\beta_\varepsilon = 60°$, and nonlinearity $\epsilon_{12} = 0.15(\varpi^1 = 0.75°)$, $\epsilon_{22} = 0.2(\varpi^2 = 4°)$. Additive noise with SNR = 30 dB was added to the signal. We randomly sampled some 1,600 points on the moiré image and applied nonlinML to estimate the phase steps $\{\alpha_\varepsilon, \beta_\varepsilon\}$ and nonlinear coefficients $\{\varpi^1, \varpi^2\}$ at each of those pixels and plotted the histogram with the retrieved values.

Figure 7.11 shows a typical simulated moiré fringe pattern. Figures 7.12a and 7.12b show the histograms of the retrieved phase steps α_ε and β_ε, and Figures 7.12c and 7.12d show the histograms of the retrieved nonlinear coefficients $\{\varpi^1, \varpi^2\}$. The maximums in the histograms are found at $\alpha_{\varepsilon\,max} = 29.6°$, $\beta_{\varepsilon\,max} = 60.1°$, $\varpi^1_{max} = 0.758°$, and $\varpi^2_{max} = 4.04°$. These simulations showed that the values were accurately estimated. Figure 7.13a shows the actual and estimated phase step responses of α'_n and β'_n. From the plot, it is evident that the phase steps were accurately estimated. Once α'_n and β'_n were estimated, the subsequent step was to estimate the phases φ_1 and φ_2. This was done by first computing the matrix $\mathbf{C}$, then the phases φ_1 and φ_2 were estimated from the arguments of c_{11} and c_{12}, respectively, because $c_{11} = I_{dc1}\gamma_{11} \exp(j\varphi_1)$, $c_{21} = I_{dc2}\gamma_{21} \exp(j\varphi_2)$. Figures 7.13b and 7.13c show the retrieved phase patterns φ_1 and φ_2, respectively. Figure 7.13d

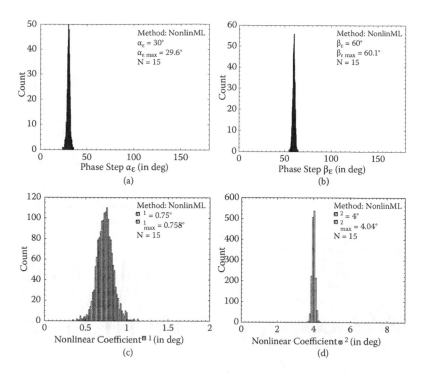

FIGURE 7.12 Histograms of retrieved parameters (a) α_ϵ, (b) β_ϵ, (c) ϖ^1, and (d) ϖ^2 with $N = 15$ and SNR = 25 dB.

shows a plot of typical errors that occurred during the estimation of the interference phase φ_1 along a row at SNR = 30 dB. The error in computation of φ_2 was of the same order as that of φ_1.

7.8 SUMMARY OF SIGNAL-PROCESSING METHODS

This chapter discussed a set of flexible and efficient high-resolution methods that are capable of extracting multiple phase information simultaneously in an optical configuration. Moreover, these methods overcome the limitations exhibited by the conventional linear phase-shifting algorithms, such as optimal performance only with a fixed phase step, inability to handle higher-order harmonics, detector non-linearity, and noise. Simulation and experimental results (see Figures 7.2, 7.3, and 7.8 and related discussion in Boxes 7.1–7.3) proved the efficacy of these signal-processing methods for cases involving single and dual PZTs in an optical setup.

This chapter also addressed the measurement of phase in the presence of a nonlinear response of the piezoelectric device to the applied voltage.

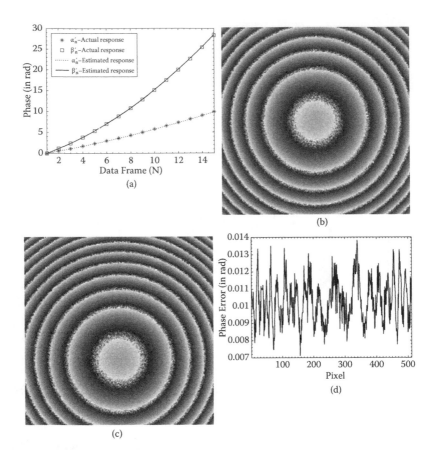

FIGURE 7.13 (a) Comparison of actual and estimated values of α'_n β'_n. (b) Wrapped phase φ_1. (c) Wrapped phase φ_2. (d) Error in estimation of phase φ_1 along a row with SNR = 30 dB.

Because the methods mentioned in Table 7.1 are sensitive to the nonlinear response of the PZT, a nonlinML method appeared to be a suitable candidate for the estimation of the phase information. In addition, under the hypothesis that the response of the PZT is monotonically nonlinear, the measurement of phase can be performed without the need of any calibration of the device. Although in this chapter we did not discuss statistical analysis of these methods, ample literature [27, 34] could be found to understand the efficacy of the signal-processing methods in phase-shifting interferometry.

REFERENCES

1. M. Takeda, H. Ina, and S. Kobayashi. Fourier-transform method of fringe-pattern analysis for computer-based topography and interferometry. *Journal of the Optical Society of America* **72**, 156–160, 1982.

2. J. H. Massig and J. Heppner. Fringe-pattern analysis with high accuracy by use of the Fourier-transform method: Theory and experimental tests. *Applied Optics* **40**, 2081–2088, 2001.

3. L. R. Watkins, S. M. Tan, and T. H. Barnes. Determination of interferometer phase distributions by use of wavelets. *Optics Letters* **24**, 905–907, 1999.

4. K. Qian, H. S. Seah, and A. Asundi. Instantaneous frequency and its application in strain extraction in moirë interferometry. *Applied Optics* **42**, 6504–6513, 2003.

5. K. Qian, Windowed Fourier transform for fringe pattern analysis. *Applied Optics* **43**, 2695–2702, 2004.

6. K. Qian. Windowed Fourier transform for demodulation of carrier fringe patterns. *Optical Engineering* **43**, 1472–1473, 2004.

7. A. Patil and P. Rastogi. Moving ahead with phase: Guest editorial. *Optics and Lasers in Engineering* **45**, 253–257, 2007.

8. Y. Surrel. Design of algorithms for phase measurements by the use of phase stepping. *Applied Optics* **35**, 51–60, 1996.

9. K. Hibino, B. F. Oreb, D. I. Farrant, and K. G. Larkin. Phase shifting for nonsinusoidal waveforms with phase-shift errors. *Journal of the Optical Society of America A* **12**, 761–768, 1995.

10. P. K. Rastogi. Phase shifting applied to four-wave holographic interferometers. *Applied Optics* **31**, 1680–1681, 1992.

11. P. K. Rastogi. Phase-shifting holographic moiré: phase-shifter error-insensitive algorithms for the extraction of the difference and sum of phases in holographic moirë. *Applied Optics* **32**, 3669–3675, 1993.

12. P. K. Rastogi, M. Spajer, and J. Monneret. In-plane deformation measurement using holographic moiré. *Optics and Lasers in Engineering* **2**, 79–103, 1981.

13. A. Patil and P. Rastogi. Phase determination in holographic moiré in presence of nonsinusoidal waveforms and random noise. *Optics Communications* **257**, 120–132, 2006.

14. A. Patil and P. Rastogi. Rotational invariance approach for the evaluation of multiple phases in interferometry in presence of nonsinusoidal waveforms and noise. *Journal of the Optical Society of America A* **22**, 1918–1928, 2005.

15. P. Stoica and R. Moses. *Introduction to Spectral Analysis*. Englewood Cliffs, NJ: Prentice Hall, 1997.

16. R. O. Schmidt. Multiple emitter location and signal parameter estimation. *Proceedings RADC, Spectral Estimation Workshop, Rome, NY*, 243–258, 1979.

17. G. Bienvenu. Influence of the spatial coherence of the background noise on high resolution passive methods. *Proceedings of the International Conference on Acoustics, Speech, and Signal Processing, Washington, DC*, 306–309, 1979.

18. R. Langoju, A. Patil, and P. Rastogi. Super-resolution Fourier transform method in phase shifting interferometry. *Optics Express* **13**, 7160–7173, 2005.

19. R. Roy and T. Kailath. ESPRIT-estimation of signal parameters via rotational invariance techniques. *IEEE Transactions on Acoustics, Speech, and Signal Processing* **37**, 984–995, 1989.

20. P. Carré. Installation et utilisation du comparateur photoëelectrique et inter-fërentiel du bureau international des poids et mesures. *Metrologia* **2**, 13–23, 1966.

21. P. Hariharan, B. F. Oreb, and T. Eiju. Digital phase-shifting interferometry: a simple error-compensating phase calculation algorithm. *Applied Optics* **26**, 2504–2506, 1987.

22. J. Schmit and K. Creath. Some new error-compensating algorithms for phase-shifting interferometry. *Optical Fabrication and Testing Workshop*, OSA Technical Digest Series, Optical Society of America, Washington, DC., **13**, PD-4, 1994.

23. K. G. Larkin and B. F. Oreb, Design and assessment of symmetrical phase-shifting algorithms. *Journal of the Optical Society of America A*, **9**, 1740–1748, 1992.

24. P. de Groot. Long-wavelength laser diode interferometer for surface flatness measurement. *Proceedings SPIE: Optical Measurements and Sensors for the Process Industries* **2248**, 136–140, 1994.

25. B. Gutmann and H. Weber. Phase-shifter calibration and error detection in phase-shifting applications: A new method. *Applied Optics* **32**, 7624–7631, 1998.

26. D. C. Ghiglia and M. D. Pritt. *Two-Dimensional Phase Unwrapping: Theory, Algorithms, and Software*. New York: Wiley, 1998.

27. A. Patil, R. Langoju, R. Sathish, and P. Rastogi. Statistical study and experimental verification of high resolution methods in phase shifting interferometry. *Journal of the Optical Society of America A* **24**, 794–813, 2007.

28. K. Hibino, B. F. Oreb, D. I. Farrant, and K. G. Larkin. Phase-shifting algorithms for nonlinear and spatially nonuniform phase shifts. *Journal of the Optical Society of America A* **4**, 918–930, 1997.

29. J. Schmit and K. Creath. Extended averaging technique for derivation of error compensating algorithms in phase shifting interferometry. *Applied Optics* **34**, 3610–3619, 1995.

30. R. Langoju, A. Patil, and P. Rastogi. Phase-shifting interferometry in the presence of nonlinear phase steps, harmonics, and noise. *Optics Letters* **31**, 1058–1060, 2006.

31. B. Raphael and I. F. C. Smith. A direct stochastic algorithm for global search. *Applied Mathematics and Computation* **146**, 729–758, 2003.

32. J. E. Grievenkamp. Generalized data reduction for heterodyne interferometry. *Optical Engineering* **23**, 350–352, 1984.

33. R. Langoju, A. Patil, and P. Rastogi. Estimation of multiple phases in interferometry in the presence of non-linear arbitrary phase steps. *Optics Express* **14**, 7686–7691, 2006.

34. R. Langoju, A. Patil, and P. Rastogi. Statistical study of generalized nonlinear phase step estimation methods in phase-shifting interferometry. *Applied Optics* **46**, 8007–8014, 2007.

CHAPTER 8

Phase Unwrapping

David R. Burton

Liverpool John Moores University
General Engineering Research Institute
Liverpool, United Kingdom

8.1 INTRODUCTION

Commonly, phase determination systems employed in interferometric analysis effectively conclude with an expression of the form

$$\phi = arctan\left(\frac{f_1}{f_2}\right) \tag{8.1}$$

where f_1 and f_2 are functions of the gray levels within the original fringe pattern image, and ϕ is the phase value within a single fringe.

The exact nature of the terms f_1 and f_2 will depend on the fringe analysis technique employed. For example, if phase shifting has been the basis of the technique, then the terms will be a series of intensity values taken from the same pixel location in a series of images, whereas if Fourier methods were the basis, then they will be respectively the imaginary and the real parts of the complex resulting from the inverse Fourier transform of the filtered spectrum. Obtaining the values for f_1 and f_2 has been the topic of the previous chapters in this volume. Whatever their origins, these values form a quotient, and we take the arctangent of this quotient to determine the phase.

The resulting value for the phase ϕ is described in the first paragraph as "the value within a single fringe"; it is effectively a localized value. It tells us how far through that particular fringe we are. The arctangent function acting on a quotient can deliver no more than this; it is bound within the values $\pm\pi$. Indeed, if we make the elementary error of determining the numerical value of the quotient prior to evaluating the arctangent function, matters will be even worse, with the function now only defined between plus $\pi/2$ and minus $\pi/2$. This existence of the phase in these separate elements, one for each fringe, leads to what we call "wrapped phase"; that is, the continuous phase distribution that we seek as the outcome of our analysis has effectively been "cut" into separate sections.

The practical, physical, manifestation of this effect is that we see the phase progress through each fringe, but each exists as an apparently separate function bounded on either side by a discontinuity. This is shown in Figure 8.1. In Figure 8.1a, we see a fringe pattern that we wish to analyze; in Figure 8.1b, we see the wrapped phase of that pattern as it would be obtained from Equation (8.1). Finally, in Figure 8.1c we see the output we really desire, the continuous

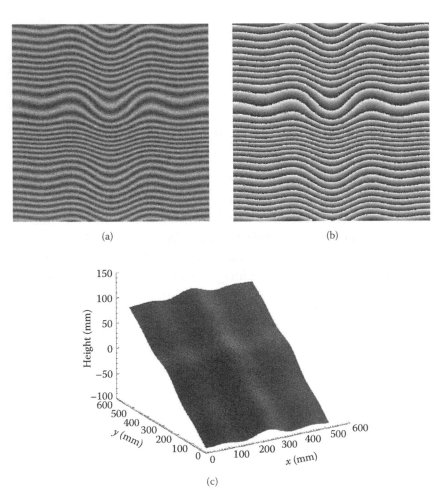

FIGURE 8.1   The fringe analysis process, including phase unwrapping.

or unwrapped phase. Phase unwrapping is the algorithmic process of taking the data from the state of Figure 8.1b to the state of Figure 8.1c.

This limitation of analysis techniques, producing only wrapped phase, is fundamental to computer-aided fringe phase determination. That is, it is not a peculiar artifact of the method used to determine f_1 and f_2; it is intrinsic to the problem of determining phase from an image using a computer. Indeed, the earliest examples of the computer analysis of interferometric patterns revealed this limitation long before the methods described previously in the book were devised. These early methods commonly skeletonized fringes, thinning them to a single line [1]. The problem then was to ascribe a correct sequence number, or order, to these lines. This was commonly referred to at the time as the "fringe-numbering problem," and it was the first realization that when we, as human beings, look at a fringe pattern and clearly see how one fringe relates to its neighbors and so on, we are in fact carrying out a high-level cognitive process. It is difficult to abstract this process and describe it in a computer algorithm.

Another way of appreciating this point is as follows: Consider Figure 8.2a, which shows a fairly complex object. The result from Equation (8.1) of analyzing a fringe pattern on this object is shown in Figure 8.2b. Despite the complexity, we, as human beings, can easily interpret this wrapped phase pattern, but in doing so we are subconsciously using a great deal of contextual information. We know what the surface under this fringe pattern is, or at least we can make an intelligent guess. This means we make use of

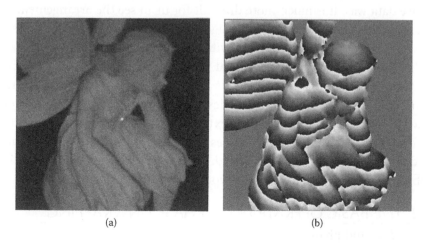

(a)   (b)

FIGURE 8.2   (See color insert.) A complex wrapped phase distribution with context.

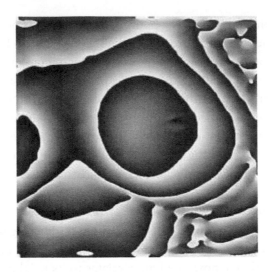

FIGURE 8.3   A noncontextual wrapped phase distribution.

layers of prior knowledge to interpret the image; in this way, we have little difficulty in guiding our eye along the discontinuities between the phase wraps and correctly assigning both connectivity and order to them. The computer does not, generally, have access to this a priori contextual knowledge; it will find the problem of correctly reassembling this distribution much more difficult. To grasp the problem from the perspective of a computer algorithm, examine Figure 8.3. Now, we have no contextual data; we do not know, a priori, what the physical field that originated this wrapped phase data was. It is much more difficult for us to see the arrangement of connections that will render this image optimally continuous. Our phase-unwrapping algorithms must cope with this problem.

This leads to the paradox that is at the heart of phase unwrapping. On first examination, the problem looks trivially simple; it is only when we come to construct a generic set of rules, an algorithm to guide a computer, that we find the problem is indeed extremely difficult without contextual information. As we shall see, successful phase-unwrapping techniques and algorithms have been developed. But, as yet the completely generic, robust, and fast solution still lies beyond our grasp.

8.2 THE BASIC OPERATION OF PHASE UNWRAPPING

To understand phase unwrapping, it is perhaps best to start by illustrating the process at its simplest manifestation. Consider Figure 8.4, which shows a simulated one-dimensional wrapped phase distribution.

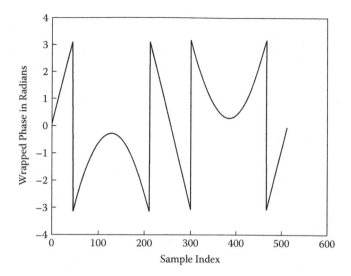

FIGURE 8.4   A simple one-dimensional wrapped phase distribution.

The method we will use is basically, albeit somewhat indirectly, attributed to Oppenheim and Shaffer [2]. Assuming that we have the function shown in Figure 8.4 as a series of digital samples in an array, then the method proceeds as shown in Box 8.1. The progress of this algorithm on the data displayed in Figure 8.4 can be seen in Figure 8.5. Figure 8.5a shows the original data; Figures 8.5b to 8.5d show the progress of the

---

BOX 8.1   UNWRAPPING PROCESS IN ONE DIMENSION

Step 1: Start with the second sample from the left in the wrapped phase signal.

Step 2: Calculate the difference between the current sample and its directly adjacent left-hand neighbor.

Step 3: If the difference between the two is larger that $+\pi$, then subtract 2π from this sample and from all the samples to the right of it.

Step 4: If the difference between the two is smaller than $-\pi$, then add 2π to this sample and to all the samples to the right of it.

Step 5: Have all the samples in the data line been processed? If not, then advance one sample to the right and go back to step 2. If they have, then stop.

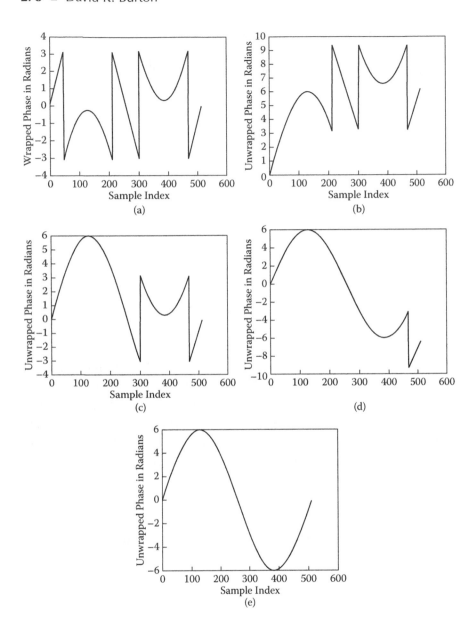

FIGURE 8.5 The progression of a simple unwrapping algorithm in one dimension.

algorithm as the wraps are progressively encountered and resolved. If the algorithm finds no substantive difference between pixels, it simply passes on to the next pixel. However, if it encounters a phase jump, which will be represented by a discontinuity in the data approaching 2π in value, it will shift the following data either up or down as appropriate.

Following the progression of the algorithm in Figure 8.5, we see the input data in Figure 8.5a, then the state when the first discontinuity has been encountered in Figure 8.5b, the second discontinuity in Figure 8.5c, and so on until we arrive at the completely continuous unwrapped phase in Figure 8.5e.

Although shown in one dimension, this algorithm can be easily extended to operate on two-dimensional data in images. To do this, we simply apply the algorithm individually to each row of pixels in the image and then finally apply it once to a single column of the data resulting from all of the row operations.

So, if this algorithm can unwrap phase, why not simply use it? The reality is that this algorithm only works in extremely simple cases. To be successful, the data must be noise free, full field, well sampled, and continuous in the sense that there must be no holes or gaps in the data.

This algorithm is extremely vulnerable to errors. If we consider it operating to unwrap data in a 512 x 512 image, then steps 3 and 4 in the algorithm must be executed over 250,000 times, and every single decision must be correct. If a single noise spike is mistaken for a phase wrap, then not only is that point wrong but also every point from there to the right and below it in the image will be incorrect. In practice, this algorithm is of little value, but it is useful as a vehicle for understanding the problem better; as we shall see, it is effectively the basis of all phase unwrappers.

8.3 PHASE UNWRAPPING: THE PRACTICAL ISSUES AND CHALLENGES

It is departures from the simplicity demanded of the basic algorithm that make robust and reliable phase unwrapping difficult. Basically, we can summarize the main issues as follows:

1. **Noise in the data:** The basic issue here is the unwrapping algorithm confusing what is in reality a noise spike for a phase wrap and applying an unwarranted correction. In practice, all fringe patterns contain noise. In some, such as those produced by speckle techniques, the noise level can be high indeed. But, even under close-to-ideal laboratory conditions, noise levels in fringe patterns are such that they usually require an unwrapping algorithm with built-in defenses to noise corruption. Unwrapping is in essence a differencing operation, so it is basically a differentiation of the data and as such can considerably amplify noise.

2. **Discontinuities:** The issue here is that the underlying field variable we are seeking to measure using fringe analysis may contain actual real discontinuities itself. The danger is that the phase-unwrapping algorithm will incorrectly attempt to resolve and remove such steps in the data, believing them to be phase wraps rather than physical attributes of the variable being measured.

3. **Areas of invalid data:** Here, we face the problem that there may be zones within the fringe image that do not contain fringe information. Holes and surrounding background on non-full-field images are the main examples we encounter.

4. **Undersampling:** Because of its susceptibility to noise, the phase-unwrapping process runs into difficulties caused by poor sampling long before the theoretical Nyquist limit. Indeed, in practical implementations if the fringe-sampling rate falls below around 10 pixels per fringe, we can expect to have to incorporate within our unwrapping software some provision against errors triggered by undersampling.

8.4 PHASE UNWRAPPING AND DEFENSIVE PROGRAMMING

All modern attempts to write robust and reliable phase unwrappers can be thought of as exercises in defensive algorithmic design. The approach of most algorithms to the actual unwrapping process itself is basically just that described previously: a series of differencing operations aimed at detecting wraps and resolving them. But, this simple kernel is usually surrounded by layers of often highly complex and sophisticated checks and strategies designed to work around the four basic issues of noise, discontinuities, invalid data, and undersampling. Indeed, effectively all of the originality in any unwrapping technique developed today lies in creating new, more efficient and effective defensive strategies.

This approach has led to a plethora of algorithms. Some are effective, but slow; others perhaps are slightly less successful but much faster in operation. One concept important to consider in the context of phase unwrappers is that of robustness. We can define robustness in this instance to mean the capability of an algorithm to successfully produce an unwrapped data set with a high confidence level in its correctness. This can include algorithms that seek to produce a maximally correct output. That is, they will unwrap the distribution as far as is possible, marking any areas where the algorithm determines that the confidence level in attempting to produce a correct unwrap is low. These areas are effectively isolated or cordoned off; the unwrapping operation will not take place in these zones. Therefore, although such algorithms do not necessarily unwrap the whole of the data in the image, the user can have high confidence that those areas that are unwrapped have been processed correctly and that the zones that have not been unwrapped are clearly identified.

8.5 PHASE-UNWRAPPING ALGORITHMS

We now consider the operation of modern phase-unwrapping algorithms. Currently, most successful phase unwrappers can be divided into two main groups:

- Algorithms that seek to guide the path of the unwrapping operation around the image in such a manner as to avoid areas of inconsistency.

- Algorithms based on area techniques in which zones where the data are believed to be consistent as to unwrapped first and then these nuclei are grown.

So, the first are fundamentally line-based techniques, and the second are based on areas. Put another way, those algorithms in the first group are built on the relationship between one pixel and its neighbor as we travel along an essentially one-dimensional path; the second consider groups of pixels that make up an area. Both of these approaches have demonstrated considerable success; both have not only their advantages but also their disadvantages.

During the discussion of these methods, it should be kept in mind that at the heart of both of them the basic unwrapping operation remains exactly the same and is no more than the process described previously for the so-called simple unwrapper. Both methods actually unwrap the image via the simple addition and subtraction of multiples of 2π. These algorithms differ purely in terms of the defensive programming that is employed to deal with areas of data affected with issues that prevent successful unwrapping. Neither of these approaches seeks to recover poor data; they merely look to ensure that these data do not corrupt the final outcome.

8.5.1 Path-Guiding Unwrapping Algorithms

The path-guiding unwrapping algorithms were established early in the development of unwrapping algorithms; one of the earliest examples

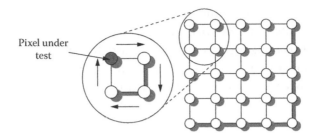

FIGURE 8.6 Defining a phase residue.

would be branch-cut methods developed by the likes of Huntley [3]. The concept behind these methods is to place lines, or "cuts," in the image. These cuts mark boundaries that the unwrapper must not cross if it is to achieve a solution in which inconsistencies are absent. To do this, these techniques must address two main issues: The first is how to decide where the cuts should be placed, and the second is how to steer the unwrapping path in such a way that the cuts are not crossed.

Probably the most successful method for delineating where the cuts should go is the residue method. Consider a small area around a single pixel as shown in Figure 8.6. In Figure 8.6, we are considering 4 pixels that constitute a small neighborhood. Designating the phase value at the upper left-hand pixel as $\phi_{i,j}$, then the others in the neighborhood loop are $\phi_{i+1,j}$, $\phi_{i+1,j+1}$, and $\phi_{i+1,j}$, respectively. We can now calculate the residue value for the top left-hand pixel as

$$r_{i,j} = \mathcal{M}\left[\frac{\phi_{i,j} - \phi_{i+1,j}}{2\pi}\right] + \mathcal{M}\left[\frac{\phi_{i+1,j} - \phi_{i+1,j+1}}{2\pi}\right]$$

$$+ \mathcal{M}\left[\frac{\phi_{i+1,j+1} - \phi_{i,j+1}}{2\pi}\right] + \mathcal{M}\left[\frac{\phi_{i,j+1} - \phi_{i,j}}{2\pi}\right] \tag{8.2}$$

where the operator $\mathcal{M}[x]$ returns a nearest-integer result. The possible values of the residual r are 0, +1, or −1. A value of 0 implies that the point (i, j) is a "good" point, and there are no potential unwrapping issues in the immediate neighborhood. However, if r has a value of ±1, then this identifies (i, j) as a source of possible inconsistencies in the phase distribution.

These residues commonly occur within the image as pairs, positive and negative, basically marking the beginning and end of a fault line within the wrapped phase distribution. We call such pairs "dipoles."

One specific type of residue pair is the so-called localized phase-noise-generated dipole residue. These pairs are caused by the random fluctuation of phase because of noise, which results in the wrapped phase gradient exceeding π. In this type of dipole, the residues tend to lie very close together, often within a single pixel, this makes this type of residue relatively easy to spot and isolate.

The second kind of residue is the dipole residue, which results from undersampling of the phase distribution. These residues are generated by the violation of Shannon's sampling theory, for which the phase is not represented with sufficient spatial resolution to correctly represent the contiguous phase. This results in spatial undersampling steps greater than $\pm\pi$. This type of residue is characterized by generating dipoles that tend to be well separated when Shannon's law is broken, which makes them hard to identify.

A further type of residue dipole is that caused by object discontinuity. Wrapped phase maps frequently contain objects that are discontinuous by nature, such as holes, sharp edges, cracks, or fluids of discontinuously varying refractive index [3, 4]. These discontinuous objects usually generate dipole residues that lie along the discontinuity edges. The existence of these dipole residues depends on the discontinuity size of the object. If the object discontinuity exceeds π when wrapped, then residues will occur. However if the phase jump caused by the discontinuity is smaller than π, then such edges are effectively invisible to the unwrapper and will not result in residue formation. It is frequently the case in practice that discontinuities follow each of these conditions at different points along their length—causing the residues to appear only over parts of the discontinuity. Object discontinuity dipole residues are characterized by generating dipoles that tend to be well separated depending on the nature of the discontinuity, which makes them difficult to identify.

The number of opposite-polarity residues in the image does not have to be equal, leading to the existence of monopoles. Such monopoles exist in two distinct types: so-called dipole-split monopoles and real monopoles. Dipole-split monopoles occur close to the borders of the image; these are really one-half of a dipole where the other, opposite-polarity residue happens to lie outside the field of view. On the other hand, real monopoles may lie anywhere within the image, generally occurring in regions of high phase gradient or areas where the phase map contains true object discontinuities [5]. However, it is difficult to locate a monopole in a wrapped phase map with a large number of residues. It is assumed that any residue

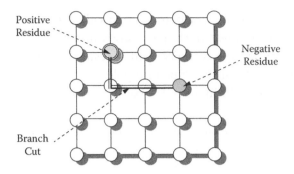

Positive Residue

Negative Residue

Branch Cut

FIGURE 8.7 (See color insert.) A branch cut placed between two residues that form a dipole pair.

has a high probability of being a monopole if the boundary lies closer than twice the distance between the residue and its closest opposite-polarity residue. The theory for approximating the number of monopoles in a wrapped phase map was presented by Gutmann [5].

How do we use these dipoles to guide the phase-unwrapping process? Put simply, we join a positive and negative residue dipole pair with a line. This line is a cut line, more generally referred to as a branch cut. Figure 8.7 shows such a cut line in place.

To achieve a correctly unwrapped phase map, the algorithm must follow a path that avoids crossing this line. So, Figure 8.8a shows an unwrapping path that, if it were followed, would result in an inconsistent unwrapped phase containing errors that would propagate throughout the image. This unwrapping path crosses a branch cut and as such will fail. In contrast, Figure 8.8b shows the unwrapper now following a correct path, around the branch cut, and this will now produce an unwrapped phase map that is at least consistent.

Figure 8.9 shows an arrangement of branch cuts that arose by analyzing the residues of a fairly simple wrapped phase distribution. In Figure 8.9 we can see that as well as the cuts we might expect to see toward the center of the image, there are many cuts that simply lead directly from one residue to an opposite-polarity residue on the border. These are effectively synthetic cuts. They have arisen because some residues located close to the border have no apparent dipole partner—probably because the actual partner lies outside the image. These are the so-called dipole-split monopoles referred to previously. To solve this, the method has simply added new residues at the adjacent border, either horizontally or vertically as appropriate.

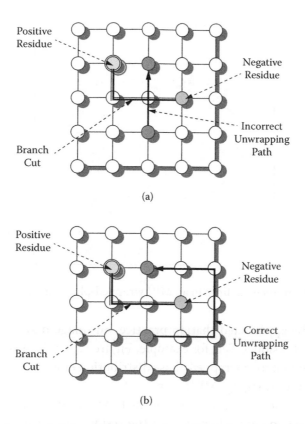

FIGURE 8.8 Incorrect and correct unwrapping paths in the presence of a branch cut.

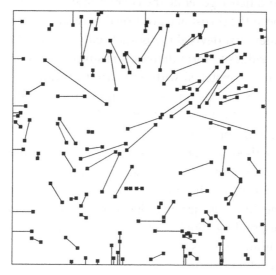

FIGURE 8.9 A branch cut arrangement resulting from analyzing a wrapped phase distribution.

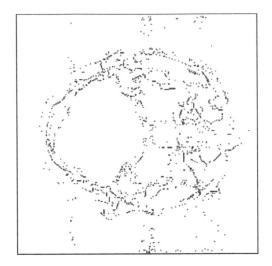

FIGURE 8.10   Branch cut lines in an MRI wrapped-phase image.

It must be appreciated that, in practice, an image may contain many thousands of residues. See, for example, Figure 8.10, which is the branch-cut arrangement for a wrapped phase distribution emanating from a magnetic resonance imaging (MRI) scan.

This raises the obvious question at the heart of this technique: How do we know how to correctly match up the residues to form correct dipole partnerships and hence generate the correct branch cuts? Answering this question has been the thrust of a great deal of recent research.

When Huntley first devised the branch-cut method [3], his algorithm for matching residues to form dipoles was simply nearest neighbor. That is, each residue was matched with its nearest residue of opposite polarity. However, although this approach is simple, regrettably it is not always successful. Since that time, several authors have developed new methods aimed at solving this problem.

Huntley's method has been improved by using more sophisticated search strategies to locate the pairs, such as improved nearest neighbor, simulated annealing, minimum-cost matching, stable marriages, reverse simulated annealing, and genetic algorithms. These algorithms seek to find the corresponding dipoles within the minimum total connection length for the branch cuts (see, e.g., [5–10]).

The tree branch cut technique was first introduced by Goldstein et al. [11]. In this technique, the pixels are utilized to create the branch cuts, such that any pixel can be marked as a part of a branch cut that connects two different residues. The major drawback of this algorithm is that the branch cuts may be placed on good pixels, which may cause error propagation. To overcome this problem, Flynn introduced the mask cut algorithm, in which the branch cut placement relies on a quality map. In his method, the branch cuts are placed on bad-quality pixels to connect residues [12]. Ghiglia and Pritt suggested removing the dipoles as a preprocessing step to enhance the Goldstein algorithm [13]. The dipoles were removed by connecting them using the nearest-neighbor procedure proposed by Huntley.

Despite a great deal of work, this problem remains an issue for algorithms based on branch cutting.

8.5.2 Area-Based Unwrapping Algorithms

The area-based unwrapping algorithm methods offer a radically different approach to that employed by the path-following variants. Instead of focusing on the relationships between individual pixel pairs, these techniques seek a solution that is, in some sense, globally optimum—based on the pattern in its entirety.

Many of these methods are based on minimization techniques, such as least squares. We can think of these methods as finding the unwrapped phase such that a global error function of the form given in Equation (8.3) is minimized.

$$\varepsilon^P = ||\text{solution} - \text{problem}||^P \qquad (8.3)$$

Providing the wrapped phase distribution is not undersampled, the local gradient of the unwrapped phase must equal the local gradient of the wrapped phase. So, we can write that

$$\Delta\phi(x,y) = w\{\Delta\phi(x,y)\} \qquad (8.4)$$

where the operator Δ signifies differentiation or differencing, and w is the wrapping operator.

So, combining Equations (8.3) and (8.4), we have the full statement of these minimizing methods:

$$\varepsilon^P = \sum_{i=0}^{M-2}\sum_{j=0}^{N-1}|\Delta^x\phi(i,j) - w\{\Delta^x\phi(i,j)\}|^P + \sum_{i=0}^{M-1}\sum_{j=0}^{N-2}|\Delta^y\phi(i,j) - w\{\Delta^y\phi(i,j)\}|^P$$

$$(8.5)$$

where $\Delta^x\phi(i,j)$ and $\Delta^y\phi(i,j)$ are the unwrapped phase gradients in the x and y directions, respectively, given by

$$\Delta^x\phi(i,j) = \phi(i+1,j) - \phi(i,j) \qquad (8.6)$$

and

$$\Delta^y\phi(i,j) = \phi(i,j+1) - \phi(i,j) \qquad (8.7)$$

The wrapped phase derivatives are similarly defined on the wrapped data.

From this, it can be seen that, although seemingly a different approach than those we have seen previously, in reality this, at its most fundamental level, is a differencing operation like the original algorithm we defined.

The exact form of the minimizing equation depends on which minimization technique is employed. Most success has been achieved using one of three major approaches:

- **Unweighted least squares:** First introduced by Hunt in 1979 [14], in this technique $p = 2$. The technique was further developed by Ghiglia and Pritt 1998 [13], who used Gauss-Seidel techniques and Fourier transform methods to solve a simplified set of equations resulting from the minimization equation.

- **Weighted least squares:** In this method, each pixel is assigned a weighting value based on an estimation of its quality [14,15]. This method is therefore linked with the quality maps concept, which is discussed further in the chapter. At this stage, we can say that the value of such weightings depends on factors such as the estimated local noise levels, whether the pixel has a residual, and so on.

- **The L^p-norm criteria:** This is a more advanced method of minimization proposed originally by Ghiglia and Romero [16] and later further developed by other workers. The L^p-norm algorithm is actually similar to that of weighted least squares, the only difference being in how the weighting factors are derived. In the weighted least-squares method, the weighted factors are defined independently from a quality map, whereas in the L^p-norm method these factors are extracted from the wrapped phase data. This means that solving the L^p-norm formulation of the problem can use the same algorithms as employed for the weighted least-squares case.

Alongside these more mainstream methods for minimization, other techniques have been employed, such as minimum energy/entropy and simulated annealing.

Brief mention was given previously to the concept of quality maps. A quality map is a series of values that in some way attempts to quantify the quality and robustness of data in the wrapped phase distribution. It is usually an array of the same dimensions as the original image ascribing a quality score to each pixel, although occasionally a lower-resolution map is employed in which pixels in a set area are given a quality score.

The use of quality maps as described was the basis for assigning weights for the weighted least-squares minimization method. But, these have also led to the development of another area-based unwrapping method known as region growing [17].

The principle of operation of the region-growing techniques is that first a quality map is created in which each pixel is scored. Unwrapping then commences from the highest scoring pixels outward into the immediate neighborhood—attempting to successfully unwrap the largest possible area. This expansion of the region continues until encountering inconsistencies, possible sources of error, or another successfully unwrapped area. When the possibilities of this operation are exhausted, the algorithm switches to the next-highest-quality scoring pixel still in an unwrapped area.

The obvious issue with this technique is related to the basis used to score the data quality: In other words, what is the form of the mathematical quality function? Several such functions have been the subject of experiments, and as with minimization methods, some formulations have proved more successful than others.

Common methods for formulating the quality map include gradient following [18], maximum correlation [19], second derivative [20], and weighted windows [21]. These methods offer the major advantage that they can be memory and time efficient. However, the obvious drawback is that they are only as good as the quality criteria they are based on, and it is difficult to build more sophisticated weighted methods into the formulation.

8.5.3 Other Methods of Phase Unwrapping

The techniques described, both the path-following and the area methods, have so far proved to be among the most successful concepts on which to base an unwrapping algorithm. But, it should not be thought that they are the only methods because this would be a long way short of the truth. In the course of the last 20 years or so, experiments have been performed with many other methods, and these methods have been extensively developed.

One different method that does seem to show some potential is the use of polynomial techniques to fit the data [22]. An example of a significant extension would be Karout et al.'s development of the residue vector concept [23], which goes a significant way toward explaining why unwrapping algorithms often fail.

8.6 ONLINE SOURCES OF UNWRAPPING CODES

There are a number of highly sophisticated unwrapping algorithms both in numerical libraries and available for download. Numerical libraries that include unwrapping code include the SciPy library (http://www.scipy.org/) and the ImageJ library (http://imagej.en.softonic.com/).

Downloads of individual routines that implement algorithms published in the scientific literature are available (see http://www.ljmu.ac.uk/GERI/120583.htm). This page contains 11 algorithms based on the work in Herráez et al. [24, 25] and Abdul-Rahman et al. [26, 27].

8.7 CONCLUSION

Every year, the phase-unwrapping literature expands, and it has been one of the strongest developing areas concerning interferometry for some time. Its popularity as a research field stems from its significance. It is used not only in interferometry but also in areas such as medical imaging (MRI) and satellite data analysis (synthetic aperture radar [SAR] techniques). This wide field of application means that there is substantial interest in constructing accurate, fast, and robust phase-unwrapping algorithms.

This constant drive forward makes it challenging to comprehensively catalogue and review all of the techniques that have been or are in use; such an undertaking could easily fill an entire weighty volume on its own. Instead, what we have tried to do here is to provide a foundation of knowledge and concepts that will prepare readers to select from the literature what is of interest to them and then tackle that content in such a way that they will understand and be able to interpret the work they find.

The "phase-unwrapping problem" is an intriguing one, at first glance apparently so easy, but on deeper acquaintanceship so tantalizingly difficult. We still cannot claim to have solved the problem, but as this review shows, we have made significant inroads. Will an all-purpose, totally generic phase unwrapper ever be developed? It seems unlikely unless some method for building in contextual data emerges. However, there is little doubt that the field will continue to advance ever closer to this aim.

REFERENCES

1. Robinson, D. W., and Reid, G. T. *Interferogram Analysis: Digital Fringe Pattern Measurement Techniques.* New York: Institute of Physics, 1993.
2. Oppenheim, A. V., and Schaffer, R. W. *Discrete-Time Signal Processing.* 2nd ed. Englewood Cliffs, NJ: Prentice Hall, 1989.
3. Huntley, J. M. Noise-immune phase unwrapping algorithm. *Applied Optics* **28**, 3268–3270, 1989.
4. Huntley, J. M. Three-dimensional noise-immune phase unwrapping algorithm. *Applied Optics* **40**, 3901–3908, 2001.
5. Gutmann, B., and Weber, H. Phase unwrapping with the branch-cut method: Clustering of discontinuity sources and reverse simulated annealing. *Applied Optics* **38**, 5577–5593, 1999.
6. Buckland, J. R., Huntley, J. M., and Turner, S. R. E. Unwrapping noisy phase maps by use of a minimum-cost-matching algorithm. *Applied Optics* **34**, 5100–5108, 1995.
7. Cusack, R., Huntley, J. M., and Goldrein, H. T. Improved noise-immune phase unwrapping algorithm. *Applied Optics* 34, 781–789, 1995.
8. Karout, S. A., Gdeisat, M. A., Burton, D. R., and Lalor, M. J. Two-dimensional phase unwrapping using a hybrid genetic algorithm. *Applied Optics* 46, 730–743, 2007.
9. Chavez, S., Xiang, Q., and An, L. Understanding phase maps in MRI: A new cutline phase unwrapping method. *IEEE Transactions on Medical Imaging* **21**(8), 966–977, 2002.
10. Salfity, M., Ruiz, P., Huntley, J. M., Graves, M., Cusack, R., and Beauregard, D. Branch-cut surface placement for unwrapping of under-sampled three-dimensional phase data: Application to magnetic resonance imaging arterial flow mapping. *Applied Optics* **45**, 2711–2721, 2006.
11. Goldstein, R. M., Zebker, H. A., and Werner, C. L. Satellite radar interferometry: Two-dimensional unwrapping. Radio Science **23**(4), ed. 713–729, 1988.
12. Flynn, T.J. Consistent 2-D phase unwrapping guided by a quality map. *Proceedings of the International Geoscience and Remote Sensing Symposium* **4**, 2057–2059, 1996.
13. Ghiglia, D. C., and Pritt, M. D. *Two Dimensional Phase Unwrapping: Theory, Algorithms and Software.* New York: Wiley-Interscience, 1998.
14. Hunt, B. R. Matrix formation of the reconstruction of Hase values from Hase differences. *Journal of the Optical Society of America* **5**, 416–425, 1979.
15. Flynn, T. J. Two-dimensional phase unwrapping with minimum weighted discontinuity. *Journal of the Optical Society of America A* **14**, 2692–2701, 1997.
16. Ghiglia, D. C., and Romero, L. A. Minimum Lp-norm two-dimensional phase unwrapping. *Journal of the Optical Society of America A* **13**, 1999–2013, 1996.

17. Herráez, M. A., Boticario, J. G., Lalor, M. J., and Burton, D. R., Agglomerative clustering-based approach for two-dimensional phase unwrapping. *Applied Optics* **44**, 1129–1140, 2005.
18. Lim, H., Xu, W., and Huang, X. Two new practical methods for phase unwrapping. *Proceedings of the International Geoscience and Remote Sensing Symposium* **3**, 196–98, 1995.
19. Xu, W., and Cumming, I. A region growing algorithm for InSAR phase unwrapping. *Proceedings of the International Geoscience and Remote Sensing Symposium* **4**, 2044–2046, 1996.
20. Bone, D. J. Fourier fringe analysis: The two dimensional phase unwrapping problem. *Applied Optics* **30**, 3627–3632, 1991.
21. Cho, B. Quality map extraction for radar interferometry using weighted window. *Electronics Letters* **40**, 2004.
22. Slocumb, B. J., and Kitchen, J. A polynomial phase parameter estimation phase unwrapping algorithm. *ICASSP '94 Proceedings of the Acoustics, Speech, and Signal Processing, 1994, IEEE International Conference* **4**, 129–132, 1994.
23. Karout, S. A., Gdeisat, M. A., Burton, D. R., and Lalor, M. J. Residue vector, an approach to branch-cut placement in phase unwrapping: Theoretical study. *Applied Optics* **46**, 4712–4727, 2007.
24. Herráez, M. A., Burton, D. R., Lalor, M. J., and Gdeisat, M. A. Fast two-dimensional phase-unwrapping algorithm based on sorting by reliability following a noncontinuous path. *Applied Optics* **41**, 7437–7444, 2002.
25. Herráez, M.A., Gdeisat, M.A., Burton, D.R., and Lalor, M.J. Robust, fast, and effective two-dimensional automatic phase unwrapping algorithm based on image decomposition. *Applied Optics* **41**, 7445–7455, 2002.
26. Abdul-Rahman, H., Gdeisat, M. A., Burton, D. R., and Lalor, M. J. Fast three-dimensional phase-unwrapping algorithm based on sorting by reliability following a noncontinuous path. *Applied Optics* **46**, 6623–6635, 2007.
27. Abdul-Rahman, H., Herráez, M. A., Gdeisat, M. A., Burton, D. R., Lalor, M. J., Lilley, F., Moore, C. J., Sheltraw, D., and Qudeisat, M. Robust three-dimensional best-path phase-unwrapping algorithm that avoids singularity loops. *Applied Optics* **48**, 4582–4596, 2009.

CHAPTER 9

Uncertainty in Phase Measurements

Erwin Hack

Electronics/Metrology/Reliability Laboratory
EMPA-Materials Science and Technology
Dubendorf, Switzerland

9.1 INTRODUCTION

An electromagnetic signal is described by its amplitude, state of phase, and state of polarization. Such signals are found in very different branches of science, such as electronics, telecommunication, interferometry, ultrasound, and phase-contrast x-ray tomography. Although the amplitude of the signal is easily accessible through the measurement of the intensity with a power detector or camera, retrieving the phase from intensity measurements (images) is more involved. There exist many different approaches to phase retrieval, several of which are described in previous chapters of this book.

Once the phase is measured, its value must be completed with an estimate of the measurement uncertainty. Many early publications dealt with the quality and the elimination of measurement uncertainty or error [1–8]. The aim of this chapter is to describe a unified process of estimating the uncertainty of the phase value obtained by means of a phase retrieval process, utilizing the *Guide to the Expression of Uncertainty in Measurement* (GUM) [9]. This guide has proven to be particularly helpful because it establishes general rules for evaluating and expressing uncertainty in measurement. Because it can be applied to a broad range of measurements, we use it in particular for phase measurement. Although the approach is based

BOX 9.1   INPUT QUANTITIES AND INFLUENCE QUANTITIES

Let us assume that the measurand y is modeled as a function of N quantities $y = f(x_1,\ldots,x_N)$. The x_i are called input quantities because to determine the value of the measurand y, all N values of the input quantities must be known together with their uncertainty value. In contrast, although an influence quantity is a quantity that does not enter the model f of the measurand, it might affect the result of the measurement. In general, their values are not known, and they are only taken into account in the estimation of the measurement uncertainty.

Note that influence quantities can become input quantities if the model of the measurement process is refined.

on a first-order approximation only, the results can provide both qualitative and quantitative understanding of the individual contributions to the combined measurement uncertainty. First, input and influence quantities in the measurement process must be identified (see Box 9.1), their uncertainty or variability must be assessed, and their effect on the measurement result must be calculated. A modular approach to describe the measurement process is taken based on the theory of propagation of uncertainty ("error propagation"; see Box 9.2). It is not essential whether an analytic treatment or a numerical simulation is applied; the bottom line of both approaches is casting the influence quantities into a model that describes their effect on the phase value. A Monte Carlo approach would allow for an incorporation of different probability density functions to describe the scatter of the input and influence quantities and is not limited to a first-order treatment.

The process of phase measurement can be separated into the modules of Figure 9.1. Information on the test object is coded in the phase of the signal by active illumination; its amplitude is measured once or several times; the phase is retrieved using a dedicated image-processing algorithm; finally, the value of the measurand to be reported is obtained from the phase value with an appropriate calibration. There might be additional steps of digital image processing, such as filtering or averaging, to improve image quality. We do not treat the effect of such filtering operations in detail because they create correlations between the individual pixel values.

BOX 9.2   PROPAGATION OF UNCERTAINTY

Let us assume that the measurand y is a function of N input quantities $y = f(x_1, \ldots, x_N)$. Then, a first-order Taylor expansion around the expectation values $\mu_i = E[x_i]$ yields

$$y = f(\mu_1, \ldots, \mu_N) + \sum_{i=1}^{N} \frac{\partial f}{\partial x_i}(x_i - \mu_i) \quad \text{or} \quad y - \mu_y = \sum_{i=1}^{N} \frac{\partial f}{\partial x_i}(x_i - \mu_i)$$

The variance of a quantity is commonly used to describe the scatter of a variable, The variance of y around its expectation value is

$$\sigma^2(y) = E[(y - \mu_y)^2] = E\left[\left(\sum_{i=1}^{N} \frac{\partial f}{\partial x_i}(x_i - \mu_i)\right)^2\right]$$

$$= E\left[\left(\sum_{i=1}^{N}\left(\frac{\partial f}{\partial x_i}\right)^2 (x_i - \mu_i)^2\right) + \left(\sum_{i,j=1}^{N} \frac{\partial f}{\partial x_i}\frac{\partial f}{\partial x_j}(x_i - \mu_i)(x_j - \mu_j)\right)\right]$$

$$= \sum_{i=1}^{N}\left(\frac{\partial f}{\partial x_i}\right)^2 E[(x_i - \mu_i)^2] + \sum_{i,j=1}^{N} \frac{\partial f}{\partial x_i}\frac{\partial f}{\partial x_j} E[(x_i - \mu_i)(x_j - \mu_j)]$$

which in short notation is written

$$\sigma^2(y) = \sum_{i=1}^{N}\left(\frac{\partial f}{\partial x_i}\right)^2 \sigma^2(x_i) + \sum_{i,j=1}^{N} \frac{\partial f}{\partial x_i}\frac{\partial f}{\partial x_j}\sigma^2(x_i, x_j)$$

The first term is the weighted sum of variances of the individual input quantities; the second term is the correlation term that describes the joint scatter of two input quantities. It is important to note that this result is independent of the probability density functions of the input quantities. Only in the simplest (but important) case, a Gaussian distribution is assumed.

If the variances serve as a measure of uncertainty, the identification $u^2 = \sigma^2$ is made.

Phase measurement is influenced by many parameters, the most important of which are discussed in the next section. Because images I_k are the basic input quantities for all phase measurement algorithms, we base the discussion of the influence quantities on the effect they have on the images. An image I_k is composed of pixel (picture element) values $I_k(m, n)$ with uncertainty values $u(I_k(m, n))$.

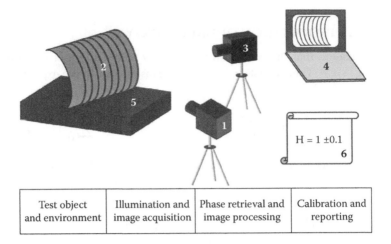

| Test object and environment | Illumination and image acquisition | Phase retrieval and image processing | Calibration and reporting |

FIGURE 9.1 Modularization of a typical experiment and the phase measurement process: 1, Illumination; 2, Test object; 3, Image acquisition; 4, Image processing and phase retrieval; 5, Environment; 6, Calibration and reporting.

9.2 INFLUENCE QUANTITIES

9.2.1 Test Object and Environment

Changes of the test object and the environment (Figure 9.1) during measurement are more critical influences than inhomogeneous but stationary properties. Table 9.1 explains important influence quantities from the test object and the environment.

9.2.2 Illumination and Image Acquisition

Light reaching the detector is generally composed of the active illumination that codes the phase and background light that is not correlated to the useful signal but incoherently added. Important influence quantities are described in Table 9.2.

Images are acquired by a detector, the properties of which can also have notable influence effects. Because the active illumination is sometimes modulated, such as with phase-stepping methods, those manipulations are also subsumed in Table 9.3.

9.2.3 Phase Retrieval and Image Processing

The phase retrieval algorithm itself is a digital process without additional influence quantities. However, depending on its construction, it can reduce the importance of some underlying influence quantities.

TABLE 9.1   Important Influence Quantities from Test Object and Environment

Object geometry	Surface gradients and perspective can lead to a varying phase value across the pixel equivalent surface element.
Vibration	If the measured phase codes object distance, object vibrations can induce varying phase values during the measurement process.[a] This can influence the fringe contrast.
Drift and creep	These are relatively slow processes that gradually change the phase value in consecutive images.
Temperature changes	These can cause a change in optical path through a change in refractive index or elongation of optical setup or change in object dimensions.[b]
Surface changes	Processes such as corrosion, oxidation, or adsorption cause a change in the reflected light intensity or spectrum or a change in the contrast of a surface speckle pattern.
Change of background lighting	This can be caused by a change in ambient light during the measurement.

[a] If the vibration amplitude itself is the measurand, this is an input quantity but not an influence quantity.
[b] If, for example, the coefficient of thermal expansion is measured, temperature is again an input quantity.

TABLE 9.2   Important Influence Quantities from Active Illumination

Periodic variation of illumination	This can be caused by a flicker of the illumination and can severely influence the image acquisition through aliasing effects.
Change in illumination power	For active phase measurement systems based on structured light of laser interferometry, this can cause multiplicative uncertainty.
Higher harmonics	Some phase retrieval algorithms assume a sinusoidal signal, but in reality, this is not the case because of higher harmonics. Although an ideal interferometric signal leads to a sinusoidal variation in irradiance, higher harmonics are induced (e.g., by nonideal sinusoidal gratings in fringe projection, by Talbot projection in shadow moiré, by interference of multiple laser beams, by nonlinear detector response, or by signal clipping because of camera saturation).
Wavelength and pointing stability	Some quantities are related to the calibration (or scaling) of the results, such as wavelength, wavelength stability, and direction of illumination; others influence the signal quality, such as irradiance stability, number of photons, and spectral content.

TABLE 9.3   Important Influence Quantities from Image Acquisition

Detector gain	In the ideal case, detector gain is linear (i.e., the signal is proportional to the incoming irradiance), but in reality nonlinearities affect the signal values.
Shot noise	Generation of electrons from photons follows Poisson statistics, in which the variance of the average number of electrons is equal to the number of electrons.
Electronic noise	This is noise accumulated because of the electronic circuitry from the detector element to the analog-to-digital converter (ADC). In a CCD or CMOS sensor, additive signal noise is caused by the reset noise, the electronic thermal noise of the readout circuit, and the shot noise of the dark current.
Digitization jitter	The digital resolution (or "bit depth") affects the signal value.
Phase-stepping jitter	Jitter in the phase-stepping device can be caused by jitter in the driver electronics.
Phase-stepping nonlinearities	In interferometry, this can be caused by a tilting mirror, a beam that is not reflected perpendicularly from a mirror, a miscalibration of the electronic drive signal, or hysteresis effects in the phase-stepping device. Linear phase step miscalibration is often treated separately.

The final step in the measurement process—the transformation of the phase value into a value of the physical quantity of interest—introduces an additional uncertainty: the calibration uncertainty. This uncertainty can either be determined directly using a physical reference material [10, 11] or be estimated from an analysis, for example, of the sensitivity vector in digital speckle pattern interferometry (DSPI). In other words, the uncertainty of the phase value must be propagated to the final result of interest, such as deformation, shape, or index of refraction. Because this book is focused on phase measurement, we do not detail the calibration uncertainty any further here.

9.3 QUANTIFICATION OF UNCERTAINTY CONTRIBUTIONS

The importance of the influence quantities identified previously is assessed in two steps. First, the sensitivity of the measurand to the influence quantity is determined, again using partial models. Then, to inform the model, the range of values of the influence quantity and their distribution in the actual experiment are estimated. This quantification is predominantly done by estimating the values of the relevant parameters from known or available information and rarely by means of specific empirical studies (see Box 9.3).

BOX 9.3   QUANTIFICATION OF INPUT QUANTITIES
VARIANCES OF TYPES A AND B

For an input quantity X determined from n independent repeated observations x_k, the standard uncertainty $u(x)$ of its estimate $x = \overline{x} = \frac{1}{n}\sum_{k=1}^{n} x_k$ is $u(x) = s(x)$ with $s^2(x) = \frac{1}{n-1}\sum_{k=1}^{n}(x_k - \overline{x})^2$. For convenience, $u^2(x)$ and $u(x)$ are called a *type A variance* and a *type A standard uncertainty*, respectively [9].

For an estimate x of an input quantity X that has not been obtained from repeated observations, the associated estimated variance $u^2(x)$ or the standard uncertainty $u(x)$ is evaluated by means of scientific judgment based on all of the available information on the possible variability of X. The pool of information may include

- previous measurement data,
- experience with or general knowledge of the behavior and properties of relevant materials and instruments,
- manufacturer's specifications,
- data provided in calibration and other certificates, or
- uncertainty values assigned to reference data taken from handbooks.

For convenience, $u^2(x)$ and $u(x)$ evaluated in this way are called a *type B variance* and a *type B standard uncertainty*, respectively. In practice, type B uncertainties prevail.

Because of the modularity of the GUM approach, all uncertainty contributions can be consistently propagated to the combined uncertainty of the phase measurement value. However, to combine individual components, a sum of squares is usually not sufficient. In modularizing the problem, it is essential to maintain consistency, that is, to avoid double counting and to take into account correlations mediated by influence factors.

In single-pixel algorithms, phase is measured for every pixel independently. This necessitates the availability of a set of images captured sequentially (temporal phase stepping) or in parallel. Representative algorithms are arctan algorithms in which the phase is retrieved from sine and cosine parts of the signal obtained from a linear combination of intensity values (linear phase-stepping or Fourier methods) [12] or from nonlinear combinations (e.g., Carré algorithms). For a single-pixel algorithm, the input quantities are the pixel values of the set of images. Because this algorithm operates on the values of each individual pixel, no correlations with neighboring pixels are expected, but correlations among subsequent phase-stepped signals are important. For single-frame algorithms, all pixels are involved in the phase retrieval algorithm, and correlations among neighboring (or all) pixels may become important.

It is essential to keep in mind that a **target uncertainty** for the experiment at hand needs to be defined; that is, the question must be answered regarding how precise the measurement result must be. Then, the uncertainty analysis may be restricted to the appropriate level, starting with the dominant influence quantities. To help develop a feeling for the uncertainty values, uncertainty components are quantified for a target uncertainty of 1% in the phase value, which corresponds to a phase measurement uncertainty of $63\ mrad$. In addition, typical values found in a real experiment are indicated to estimate the overall phase measurement uncertainty that can be expected for a specific method. In the next section, the influence of the quantities identified previously on the uncertainty of the pixel values is detailed.

9.4 UNCERTAINTY CONTRIBUTIONS FOR IMAGING

9.4.1 Lateral and Temporal Image Resolution

Let an image be represented by an array of pixels (picture element) values $I(i,j)$, which can be thought of as the result of an integration process over space and time,

$$I(i,j;t) = \int_{t}^{t+T}\int_{x}^{x+p}\int_{y}^{y+p} i(\xi,\eta,\tau)d\xi\, d\eta\, d\tau \tag{9.1}$$

where T is the integration time, p is the physical pitch of the pixel assumed to be square to avoid subindices, and (ξ,η) are the area detector coordinates. In imaging applications, each pixel area corresponds to a projected area on the object surface of size $Mp \times Mp$, where M is the magnification of the lens.

Even assuming a purely harmonic irradiance $i = |a\cos(\frac{\phi+\beta}{2})|^2$, the parameter values may depend on space and time. If they are constant across the pixel of size p and during the acquisition time T, integrations in Equation (9.1) are easily performed:

$$I(i,j;t) = |a(i,j;t)|^2 \cos^2\left(\frac{\phi(i,j;t)+\beta}{2}\right)p^2T \;\; = \;\; I_B + I_\Gamma \cos(\phi(i,j;t)+\beta)$$

$$I_B = I_\Gamma = \tfrac{1}{2}|a(i,j;t)|^2 p^2T \tag{9.2}$$

where I_B represents the bias or background intensity and I_Γ is the modulation intensity, which are equal in this ideal case.

If a varies across the pixel, the expressions are replaced by an average value, that is,

$$I_B = I_\Gamma = \tfrac{1}{2} T \int_x^{x+p} \int_y^{y+p} |a(\xi,\eta)|^2 d\xi d\eta \qquad (9.3)$$

but the phase expression remains unchanged.

If, however, the phase varies across the pixel because of a phase gradient or the phase varies over time because of vibration or drift, for example, a resulting phase average value is obtained. For linear changes, the respective variances are given by

$$u^2(\phi,g) = \frac{1}{24} g^2 p^2 \qquad g = \frac{\partial \phi}{\partial x}$$
$$\text{with} \qquad (9.4)$$
$$u^2(\phi,d) = \frac{1}{24} d^2 T^2 \qquad d = \frac{\partial \phi}{\partial t}$$

For a target uncertainty of $u(\phi) = 63$ mrad, this limits the phase gradient to $g < \frac{0.3rad}{p}$, corresponding to one fringe over 20 pixels, or $d < \frac{0.3rad}{T}$, which corresponds to a drift of less than 1/20 of a fringe per frame.

For visualization of some results, we use a fringe map simulating the shape of a cylinder. The ideal phase distribution is given in Figure 9.2a while Figure 9.2b shows the fringe pattern. The background intensity is assumed to increase linearly from left to right, and the fringe contrast Γ increases from top ($\Gamma = 0.01$) to bottom ($\Gamma = 1.0$).

9.4.2 Signal-Independent Contributions

9.4.2.1 Electronic Noise and Digitization Jitter

Electronic noise $u_{el.noise}$ and **digitization jitter** (i.e., the uncertainty introduced by the resolution of the least-significant bit u_{LSB}) are random contributions that are uncorrelated between the pixels and independent of the signal value S. The signal uncertainty caused by the LSB is obtained from the variance of an equal (or rectangular) probability density function, the width of which is given by the signal equivalent value of the LSB, S_{LSB}, which is given by

$$u_{LSB}^2 = \frac{1}{12} S_{LSB}^2. \qquad (9.5)$$

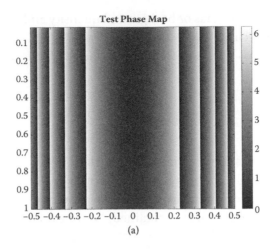

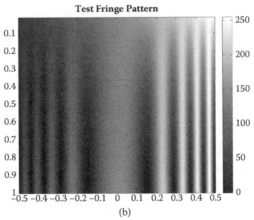

FIGURE 9.2   Simulated phase map (a) and fringe pattern (b) representing a cylinder shape measurement.

The squared contributions are added to yield

$$u^2(S, add) = u_{el.noise}{}^2 + u_{LSB}{}^2. \tag{9.6}$$

The digitization jitter or quantization noise [13,14] can normally be neglected against the electronic noise if the digitization depth is 8 bits or more [15].

9.4.2.2 Background Fluctuations

Rapid **fluctuations of the incoherent background illumination** are a noise source that is independent of the signal value but is correlated between neighboring pixels.

$$u^2(S, background) = S^2_{background}. \tag{9.7}$$

9.4.3 Signal-Dependent Contributions

9.4.3.1 Shot Noise

In a photon detector (charge-coupled device [CCD] and complementary metal-oxide semiconductor [CMOS] cameras, photon counter), the number of electrons per exposure varies according to Poisson statistics, even with constant average irradiance. The variance of the signal caused by shot noise is proportional to the signal itself [16] when signals are measured in electrons or counts [cnt]:

$$u^2(S, shot[\text{cnt}]) = S[\text{cnt}] \tag{9.8}$$

The full well depth of a pixel describes the maximum number of electrons the pixel can hold, with typical values less than 10^5. This leads to a shot noise level greater than 0.3%.

9.4.3.2 Illumination Instability and Camera Gain Fluctuations

An ideal detector converts the irradiance I linearly into an electrical signal S:

$$S = \gamma I \tag{9.9}$$

If the illuminating light source is not stable, it induces an irradiance fluctuation proportional to the signal itself. In view of the symmetry of Equation (9.9) in gain and irradiance, this has the same effect as a gain fluctuation; that is, it leads to a multiplicative noise term. Correlation between the frames can be neglected if the fluctuations are faster than the frame acquisition rate. Otherwise, the intensity is regarded as drifting (see further discussion in Section 9.5.3). The uncertainty contributions are

$$
\begin{aligned}
u^2(S, \gamma) &= I^2 u^2(\gamma) = S^2 r^2(\gamma) \\
u^2(S, I_{fluct}) &= \gamma^2 u^2(I) = S^2 r^2(I)
\end{aligned}
\tag{9.10}
$$

where γ is the detector gain, and r is the relative uncertainty.

9.4.3.3 Detector Nonlinearity

Detector nonlinearity is a systematic uncertainty that causes higher harmonics and correlates the signals in a multiframe algorithm [17]. Because of the ubiquitous use of CMOS sensors (which are slightly nonlinear and inferior to CCDs in this regard), this effect has regained attention in the last decade. Let the ideal signal, Equation (9.9), be replaced by

$$\tilde{S} = \gamma I + \gamma_2 I^2 + \cdots + \gamma_J I^J = S + r_2 S^2 + \cdots + r_J S^J \qquad (9.11)$$

where J is the highest order of detector nonlinearity considered, and r_n is the nonlinearity parameter for signal order n. Of course, if the nonlinearity is known (e.g., from a calibration curve), it can be corrected for. However, if the specific form of nonlinearity is not known exactly or not corrected completely, it must be treated as an uncertainty component given by

$$u^2(S, r_n) = S^{2n} r_n^2 \qquad u(S_j, S_k, r_n) = S_j^n S_k^n r_n^2 \qquad (9.12)$$

Second-order nonlinearity is sometimes specified in counts; for instance, Equation (9.11) is replaced by

$$\tilde{N} = N + \alpha N^2 \qquad (9.13)$$

so that the parameters are related by

$$r_2 = \alpha \frac{N_{full\,well}}{S_{sat}} \qquad (9.14)$$

9.5 UNCERTAINTY CONTRIBUTIONS FOR LINEAR PHASE-STEPPING ALGORITHMS

9.5.1 Combined Uncertainty

Let the set of phase-stepped signals be written as

$$S_k = \gamma I_k = S_B + S_\Gamma \cos(\phi + \beta_k), \qquad (9.15)$$

where S_B is the background or direct current (DC) bias, S_Γ is the modulation, and β_k is the kth phase-stepping angle, $k = 0 \ldots N - 1$. Linear phase-stepping algorithms use weighted sums of the signals S_k

to synthesize sine and cosine terms and retrieve the phase from the general expression

$$\phi+\alpha = \mathrm{atan}\left(\frac{G\sin(\phi+\alpha)}{G\cos(\phi+\alpha)}\right) = \mathrm{atan}\left(\frac{\sum_{k=0}^{N-1} b_k S_k}{\sum_{k=0}^{N-1} a_k S_k}\right) = \mathrm{atan}\left(\frac{\langle\mathbf{b}|\mathbf{S}\rangle}{\langle\mathbf{a}|\mathbf{S}\rangle}\right)$$

(9.16)

where the coefficients a_k and b_k derive from the specific N-frame algorithm. A factor G is included, which absorbs the scaling of the terms. Here, *atan* is short for the Arctan2 function, which uses the signs of numerator and denominator to map the phase over the full 2π range [18]. The offset angle α is introduced to allow for a shift in the origin of the phase angles.

Because the signal values are real, we can cast Equation (9.16) into the simple form

$$\phi+\alpha = \arg\left(\langle\mathbf{c}|\mathbf{S}\rangle\right) \quad \text{where} \quad \mathbf{c}=\mathbf{a}+i\mathbf{b} \tag{9.17}$$

Many approaches to generate an algorithm that is adapted to compensate for specific uncertainty contributions have been reported, such as using characteristic polynomials [19] or spectral filters based on a linear system engineering approach [20].

The combined phase measurement uncertainty is calculated from Equation (9.16) [21]:

$$u^2(\phi) = \frac{1}{G^2}\sum_{k=0}^{N-1}(b_k\cos\phi - a_k\sin\phi)^2 u^2(S_k)$$

$$+\frac{2}{G^2}\sum_{k=0}^{N-1}\sum_{j>k}^{N-1}(b_j\cos\phi - a_j\sin\phi)(b_k\cos\phi - a_k\sin\phi)u(S_j,S_k)$$

(9.18)

The first sum takes into account uncorrelated uncertainties; the second line contains correlations among the signals. Such correlations arise because, for example, the same measuring instrument is used or the same environmental influence applies. If the parameters g_l responsible for correlation are introduced explicitly, the resulting covariance $u(S_j, S_k)$ can be calculated according to [9]

$$u(S_j, S_k) = \sum_l \left(\frac{\partial S_j}{\partial g_l}\right)\left(\frac{\partial S_k}{\partial g_l}\right) u^2(g_l) \qquad (9.19)$$

The general linear phase-stepping algorithms are discussed in Hack and Burke [21]; here, we restrict the discussion to the classical N-bucket algorithms for which $\beta_k = k \times \frac{2\pi}{N}$ and Equation (9.18) simplifies to

$$u_N^2(\phi) = \frac{4}{N^2 S_\Gamma^2}$$

$$\left\{ \sum_{k=0}^{N-1} \sin^2(\phi + \beta_k)\, u^2(S_k) + 2\sum_{k=0}^{N-1} \sum_{j>k}^{N-1} \sin(\phi + \beta_j)\sin(\phi + \beta_k)\, u(S_j, S_k) \right\}$$

$$(9.20)$$

9.5.2 Uncertainty from Uncorrelated Influences

When signal-independent and hence uncorrelated contributions, such as from Equations (9.6) and (9.7), are inserted into Equation (9.20), this leads to [22–24]

$$u_N^2(\phi) = \frac{4}{N^2 S_\Gamma^2} \sum_{k=0}^{N-1} \sin^2(\phi + \beta_k)\, u^2(S) \quad = \quad \frac{2\, u^2(S)}{N\, S_\Gamma^2} \qquad (9.21)$$

so that u_N drops with $\sqrt{N}$ and with increasing signal modulation S_Γ. In turn, to meet a target uncertainty of 63 mrad, the acceptable relative signal uncertainty must satisfy

$$\frac{u(S)}{S_\Gamma} \leq \sqrt{\frac{N}{2}} u_N(\phi, S) = \sqrt{N} \times 0.045 \qquad (9.22)$$

In other words, for the classical four-bucket algorithm, we can accept 9% relative random signal uncertainty $u(S)/S_\Gamma$.

When signal dependent, but uncorrelated, uncertainty components are considered, their squares are added up according to Equation (9.20):

$$u_N^2(\phi) = \frac{4}{N^2 S_\Gamma^2} \sum_{k=0}^{N-1} \sin^2(\phi + \beta_k)\, u^2(S_k) \qquad (9.23)$$

For the shot noise, using Equation (9.8) we obtain

$$u_N^2(\phi, shot) = \frac{4}{N^2 S_\Gamma[cnt]^2} \sum_{k=0}^{N-1} \sin^2(\phi + \beta_k) \, S_k[cnt]$$

$$= \frac{4 \, S_B[cnt]}{N^2 S_\Gamma[cnt]^2}$$

$$\sum_{k=0}^{N-1} \sin^2(\phi + \beta_k) + \frac{4}{N^2 S_\Gamma[cnt]} \sum_{k=0}^{N-1} \sin^2(\phi + \beta_k) \, \cos(\phi + \beta_k)$$

(9.24)

The second sum over the trigonometric functions vanishes, resulting in

$$u_N^2(\phi, shot) = \frac{2 \, S_B[cnt]}{N S_\Gamma[cnt]^2} = \frac{2}{N v \, S_\Gamma[cnt]} \tag{9.25}$$

where $v = S_\Gamma/S_B$ is the fringe visibility.

For camera gain and illumination fluctuations, an analogous procedure yields

$$u_N^2(\phi) = \frac{4r^2}{N^2 S_\Gamma^2} \sum_{k=0}^{N-1} \sin^2(\phi + \beta_k) \, S_k^2$$

$$= \frac{4r^2}{N^2 S_\Gamma^2} \sum_{k=0}^{N-1} \sin^2(\phi + \beta_k) (S_B^2 + S_\Gamma^2 \cos^2(\phi + \beta_k) + 2 S_B S_\Gamma \cos(\phi + \beta_k))$$

(9.26)

where r^2 is either $r^2(\gamma)$ or $r^2(I)$. Again, using an analytic evaluation of the sums, we obtain

$$u_N^2(\phi) = \frac{2r^2}{Nv^2} + \frac{2r^2}{N^2}\left(\frac{N}{4} - \delta_{N4} \cos(4\phi)\right) \tag{9.27}$$

9.5.3 Uncertainty from Phase Stepping
9.5.3.1 Phase Step Jitter
If the phase step angle β_k in the phase-stepping procedure is randomly jittering around its expectation value with a standard deviation of $u(\beta)$, this entails phase measurement uncertainty. The same is true for random

object vibrations that induce jitter in the phase value itself. From Equation (9.15), the signal uncertainty caused by phase step (or object phase) jitter is obtained according to

$$u^2(S_k, \beta_k) = S_\Gamma^2 \sin^2(\phi + \beta_k)\, u^2(\beta_k). \tag{9.28}$$

Note that the uncertainty is proportional to the signal modulation S_Γ and depends on the phase angle ϕ, even if we assume that the phase step jitter itself is independent of k and uncorrelated, that is, $u^2(\beta_k) = u^2(\beta)$. Inserting the expression into Equation (9.20) yields

$$u_N^2(\phi, \beta) = \frac{u^2(\beta)}{2N}(3 + \delta_{N4}\cos(4\phi)) \tag{9.29}$$

so that u_N drops again with $\sqrt{N}$. It is independent of the signal modulation S_Γ but varies with the phase value in the case of $N = 4$. To meet our target uncertainty of 63 mrad on average, the phase-stepping jitter must be less than

$$u(\beta) \le \sqrt{\frac{2N}{3}} u_N(\phi, \beta) = \sqrt{N} \times 51mrad. \tag{9.30}$$

9.5.3.2 Linear Phase Step Miscalibration

If the phase step angle β does not exactly match the design phase step, there is a systematic phase angle deviation, sometimes also called detuning or slope error [25–29]. For the case of a linear phase step miscalibration, we have from Equation (9.15)

$$\tilde{S}_k = S_B + S_\Gamma \cos(\phi + k\beta(1 + \varepsilon))$$

$$\left.\frac{\partial \tilde{S}_k}{\partial \varepsilon}\right|_{\varepsilon=0} = -S_\Gamma \sin(\phi + k\beta)\,k\beta \tag{9.31}$$

where ε is the relative miscalibration error. The signal deviation increases with k and is again proportional to the modulation signal S_Γ. As a consequence, the phase measurement uncertainty caused by a linear phase step miscalibration cannot be reduced by increasing S_Γ. The miscalibration ε

introduces a systematic phase variation that is not known a priori (otherwise, one could correct for it); therefore, it has to be treated as a measurement uncertainty. Further, the signals are fully correlated, which yields the relations

$$u^2(S_k,\varepsilon) = S_\Gamma^2 \sin^2(\phi+k\beta) k^2\beta^2\varepsilon^2$$

$$u_\varepsilon(S_j,S_k) = S_\Gamma^2 \sin(\phi+j\beta)\sin(\phi+k\beta) j k \beta^2\varepsilon^2 \qquad (9.32)$$

Insertion into the general uncertainty expression for N-bucket algorithms, Equation (9.20), leaves us with [6, 30, 31]

$$u_N^2(\phi,\varepsilon) = \frac{4}{N^2} \left\{ \beta\varepsilon \sum_{k=0}^{N-1} k \sin(\phi+\beta_k) \right\}^2 \quad \text{or}$$

$$u_N(\phi,\varepsilon) = \varepsilon\frac{\pi}{N}\left((N-1) - \frac{\sin\left(2\phi-\frac{2\pi}{N}\right)}{\sin\left(\frac{2\pi}{N}\right)} \right) \qquad (9.33)$$

To meet a target uncertainty of 63 mrad on average, the linear phase step miscalibration must be less than

$$\varepsilon \le \frac{N}{(N-1)\pi} u_N(\phi,\varepsilon) \le 3\% \qquad (9.34)$$

9.5.3.3 Higher-Order Phase Step Miscalibration

The general phase step uncertainty including higher-order terms [32, 33] is written as a power series in k up to the order P:

$$\varepsilon_k = k\varepsilon\beta + k^2\varepsilon_2 + \cdots + k^P\varepsilon_P. \qquad (9.35)$$

Hence, in analogy to Equation (9.32), the deviation is again proportional to the signal modulation, and treating each order independently, we have from Equation (9.33)

$$u_N^2(\phi,\varepsilon_p) = \frac{4}{N^2} \left\{ \varepsilon_p \sum_{k=0}^{N-1} k^P \sin(\phi+\beta_k) \right\}^2 \qquad (9.36)$$

A closed-form solution is cumbersome, but the average uncertainty scales with N^{p-1}, showing that uncertainty caused by higher-order nonlinearities increases steeply with the number of frames.

9.5.3.4 Detector Nonlinearity

Detector nonlinearity is a systematic error that causes higher harmonics and correlates the individual signals in a multiframe algorithm [17, 33, 34]. The uncertainty components given in Equation (9.12) are fully correlated because, for each order n, the detector nonlinearity $r_n S_k^n$ affects all signals equally. Inserting into Equation (9.20), we obtain

$$u_N^2(\phi) = \frac{4}{N^2 S_\Gamma^2} \left\{ \sum_{k=0}^{N-1} \sin(\phi + \beta_k) \, S_k^n r_n \right\}^2 \tag{9.37}$$

Decomposing the signal into its background and modulation, the expression for the N-bucket algorithm can be evaluated analytically. It is found that the sum over the powers of the cosine vanishes, except for orders $n = N - 1$ and above. For the lowest contributing order, this yields [21]

$$u_N^2(\phi, r_{N-1}) = r_{N-1}^2 \left(\frac{S_\Gamma}{2} \right)^{2N-4} \sin^2(N\phi) \tag{9.38}$$

Note that the uncertainty increases with increasing fringe contrast. To meet our target uncertainty of 63 mrad on average using $\langle \sin^2(N\phi) \rangle = 1/2$, the contribution of the detector nonlinearity must be less than

$$\frac{r_{N-1} S_\Gamma^{N-1}}{S_\Gamma} \leq 2^{N-1.5} u_N(\phi, r_{N-1}) = 2^N \times 2.2\% \tag{9.39}$$

9.5.3.5 Signal Harmonics

Higher harmonics in the signal lead to additional terms up to the order R:

$$\tilde{S}_k = S_k + \sum_{r=2}^{R} S_r \cos(r(\phi + \beta_k)) \tag{9.40}$$

If nothing is known about the higher harmonics present in the signal (i.e., no Fourier decomposition is given), then these orders are taken to be uncorrelated and are treated individually. The sensitivity to order r is given by

$$\frac{\partial \tilde{S}_k}{\partial S_r} = \cos(r(\phi + \beta_k)) \tag{9.41}$$

which leads again to a full correlation of all signals within a given harmonic order r. Hence, the uncertainty is, in analogy to Equation (9.37),

$$u_N^2(\phi, S_r) = \frac{4S_r^2}{N^2 S_\Gamma^2} \left\{ \sum_{k=0}^{N-1} \sin(\phi + \beta_k) \cos(r(\phi + \beta_k)) \right\}^2. \tag{9.42}$$

The uncertainty is directly proportional to the ratio of the higher-order signal content to the fundamental frequency signal content. Because $\cos(r\phi)$ can be decomposed into powers of $\cos(\phi)$ up to the order r, we have a similar situation as with the detector nonlinearity discussed previously. The sum for the N-bucket algorithm vanishes again, except for the cases when $r \pm 1 = ZN$, where Z is an integer. The lowest contributing order for a given N is $r = N - 1$, and the next higher order is $r = N + 1$. Their contribution is given by the identical expression [21]

$$u_N^2(\phi, S_{N\pm1}) = \frac{S_{N\pm1}^2}{S_\Gamma^2} \sin^2(N\phi) \tag{9.43}$$

Hence, if insensitivity of the phase measurement to the signal harmonics up to the Rth order is required, the minimal number of signal values to be acquired is $N = R + 2$ [19, 30, 35, 36]. The target uncertainty of 63 mrad would allow a higher harmonic signal content of 8.9%.

9.5.3.6 Background Drift
Background irradiance can change during phase stepping [37, 38]. If a linear drift is assumed, we can write

$$\tilde{S}_k = S_B(1 + k\delta_{drift}) + S_\Gamma \cos(\phi + k\beta)$$

$$\left. \frac{\partial \tilde{S}_k}{\partial \delta_{drift}} \right|_{\varepsilon=0} = kS_B \tag{9.44}$$

The deviation increases linearly with k, but it is not modulated. Because δ_{drift} is unknown, it must be treated as an uncertainty. It leads again to full correlation among the signals:

$$u^2(S_k, \delta_{drift}) = k^2 S_B^2 \delta_{drift}^2$$

$$u_{drift}(S_j, S_k) = S_B^2 jk\delta_{drift}^2 \tag{9.45}$$

Inserting into the general expression Equation (9.20), we obtain the following after analytical summation:

$$u_N^2(\phi, \delta_{drift}) = \frac{\delta_{drift}^2}{v^2} \left\{ \frac{\cos^2\left(\phi - \frac{\pi}{N}\right)}{\sin^2\left(\frac{\pi}{N}\right)} \right\}. \tag{9.46}$$

Drift in the background irradiance is a major source of uncertainty because it is inversely proportional to the fringe contrast v. To be compatible with our target uncertainty of 63 mrad, the drift between two frames must (on average) respect

$$\delta_{drift} \le v\sqrt{2}\sin\left(\frac{\pi}{N}\right)u_N(\phi,\delta_{drift}) = v\sin\left(\frac{\pi}{N}\right) \times 0.09 \qquad (9.47)$$

9.5.3.7 Illumination Drift

Drift in the illumination source affects the entire signal (background and modulation), that is, for a linear drift

$$u^2(S_k, I_{drift}) = k^2 S_k^2 \delta_{illum.drift}^2$$

$$u_{drift}(S_j, S_k) = S_j S_k j k \delta_{illum.drift}^2 \qquad (9.48)$$

This effect occurs, for example, during wavelength shifting when the source power cannot be kept constant [37]. The contribution to the uncertainty is again fully correlated. For the N-bucket algorithm, the analytic expression after summation is given by

$$u_N^2(\phi,\delta_{illum.drift}) = \delta_{illum.drift}^2 \left\{\frac{\cos\left(\phi-\frac{\pi}{N}\right)}{v\,\sin\left(\frac{\pi}{N}\right)} + \frac{\cos\left(2\phi-\frac{2\pi}{N}\right)}{2\sin\left(\frac{2\pi}{N}\right)}\right\}^2 \qquad (9.49)$$

On average, this is

$$u_N^2(\phi,\delta_{illum.drift}) = \delta_{illum.drift}^2 \left\{\frac{1}{v^2}\frac{1}{2\sin^2\left(\frac{\pi}{N}\right)} + \frac{1}{4\sin^2\left(\frac{2\pi}{N}\right)}\right\} \qquad (9.50)$$

from which we find that the boundary imposed by our target uncertainty of 63 mrad is

$$\delta_{illum.drift} \le \frac{0.063}{\sqrt{\frac{1}{v^2}\frac{1}{2\sin^2\left(\frac{\pi}{N}\right)} + \frac{1}{4\sin^2\left(\frac{2\pi}{N}\right)}}} \qquad (9.51)$$

9.5.4 Example of Combined Uncertainty

To illustrate the influence of the individual contributions to the overall phase measurement uncertainty, we calculate the variances and covariances for the classical four-frame algorithm based on an 8-bit digitizer camera and sum them according to Equation (9.20). Table 9.4 is a

TABLE 9.4   Summary of Contributions to the Phase Measurement Uncertainty uFF of a Four-Frame, 8-Bit Algorithm, Keeping a Laser Interferometer in Mind

Component f	Uncertainty Value $u(f)$	Phase Variance for $N = 4$, $u^2(\phi, f)$	Value $u^2(\phi, f)$ (mrad²)
Electronic noise	$u_{el.noise} = 1.5$ GL (type A)	$u_{FF}^2(\phi, el.noise) = \dfrac{u_{el.noise}^2}{2\,S_\Gamma^2}$ Equation (9.21)	275
Digitization noise	$S_{LSB} = 1$ GL per definition (type B)	$u_{FF}^2(\phi, LSB) = \dfrac{S_{LSB}^2}{24\,S_\Gamma^2}$ Equation (9.21)	10
Background fluctuations[a]	$S_{background} = 1.2$ GL (type B)	$u_{FF}^2(\phi, background) = \dfrac{S_{background}^2}{2\,S_\Gamma^2}$ Equation (9.21)	200
Shot noise [b]	$S_\Gamma = 4{,}000$ cnt (type B)	$u_{FF}^2(\phi, shot) = \dfrac{1}{2\nu\,S_\Gamma[\mathrm{cnt}]}$ Equation (9.25)	250
Illumination instability	Typical laser stability rated as 1% $r(I) = 0.01$ (data sheet, type B)	$u_{FF}^2(\phi, I) = \dfrac{r^2(I)}{2\nu^2} + \dfrac{r^2(I)}{8}(1 - \cos(4\phi))$ Equation (9.27)	213
Detector gain fluctuation[c]	$r(\gamma) = 1/256$ (type B)	$u_{FF}^2(\phi, \gamma) = \dfrac{r^2(\gamma)}{2\nu^2} + \dfrac{r^2(\gamma)}{8}(1 - \cos(4\phi))$ Equation (9.27)	32
Phase step jitter[d]	$u(\beta) = 1.4°$ (type B)	$u_{FF}^2(\phi, \beta) = \dfrac{u^2(\beta)}{8}(3 + \cos(4\phi))$ Equation (9.30)	209
Linear phase step miscalibration[e] [39]	$\varepsilon = 0.005$ (type B)	$u_{FF}^2(\phi, \varepsilon) = \varepsilon^2 \dfrac{\pi^2}{16}(3 + \cos(2\phi))^2$ Equation (9.33)	147
Higher-order phase step miscalibration	Assumed negligible	$u_{FF}^2(\phi, \varepsilon_p) = \dfrac{1}{4}\varepsilon_p^2 \left\{ \sum_{k=0}^{3} k^p \sin\left(\phi + k\dfrac{\pi}{2}\right) \right\}^2$ Equation (9.36)	0

(continued)

TABLE 9.4 Summary of Contributions to the Phase Measurement Uncertainty uFF of a Four-Frame, 8-Bit Algorithm, Keeping a Laser Interferometer in Mind *(continued)*

Component f	Uncertainty Value $u(f)$	Phase Variance for $N = 4$, $u^2(\phi, f)$	Value $u^2(\phi, f)$ (mrad²)
Detector nonlinearity	Negligible[f]	$u_{FF}^2(\phi, r_3) = r_3^2 \left(\dfrac{S_\Gamma}{2} \right)^4 \sin^2(4\phi)$ Equation (9.38)	0
Higher signal harmonics[g]	Assuming total of 0.0025	$u_{FF}^2(\phi, S_{N\pm 1}) = \dfrac{S_{N\pm 1}^2}{S_\Gamma^2} \sin^2(4\phi)$ Equation (9.43)	1,250
Background irradiance drift[h]	$\delta_{drift} = 0.005$	$u_{FF}^2(\phi, \delta_{drift}) = \dfrac{\delta_{drift}^2}{v^2}(1 + \sin(2\phi))$ Equation (9.46)	100
Illumination source drift[i]	$\delta_{illum.drift} = 0.0025$	$u_{FF}^2(\phi, \delta_{illum.drift}) = \delta_{illum.drift}^2$ $\left\{ \dfrac{1}{v}(\cos(\phi) + \sin(\phi)) + \dfrac{\sin(2\phi)}{2} \right\}^2$ Equation (9.49)	27
Total		$u^2(\phi)$ (mrad²)	2,712
		$u(\phi)$ (mrad)	52

[a] Assuming 10% background irradiance on the camera with 5% fluctuation yields a background signal of $S_{background} = 1.2$ GL.

[b] Assuming a typical 16,000 e⁻ full-well capacity for a 6.5 x 6.5 µm pixel (256 GL), then $S_\Gamma = 64$ GL corresponds to 4,000 cnt.

[c] Typical gain stability is rated as 1 GL/frame: $r(\gamma) = 1/256$ (data sheet, type B).

[d] A typical position reproducibility of a high-end piezo actuator is 1 nm. For a green laser interferometer, $\lambda = 532$ nm, and a 90° phase step in reflection corresponds to $\lambda/8$ or 66 nm, so $u(\beta) = 90°/66 = 1.4°$.

[e] A literature value for a high-end piezo actuator linear miscalibration is 0.1%. A more realistic value includes angular deviations across the beam and calibration uncertainty, so we assume 0.5%.

[f] The second-order nonlinearity is dominant. For a scientific CCD, the parameter is typically $\alpha < 10^{-7}$, which, according to Equation (9.14), is $r_2 < 10^{-5}$/GL.

[g] For the worst case of a rectangular signal, the power ratios of the third and fifth harmonic content as obtained from a Fourier analysis would be 0.123 and 0.055, respectively. For a laser interferometer, the higher harmonics content is much less pronounced, for example, 5%, which results in a power ratio of 0.0025.

[h] We assume again a 10% background intensity and a drift of the same order of magnitude as the jitter above (i.e., 5%), resulting in $\delta_{drift} = 0.005$.

[i] We assume that the illumination drift over four frames is comparable to the illumination stability (i.e., 1% for a typical laser).

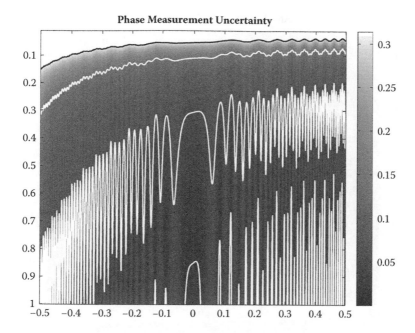

FIGURE 9.3   Total phase measurement uncertainty expected for the fringe map in Figure 9.2, based on the uncertainty contributions given in Table 9.4. For better visibility, the color bar has been restricted to 1/20 of a fringe. Contour lines are drawn (from bottom to top) at π/100, π/50, π/20, and π/10.

summary of the individual contributions and gives typical values assumed for the calculation/simulation of the combined measurement uncertainty. Table 9.4 provides a single value assuming a contrast of $v = 0.5$, $S_\Gamma = 64\ GL$, and a mean square average for the phase angles.

Summing the contributions given in Table 9.4 results in a combined phase measurement uncertainty of 52 mrad. The dominant term is caused by the higher harmonics content in the signal, which makes up half of the phase variance. The values in Table 9.4 are calculated for a fixed fringe contrast of 0.5 and are averaged over all phase values, but Figure 9.3 highlights the variation with both phase and contrast across the fringe map of Figure 9.2. In addition, Figure 9.4 shows the difference of the phase retrieved using the four-frame algorithm when each frame contains all uncertainty contributions simulated using the values given in Table 9.4. Clearly, the two approaches are consistent.

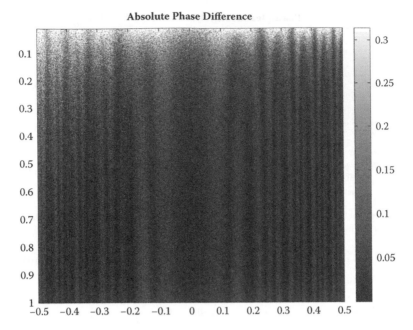

FIGURE 9.4   Phase difference simulated for the fringe map in Figure 9.2 based on the combined uncertainty contributions given in Table 9.4. For better visibility, the color bar has been restricted to 1/20 of a fringe.

9.6 PHASE MEASUREMENT UNCERTAINTY FOR CARRÉ-TYPE ALGORITHMS

Carré-type algorithms [40] are based on linearly phase-stepped images but use these images for an estimate of the phase-stepping angle β. As a result, the phase is calculated from the general expression [41–46]

$$\phi+\alpha = \mathrm{atan}\left(\frac{\sqrt{\langle S|D|S\rangle}}{\langle a|S\rangle}\right) = \mathrm{atan}\left(\frac{G\sin(\phi+\alpha)}{G\cos(\phi+\alpha)}\right) \qquad (9.52)$$

where the matrix $\mathbf{D}$ is a bilinear form. In the important cases when $\mathbf{D}$ can be written as the outer product of two vectors $\mathbf{D} = |\mathbf{b}^1\rangle\,\langle\mathbf{b}^2|$, Equation (9.52) transforms to the generalized form [47]

$$\tan(\phi+\alpha) = \sqrt{\frac{\langle S|\mathbf{b}^1\rangle\,\langle\mathbf{b}^2|S\rangle}{\langle\kappa\,a|S\rangle\,\langle\kappa^{-1}a|S\rangle}} \qquad (9.53)$$

The scaling factor κ is introduced for consistency in factorizing the denominator and is $\kappa = \tan(\beta/2)$. Equation (9.53) suggests that the Carré-type algorithm can be written as the geometric mean of two linear phase-stepping algorithms with the coefficients $(\kappa\mathbf{a}, \mathbf{b}^1)$ and $(\kappa^{-1}\mathbf{a}, \mathbf{b}^2)$, respectively. The well-known Carré and Novak [43] methods have respectively.

$$\langle\mathbf{b}^1| = (-1 \quad 3 \quad -3 \quad 1) \qquad \langle\mathbf{b}^1| = (-1 \quad 2 \quad 0 \quad -2 \quad 1)$$

$$\langle\mathbf{b}^2| = (1 \quad 1 \quad -1 \quad -1) \quad \text{and} \quad \langle\mathbf{b}^2| = (1 \quad 2 \quad 0 \quad -2 \quad -1) \qquad (9.54)$$

$$\langle\mathbf{a}| = (-1 \quad 1 \quad 1 \quad -1) \qquad \langle\mathbf{a}| = (-1 \quad 0 \quad 2 \quad 0 \quad -1)$$

In view of Equation (9.53), the measurement uncertainty of Carré-type algorithms can be expressed by the measurement uncertainty of the two linear phase-stepping algorithms [47], again using the GUM procedure consistently. If we write Equation (9.53) in the form

$$f = \tan(\phi) = \sqrt{f_1 \, f_2} \quad \text{with} \quad f_i = \frac{\langle\mathbf{b}^i|\mathbf{S}\rangle}{\langle\mathbf{a}^i|\mathbf{S}\rangle} \qquad (9.55)$$

the propagation of uncertainties of the individual linear algorithms f_i to the uncertainty of the tangent and the phase itself yields

$$u^2(f) = \frac{f_2^2}{4f^2}u^2(f_1) + \frac{f_1^2}{4f^2}u^2(f_2) + \frac{1}{2}u(f_1, f_2)$$

$$u^2(\phi) = \frac{u^2(f)}{(1+f^2)^2} \qquad (9.56)$$

Therefore, the expressions found for the linear phase-stepping procedures can be used directly. The covariance term is discussed further in Hack [47].

9.7 PHASE MEASUREMENT UNCERTAINTY FOR SINGLE-FRAME ALGORITHMS

9.7.1 Relation to Linear Phase-Stepping Algorithms

In single-frame algorithms, the phase value of a pixel is retrieved from a single frame. Therefore, single-frame algorithms involve neighboring pixels (e.g., spatial phase-stepping algorithms and pixelated phase-mask methods [48]) or even the entire frame (e.g., Fourier transform algorithms) [49]. To be specific, we look at the Fourier technique for phase retrieval as described in Chapter 1 of this book. Let the fringe signal including a carrier frequency be written as

$$S(\mathbf{x}) = \gamma I(\mathbf{x}) = S_B(\mathbf{x}) + S_\Gamma(\mathbf{x})\cos(\phi(\mathbf{x}) + \mathbf{f} \cdot \mathbf{x}), \qquad (9.57)$$

where $\mathbf{f}$ is a frequency vector adding a linear carrier to the phase ϕ, and $\mathbf{x}$ is short for (x, y).

The Fourier transform of this fringe pattern is

$$F[S(\mathbf{x})] = F[S_B(\mathbf{x})] + \tfrac{1}{2}F[S_\Gamma(\mathbf{x})e^{i(\phi(\mathbf{x}) + \mathbf{f} \cdot \mathbf{x})}] + \tfrac{1}{2}F\left[S_\Gamma(\mathbf{x})e^{-i(\phi(\mathbf{x}) + \mathbf{f} \cdot \mathbf{x})}\right] \quad (9.58)$$

The first term, which describes the background signal, is found in the center of the Fourier plane around low frequencies. Figure 9.5 shows a test fringe pattern with a carrier in the y-direction of around 96 fringes across the image. The Fourier transform of the real modulation signal gives rise to two symmetric peaks centered on $+\mathbf{f}$ and $-\mathbf{f}$ (see Figure 9.5c). Because these terms are separated from the first term in the Fourier plane, the background peak and the area around the negative frequency $-\mathbf{f}$ can be filtered out. Guidelines on how to perform these filtering steps are found in Chapters 1 and 2. The back-transformation of the area around the positive frequency $+\mathbf{f}$ alone yields the complex quantity

$$I(\mathbf{x}) = F^{-1}[F[S_\Gamma(\mathbf{x})\exp(i\phi(\mathbf{x}))]] = S_\Gamma(\mathbf{x})\exp(i\phi(\mathbf{x})) \qquad (9.59)$$

after subtraction of the carrier phase. Finally, the phase constituting the argument of I is retrieved from the real and imaginary part using the arctan function (see Figure 9.5d):

$$\phi(\mathbf{x}) = \mathrm{atan}\frac{\Im[I(\mathbf{x})]}{\Re[I(\mathbf{x})]} \qquad (9.60)$$

This is of exactly the same form as Equation (9.16) for linear phase-stepping algorithms, so we can use the expressions derived above *mutatis mutandis*. First, we have to express I as a function of the original signal S. Because in single-frame algorithms there is only one frame, I will be expressed as a sum of different pixel values $S(m,n)$ from the single-frame S rather than from sequential frames.

Test Phase Map

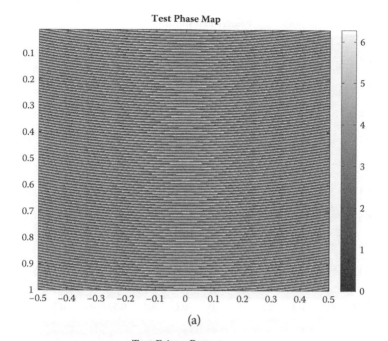

(a)

Test Fringe Pattern

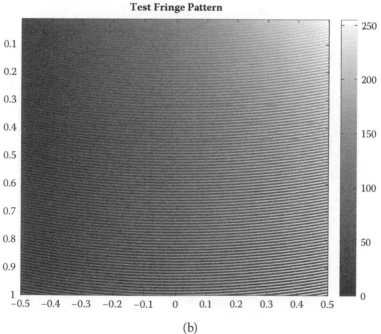

(b)

FIGURE 9.5 Simulated phase map (a) and fringe pattern (b) based on Figure 9.2 with additional carrier in *y*-direction. Fourier transform amplitude (c) and retrieved object phase map (d). *(continued)*

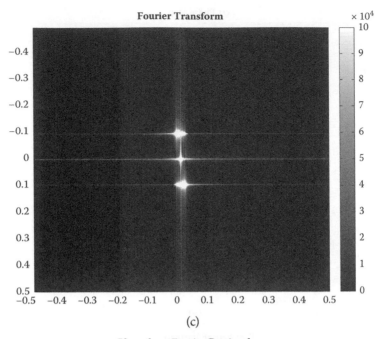

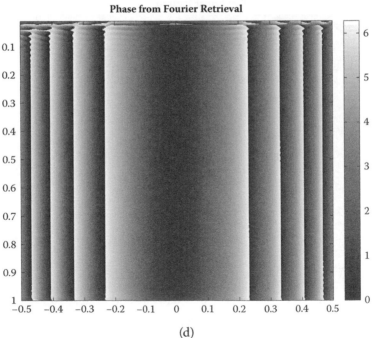

FIGURE 9.5 *(continued)* Simulated phase map (a) and fringe pattern (b) based on Figure 9.2 with additional carrier in *y*-direction. Fourier transform amplitude (c) and retrieved object phase map (d).

9.7.2 Combined Phase Measurement Uncertainty

For an array of $M \times N$ pixels (M and N being even numbers for convenience), the discrete Fourier transform is applied, and the processes described previously are in mathematical notation:

1. Fourier transformation: Equation (9.58) yields the spectral components in $(\omega_x, \omega_y) = (2\pi v_x, 2\pi v_y)$:

$$T(v_x, v_y) = \frac{1}{MN} \sum_{n=-N/2}^{N/2-1} \sum_{m=-M/2}^{M/2-1} S(m,n) \, e^{-2\pi i \frac{v_x}{M} m} \, e^{-2\pi i \frac{v_y}{N} n}$$

$$T(\omega) = \frac{1}{MN} \sum_{m} S(m) \exp(-i\omega \cdot m) \qquad (9.61)$$

$$S(m) = S_B(m) + \tfrac{1}{2} S_\Gamma(m) e^{i(\phi(m)+f \cdot m)} + \tfrac{1}{2} S_\Gamma(m) e^{-i(\phi(m)+f \cdot m)}$$

where $m = (m,n)$, $k = (l,k)$, $\omega = (\omega_x/M, \omega_y/N)$.

2. Filtering in the Fourier domain and shifting by $-f$ selects the appropriate coefficients $T(\omega)$, $\omega \in D(+f)$ from the filtered domain D around the carrier frequency $+f$. If we denote the real Fourier filter function by $H(\omega, f)$ to highlight that it is centered on f, then filtering corresponds to a multiplication of the Fourier spectrum with H (see also Box 1.2 in Chapter 1). This step as such does not introduce additional uncertainties, but of course the performance depends on the choice of H (see Chapter 2).

3. Back-transformation is now formally written as

$$I(p) = \sum_{\omega} e^{i\omega \cdot p} T(\omega) H(\omega, f)$$

$$h(p) = \sum_{\omega} e^{i\omega \cdot p} H(\omega, f) \qquad (9.62)$$

Because the signal values are the primary input quantities, we insert Equation (9.61) into Equation (9.62) to obtain

$$I(p) = \frac{1}{MN} \sum_{\omega} \sum_{m} S(m) e^{i\omega \cdot (p-m)} H(\omega, f) = \sum_{m} c(m) S(m) \quad (9.63)$$

This is the sought expression, that is, I written as a function of pixel values $S(m)$, which is the convolution $I = S*h$ of the original signal S with the inverse Fourier transform h of the Fourier filter function H. The "equivalent phase-stepping coefficients" $\mathbf{c} = \mathbf{a} + i\mathbf{b}$ of Equation (9.17) can thus be identified with

$$c(\mathbf{m}) = \frac{1}{MN} \sum_{\omega} H(\omega, \mathbf{f}) \exp(i\omega \cdot (\mathbf{p} - \mathbf{m})) = \frac{1}{MN} h(\mathbf{p} - \mathbf{m}) \quad (9.64)$$

For a rectangular window H around the carrier frequency peak, the coefficients c are sync-type functions. The combined uncertainty of φ in pixel $\mathbf{p}$ is, in analogy to Equation (9.18), using Equation (9.59) [50],

$$u^2(\phi(\mathbf{p})) = \frac{1}{S_\Gamma(\mathbf{p})^2} \sum_{\mathbf{m}} (b(\mathbf{m})\cos\phi(\mathbf{p}) - a(\mathbf{m})\sin\phi(\mathbf{p}))^2 u^2(S(\mathbf{m}))$$

$$+ \frac{2}{S_\Gamma(\mathbf{p})^2} \sum_{\mathbf{m}} \sum_{k>m}^{N-1} (b(\mathbf{k})\cos\phi(\mathbf{p}) - a(\mathbf{k})\sin\phi(\mathbf{p})) (b(\mathbf{m})\cos\phi(\mathbf{p})$$

$$- a(\mathbf{m})\sin\phi(\mathbf{p})) u(S(\mathbf{k}), S(\mathbf{m}))$$

$$(9.65)$$

The covariances $u(S(\mathbf{k}), S(\mathbf{m}))$ now concern the signal values in different pixels rather than in subsequent images as is the case with the temporal phase-stepping algorithms, Equation (9.19).

Inserting $a(\mathbf{m})$ and $b(\mathbf{m})$ from Equation (9.64) into Equation (9.65) and using trigonometric identities, we obtain

$$u^2(\phi(\mathbf{p})) = \frac{1}{S_\Gamma(\mathbf{p})^2} \frac{1}{M^2 N^2} \sum_{\mathbf{m}} \left(\sum_{\omega} H(\omega, \mathbf{f})\sin(\omega \cdot (\mathbf{p} - \mathbf{m}) - \phi) \right)^2 u^2(S(\mathbf{m}))$$

$$+ \frac{2}{S_\Gamma^2} \frac{1}{M^2 N^2} \sum_{\mathbf{m}} \sum_{k>m}^{N-1} \left(\sum_{\omega} H(\omega, \mathbf{f})\sin(\omega \cdot (\mathbf{p} - \mathbf{k}) - \phi) \right)$$

$$\times \left(\sum_{\omega} H(\omega, \mathbf{f})\sin(\omega \cdot (\mathbf{p} - \mathbf{m}) - \phi) \right) u(S(\mathbf{k}), S(\mathbf{m}))$$

$$(9.66)$$

Replacing the covariance terms by Equation (9.19), we can formally write these sums using the convolutions

$$u^2(\phi) = \frac{1}{S_\Gamma^2 M^2 N^2} \left\{ \tilde{h}^2 * u^2(S) + \sum_l \left[\left(\tilde{h} * \frac{\partial S}{\partial g_l} \right)^2 - \tilde{h}^2 * \left(\frac{\partial S}{\partial g_l} \right)^2 \right] u^2(g_l) \right\}$$

$$\tilde{h}(\mathbf{p}) = \sum_\omega H(\omega, \mathbf{f}) \sin(\omega \cdot \mathbf{p} - \phi) = \mathfrak{J}[e^{-i\phi} h(\mathbf{p})]$$

$$(9.67)$$

9.7.3 Uncertainty from Uncorrelated Influences

When contributions that are uncorrelated among pixels but have identical variance such as electron noise are inserted into Equation (9.67), this leads to

$$u^2(\phi, uncorr) = \frac{1}{S_\Gamma^2} \frac{u^2(S)}{M^2 N^2} \{ \tilde{h}^2 * 1 \} = \frac{1}{S_\Gamma^2} u^2(S) \{ FT[\tilde{h}^2](0) \} \quad (9.68)$$

where we have used the fact that convolution of a function with the unit function is proportional to the zero-frequency Fourier component of the function. Because $H(\omega, \mathbf{f})$ is centered on $\mathbf{f}$, the sum of $\tilde{h}^2$ over $\mathbf{p}$, Equation (9.67) is

$$u^2(\phi, uncorr) = \frac{1}{S_\Gamma^2} \frac{u^2(S)}{4MN} \sum_\omega H^2(\omega, \mathbf{f}) \quad (9.69)$$

For a mask of unit height, the last sum in Equation (9.69) is just the area of the mask in the Fourier plane, whereas MN is the total area of the Fourier plane. In other words, to meet our target uncertainty of 63 mrad, the acceptable relative signal uncertainty must satisfy

$$u(\phi, uncorr) = \frac{u(S)}{2S_\Gamma} \sqrt{\text{area ratio}} \le 0.063 \quad (9.70)$$

so that the uncertainty drops with increasing signal modulation S_Γ.

Shot noise is uncorrelated between pixels, but proportional to the signal, so that Equation (9.67) transforms into

$$u^2(\phi, shot) = \frac{1}{S_\Gamma^2 M^2 N^2} \{ \tilde{h}^2 * u^2(S) \} = \frac{1}{S_\Gamma^2 M^2 N^2} \{ \tilde{h}^2 * S \} \quad (9.71)$$

9.7.4 Uncertainty from Correlated Influences

Correlations between the signal values at different pixels are created when influence quantities affect pixels in a predictable way. Background fluctuations and drift, Equation (9.7), are additive constants that contribute to the zero-frequency peak and hence do not affect the phase measurement. Inserting such a fully correlated effect into Equation (9.67) yields

$$u^2(\phi) = \frac{u^2(g_l)}{S_\Gamma^2 M^2 N^2} \left(\tilde{h} * \frac{\partial S}{\partial g_l} \right)^2 \tag{9.72}$$

For **background fluctuations**, the derivative equals 1, and the convolution in Equation (9.72) vanishes.

For **illumination instability and camera gain fluctuations**, Equation (9.9), the derivative in Equation (9.72) is S/γ; therefore,

$$u^2(\phi, \gamma) = \frac{r^2(\gamma)}{S_\Gamma^2 M^2 N^2} (\tilde{h} * S)^2 \tag{9.73}$$

Likewise, nth order **detector nonlinearity** from Equation (9.11) leads to the correlations

$$u^2(\phi, r_n) = \frac{r^2(r_n)}{S_\Gamma^2 M^2 N^2} (\tilde{h} * S^n)^2 \tag{9.74}$$

Higher harmonics affect all pixels but are separated in the Fourier plane when the carrier frequency is chosen carefully (see Chapter 2). Hence, they presumably do not influence the retrieved phase.

Pixel-to-pixel **gain jitter** again causes high-frequency contributions that are filtered out. An overall **gain change** would affect the modulation intensity S_Γ. This would indirectly affect the phase value because the fringe contrast changes. Because Fourier transform is a linear operation, scaling of the input image does not affect the phase arguments.

To illustrate the combined effect of the various sources of uncertainty, the fringe pattern with carrier shown in Figure 9.5b is exposed to the same uncertainty contributions of Table 9.4 as was done for the four-frame algorithm. The retrieved phase is displayed in Figure 9.6 and compared to the phase retrieved from the corresponding noiseless fringe pattern. Although

Phase from Fourier Retrieval

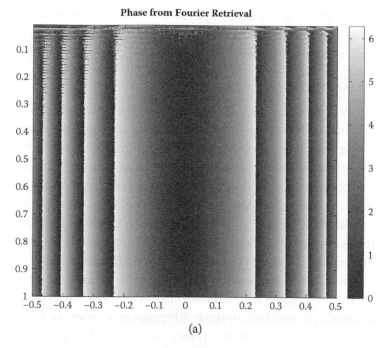

(a)

Absolute Phase Difference

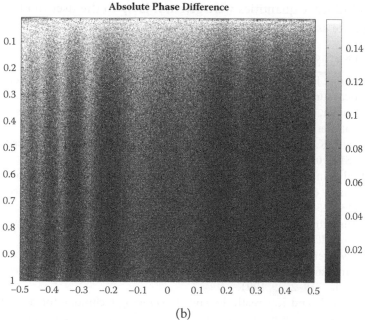

(b)

FIGURE 9.6   Phase map retrieved with the Fourier transform method from a noisy intensity map (a) and absolute phase difference compared to a noiseless Fourier phase retrieval (b).

the retrieved phase is less accurate than from the four-frame algorithm, when compared to the original phase map of Figure 9.2a, it is more robust against noise relative to the Fourier retrieval from a noiseless fringe pattern.

9.8 SUMMARY

This chapter presented a methodology based on the *Guide to the Expression of Uncertainty in Measurement* [9] to calculate the measurement uncertainty for a variety of phase retrieval techniques. It was shown that uncertainty related to Carré and Fourier methods can be reduced to the general expression of uncertainty for linear phase-stepping algorithms. This opens the way to compare the methods and, once the experimentalist has decided on a phase retrieval method, to calculate the combined phase measurement uncertainty. Although the approach is based on first-order propagation of uncertainties, the basic equations can also be used in a numerical simulation to assess the phase measurement uncertainties. Depending on the application, variations of the phase uncertainty with the phase itself, the contrast, or the brightness can be more or less important. Even though some estimated values to quantify the upper bounds of several influence quantities were given, it is up to the user to obtain the appropriate estimates for an experimental setup.

REFERENCES

1. Bruning, J. H., et al. Digital wavefront measuring interferometer for testing optical surfaces and lenses. *Applied Optics* **13**(11), 2693–2703, 1974.
2. Schwider, J., et al. Digital wave-front measuring interferometry—some systematic-error sources. *Applied Optics* **22**(21), 3421–3432, 1983.
3. Creath, K. Comparison of phase measurement algorithms. Proceedings of the SPIE **0680**, 19–28, 1986.
4. Hariharan, P., B. F. Oreb, and T. Eiju. Digital phase-shifting interferometry—a simple error-compensating phase calculation algorithm. *Applied Optics* **26**(13), 2504–2506, 1987.
5. Kinnstaetter, K., et al. Accuracy of phase-shifting interferometry. *Applied Optics* **27**(24), 5082–5089, 1988.
6. Vanwingerden, J., H. J. Frankena, and C. Smorenburg. Linear-approximation for measurement errors in phase-shifting interferometry. *Applied Optics* **30**(19), 2718–2729, 1991.
7. Schmit, J., and K. Creath. Extended averaging technique for derivation of error-compensating algorithms in phase-shifting interferometry. *Applied Optics* **34**(19), 3610–3619, 1995.
8. Zhang, H., M. J. Lalor, and D. R. Burton. Error-compensating algorithms in phase-shifting interferometry: A comparison by error analysis. *Optics and Lasers in Engineering* **31**(5), 381–400, 1999.

9. ISO/IEC. *Guide to the Expression of Uncertainty in Measurement (GUM)*. Geneva, ISO, 2008.

10. Davighi, A., et al. The development of a reference material for calibration of full-field optical measurement systems for dynamic deformation measurements. *Advances in Experimental Mechanics VIII* **70**, 33–38 (2011).

11. Patterson, E. A., et al. Calibration and evaluation of optical systems for full-field strain measurement. *Optics and Lasers in Engineering* **45**(5), 550–564, 2007.

12. Schreiber, H. A. B., and J. H. Bruning. Phase shifting interferometry. In *Optical Shop Testing*. Edited by D. Malacara. New York: Wiley, 2007, 547–666.

13. Zhao, B., and Y. Surrel. Effect of quantization error on the computed phase of phase-shifting measurements. *Applied Optics* **36**(10), 2070–2075, 1997.

14. Skydan, O. A., et al. Quantization error of CCD cameras and their influence on phase calculation in fringe pattern analysis. *Applied Optics* **42**(26), 5302–5307, 2003.

15. Helmers, H., and M. Schellenberg. CMOS vs. CCD sensors in speckle interferometry. *Optics and Laser Technology* **35**(8), 587–595, 2003.

16. Loudon, R. *The Quantum Theory of Light*. 3rd ed. New York: Oxford University Press, 2000, ix.

17. Schodel, R., A. Nicolaus, and G. Bonsch. Phase-stepping interferometry: Methods for reducing errors caused by camera nonlinearities. *Applied Optics* **41**(1), 55–63, 2002.

18. Larkin, K. G., and B. F. Oreb. Propagation of errors in different phase-shifting algorithms—a special property of the arc tangent function. *Interferometry: Techniques and Analysis* **1755**, 219–227, 1993.

19. Surrel, Y. Design of algorithms for phase measurements by the use of phase stepping. *Applied Optics* **35**(1), 51–60, 1996.

20. Servin, M., J. C. Estrada, and J. A. Quiroga. The general theory of phase shifting algorithms. *Optics Express* **17**(24), 21867–21881, 2009.

21. Hack, E., and J. Burke. Invited review article: Measurement uncertainty of linear phase-stepping algorithms. *Review of Scientific Instruments* **82**(6), 061101, 2011.

22. Ohyama, N., et al. Accuracy of phase determination with unequal reference phase-shift. *Journal of the Optical Society of America A, Optics Image Science and Vision* **5**(12), 2019–2025, 1988.

23. Hibino, K. Susceptibility of systematic error-compensating algorithms to random noise in phase-shifting interferometry (vol 36, pg 2084, 1997). *Applied Optics* **36**(22), 5362–5362, 1997.

24. Malacara, D., M. Servín, and Z. Malacara. *Interferogram Analysis for Optical Testing*. 2nd ed. Boca Raton, FL: Taylor & Francis, 2005, xv.

25. Larkin, K. G., and B. F. Oreb. Design and assessment of symmetrical phase-shifting algorithms. *Journal of the Optical Society of America A, Optics Image Science and Vision* **9**(10), 1740–1748, 1992.

26. Surrel, Y. Phase-stepping—A new self-calibrating algorithm. *Applied Optics* **32**(19), 3598–3600, 1993.

27. Estrada, J. C., M. Servin, and J. A. Quiroga. A self-tuning phase-shifting algorithm for interferometry. *Optics Express* **18**(3), 2632–2638, 2010.
28. Malacara-Doblado, D., and B. V. Dorrio. Family of detuning-insensitive phase-shifting algorithms. *Journal of the Optical Society of America A, Optics Image Science and Vision* **17**(10), 1857–1863, 2000.
29. Bi, H. B., et al. Class of 4+1-phase algorithms with error compensation. *Applied Optics* **43**(21), 4199–4207, 2004.
30. Huntley, J. M. Automated analysis of speckle interferograms. In *Digital Speckle Pattern Interferometry and Related Techniques*. Edited by P. K. Rastogi. New York: Wiley, 2001, 59–139.
31. Cordero, R. R., et al. Uncertainty analysis of temporal phase-stepping algorithms for interferometry. *Optics Communications* **275**(1), 144–155, 2007.
32. Schwider, J., T. Dresel, and B. Manzke. Some considerations of reduction of reference phase error in phase-stepping interferometry. *Applied Optics* **38**(4), 655–659, 1999.
33. Ai, C., and J. C. Wyant. Effect of piezoelectric transducer nonlinearity on phase-shift interferometry. *Applied Optics* **26**(6), 1112–1116 (1987).
34. Ohyama, N., et al. An analysis of systematic phase errors due to nonlinearity in fringe scanning systems. *Optics Communications* **58**(4), 223–225, 1986.
35. Hibino, K., et al. Phase-shifting for nonsinusoidal wave-forms with phase-shift errors. *Journal of the Optical Society of America A, Optics Image Science and Vision* **12**(4), 761–768, 1995.
36. Han, C. W., and B. T. Han. Error analysis of the phase-shifting technique when applied to shadow moire. *Applied Optics* **45**(6), 1124–1133, 2006.
37. Onodera, R., and Y. Ishii. Phase-extraction analysis of laser-diode phase-shifting interferometry that is insensitive to changes in laser power. *Journal of the Optical Society of America A, Optics Image Science and Vision* **13**(1), 139–146, 1996.
38. Surrel, Y. Design of phase-detection algorithms insensitive to bias modulation. *Applied Optics* **36**(4), 805–807, 1997.
39. Li, Y. Q., Z. Y. Zhu, and X. Y. Li. Elimination of reference phase errors in phase-shifting interferometry. *Measurement Science & Technology* **16**(6), 1335–1340, 2005.
40. Carré, P. Installation et utilisation du comparateur photoélectrique et inter-férentiel du BIPM. *Metrologia* **2**, 13–23, 1966.
41. Stoilov, G., and T. Dragostinov. Phase-stepping interferometry: Five-frame algorithm with an arbitrary step. *Optics and Lasers in Engineering* **28**(1), 61–69, 1997.
42. Qian, K. M., F. J. Shu, and X. P. Wu. Determination of the best phase step of the Carre algorithm in phase shifting interferometry. *Measurement Science & Technology* **11**(8), 1220–1223, 2000.
43. Novak, J. Five-step phase-shifting algorithms with unknown values of phase shift. *Optik* **114**(2), 63–68, 2003.
44. Chen, L. J., C. G. Quan, and C. J. Tay. Simplified Carre's method for phase extraction. *Third International Conference on Experimental Mechanics and Third Conference of the Asian-Committee-on-Experimental-Mechanics, Pts 1 and 2* **5852**, 198–202, 2005.

45. Novak, J., P. Novak, and A. Miks. Multi-step phase-shifting algorithms insensitive to linear phase shift errors. *Optics Communications* **281**(21), 5302–5309, 2008.

46. Magalhaes, P. A. A., P. S. Neto, and C. S. de Barcellos. Generalization of Carre equation. *Optik* **122**(6), 475–489, 2011.

47. Hack, E. Measurement uncertainty of Carre-type phase-stepping algorithms. *Optics and Lasers in Engineering* **50**(8), 1023–1025, 2012.

48. Kimbrough, B. T. Pixelated mask spatial carrier phase shifting interferometry algorithms and associated errors. *Applied Optics* **45**(19), 4554–4562, 2006.

49. Takeda, M., H. Ina, and S. Kobayashi. Fourier-transform method of fringe-pattern analysis for computer-based topography and interferometry. *Journal of the Optical Society of America* **72**(1), 156–160, 1982.

50. Betta, G., C. Liguori, and A. Pietrosanto. Propagation of uncertainty in a discrete Fourier transform algorithm. *Measurement* **27**(4), 231–239, 2000.

Index